The Challenge of Rural Electrification

Strategies for Developing Countries

Edited by

Douglas F. Barnes

RFF PRESS
RESOURCES FOR THE FUTURE

New York • London

ENERGY SECTOR MANAGEMENT ASSISTANCE PROGRAM
Washington, DC, USA

An RFF Press book

For a full list of publications please contact:
Earthscan
4 Park Square, Milton Park, Abingdon, Oxon OX14 4RN
605 Third Avenue, New York, NY 10017

Earthscan is an imprint of the Taylor & Francis Group, an informa business

Library of Congress Cataloging-in-Publication Data

The challenge of rural electrification : strategies for developing countries / Douglas F. Barnes, editor. — 1st ed.
    p. cm.
  Includes bibliographical references and index.
  ISBN 978-1-933115-44-3 (hardcover : alk. paper) — ISBN 978-1-933115-43-6 (pbk. : alk. paper)
  1. Rural electrification—Developing countries.   I. Barnes, Douglas F.
  HD9688.D44C53 2007
  333.793'2—dc22   2007004676

The paper in this book meets the guidelines for permanence and durability of the Committee on Production Guidelines for Book Longevity of the Council on Library Resources. This book was typeset by Integrated Book Technology. It was copyedited by Paula Berard. The cover was designed by Nancy Bratton.

The findings, interpretations, and conclusions offered in this publication are those of the contributors. They do not necessarily represent the views of Resources for the Future, its directors, or its officers.

ISBN 978-1-933115-44-3 (hbk)
ISBN 978-1-933115-43-6 (pbk)

About Resources for the Future *and* RFF Press

Resources for the Future (**RFF**) improves environmental and natural resource policy-making worldwide through independent social science research of the highest caliber. Founded in 1952, RFF pioneered the application of economics as a tool for developing more effective policy about the use and conservation of natural resources. Its scholars continue to employ social science methods to analyze critical issues concerning pollution control, energy policy, land and water use, hazardous waste, climate change, biodiversity, and the environmental challenges of developing countries. RFF Press supports the mission of RFF by publishing book-length works that present a broad range of approaches to the study of natural resources and the environment. Its authors and editors include RFF staff, researchers from the larger academic and policy communities, and journalists. Audiences for publications by RFF Press include all of the participants in the policymaking process—scholars, the media, advocacy groups, NGOs, professionals in business and government, and the public.

About ESMAP

The **Energy Sector Management Assistance Program (ESMAP)** is a global technical assistance program that promotes the role of energy for poverty reduction and economic growth. It helps build consensus and provides policy advice on sustainable energy development to governments of developing countries and economies in transition. ESMAP was established in 1983 under the joint sponsorship of the World Bank and the United Nations Development Programme as a partnership with other donors in response to global energy crises. Since its creation, ESMAP has operated in over 100 different countries, through approximately 450 activities covering a broad range of energy issues.

To my family—Mary Ann, Andrea and Chris

Contents

Figures and Tables

Preface

Most people in the world just have to flip a switch and they can use lights, televisions, computers, machines, tools, and many other appliances that enable them to read, obtain news, be entertained, and participate in productive work. Rural electrification had been quite successful in stimulating development in all developed and many developing countries, so this study was initiated to see why it has been difficult to reach the approximately 1.6 billion people who still do not have electricity. In this day and age it seems almost unthinkable that in Africa 90 percent of people in rural areas are without electricity. Even in growing economic powers such as India there are close to 400 million rural people without access.

The Challenge of Rural Electrification describes how to develop effective institutions, provide efficient and enabling subsidies, and keep distribution utilities afloat financially during a period of rapid expansion, all in ways that are politically acceptable to a wide range of constituents. The contributors to this volume document how a diverse set of countries in the developing world has met and conquered these challenges. This book can serve as a guide for countries embarking on the road toward implementing programs that provide electricity to difficult-to-reach populations.

This book is of particular interest to a wide range of people, including policymakers, electricity distribution specialists, economists, and politicians, who have not yet achieved high rates of providing electricity to people in rural areas. In academic circles, those interested include historians, economists, political scientists, and sociologists, as the case studies touch on many aspects of these various disciplines. In fact, the authors of these case studies cover a wide range of disciplines, including engineering, energy, social science, and economic history.

When we first started this project, rural electrification was falling out of favor with many international financial agencies, in part because they perceived large public projects as going against the trend toward privatization of energy services. This was compounded by problems encountered in their own portfolio of rural electrification projects. In the private sector, approaches to rural electrification never seemed to take off because it is a difficult business. In its beginning stages, rural electrification is not a profitable business, and few private sector and sometimes even public sector companies are interested in making the investments needed to serve mainly poor rural people. As a consequence, there must be subsidies in such programs that encourage the development of commercially viable service companies. One should not be surprised that this sometimes leads to political interference, poor subsidy design, and consequently poor performance of many distribution companies. Most recently, rural electrification and energy access in general have come back into favor with a somewhat different orientation. Governments are involved in promoting rural electrification in almost all instances. Today we also are seeing decentralized electricity services in many remote areas that were not possible many years ago.

In the earliest stages of this study, in a time when rural electrification seemed to be quite unpopular among development agencies, we decided to complete two case studies of successful programs and document the ways in which they were organized and implemented. After completing the first two case studies in Thailand and Costa Rica, it was clear that despite significant differences—Thailand has a publicly run program and Costa Rica involves distribution cooperatives—they faced similar problems. Also, during the preparation of these case studies, we found that there had been many successful programs around the world that were quietly and successfully providing electricity to rural customers, some of which were not even relying on any kind of financing from outside international agencies. Thus, we decided to document a broader number of successful programs to determine whether lessons could be learned from them. The project expanded slowly until we had documented enough programs—ten in all—so that we could be fairly certain of the reasons for successful rural electrification in the face of so many that had failed.

The case studies in this book provide evidence that there certainly are no simple solutions or magic formulas for successful rural electrification programs, as this study found a variety of models that can work, including those from public sector companies, cooperatives, and private firms. Instead, to be successful, countries must follow a set of fairly well-defined but rather flexible principles, including ways to approach subsidies, a clear path toward financial viability, cooperation with local communities, adoption of appropriate standards to achieve low cost electricity distribution, and an arms-length relation with the government.

Acknowledgments

This book is the culmination of the work on rural electrification in developing studies financed and supported by the Energy Sector Management Assistance Program of the World Bank. We owe much debt and gratitude to the dedicated staff of ESMAP who over the years have encouraged us to prepare this book. Jamal Saghir and Ede Ijjasz-Vasquez have provided valuable support to bring this work to completion. We also are especially appreciative of the support of Dominique Lallement and Karl Jechoutek, both of whom have a keen interest in alleviating energy poverty and promoting energy development around the world. Richard Stern and Dennis Anderson must be given credit for starting the World Bank efforts on rural energy. Also, the innovative rural energy work of Willem Floor and Robert van der Plas provided a significant background for many of the findings in this book.

In addition, we would like to acknowledge the contributions of the original ESMAP support staff and the case study editors. The publications staff, including Heather Worley and Marjorie Araya, has been instrumental in assisting in the development of this joint publication with RFF Press. The support staff that coordinated and administered much of the research contained in this book includes Nyra Wallace and Maureen Cuffley. Norma Adams tackled the unenviable task of editing many of the original case studies. We thank Paula Berard for the editing and preparation of the manuscript. Finally, we also wish to thank Grace Hill and Miriam Dowd for the production and marketing of this book. They were helpful beyond expectations in completing this project in a timely manner.

I truly appreciate the comments and insights of those who provided peer review comments on an earlier version of this study. Karl Jechoutek provided a review that assisted in setting the framework for the book. Anton Eberhard provided advice for the conclusion and other parts of the work. A long time colleague, Andrew Barnett, not only gave constant support for the project in his role as an advisor to ESMAP, but also provided useful insights for the final editing of the manuscript. This work would have been impossible to do by myself. Within the World Bank, excellent comments were provided by Demetrios Papathanasiou and Fanny Missfeldt-Ringius. Also, I wish to thank all of the authors who not only wrote their chapters, but had to put up with my constant advice and recommendations for revising them for what seemed like a million times.

The research was financed by the World Bank. However, the findings, interpretation, and conclusions expressed in this book are entirely our own and should not be attributed in any manner to the World Bank, to its affiliated organizations, or to members of its Board of Executive Directors or the governments they represent.

Contributors

DOUGLAS F. BARNES is a senior energy specialist within the Energy and Water Department of the World Bank. Prior to joining the Bank, he worked at the Center for Energy Policy Research at Resources for the Future. He has been involved in social aspects of development for both rural and urban areas for more than 25 years, leading efforts to develop a strategy for rural energy for the World Bank Group, which was published as *Rural Energy and Development: Improving Energy Supplies for Two Billion People*. He published a book entitled *The Urban Energy Transition: Energy, Poverty and the Environment in the Developing World* and was a coauthor of *Environmental Health and Traditional Fuel Use in Developing Countries*. He is also the author of *Electric Power for Rural Growth: How Electricity Affects Rural Life in Developing Countries*.

MONCEF AISSA is a senior electrical engineer. He is an international consultant on low cost system designs for rural electrification after having had a 33-year career with *Societe Tunisienne de l'Electricite et du Gaz*, the Tunisian public utility for electricity and gas. There he headed the Department of Electrical Distribution Network Planning for more than 20 years and now heads the Department of Electrical Equipment Standards and Technical Specifications.

ELIZABETH CECELSKI has worked for more than 25 years on rural electrification and rural development. She was an energy economist at Resources for the Future and later worked in the Rural Employment Policies Branch of the International Labour Organisation in Geneva. She is a founding member of the Advisory Group and technical adviser for ENERGIA, the International Network on Gender and Sustainable Energy. She serves on the 3-member Technical Advisory Group for the Energy Trust Funded Programs of the World Bank.

JOY DUNKERLEY is a consultant on rural energy supplies in India and Tunisia for the World Bank; global prospects for civilian nuclear power for the Atlantic Council; energy policies and economic development for the United Nations Educational, Scientific and Cultural Organization; and a wide range of energy issues for the African Energy Policy Research Network. At Resources for the Future, she developed an economics-based policy research program on energy in developing countries, and co-authored *Energy Strategies for Developing Nations.* She received the 2000 U.S. Association for Energy Economics Adelman-Frankel Award.

GERALD FOLEY has written widely on energy issues and has spent many years as a consultant specializing in rural energy issues in the developing world. During his work with the Panos Institute in London, he wrote a study of rural electrification, *Energy for Rural People.* As a partner in the Nordic Consulting Group in Copenhagen he undertook a variety of energy consultancy projects for the World Bank, United Nations Development Programme, the European Union, and the overseas development agencies of Denmark, Norway, and other countries.

LUIS E. GUTIERREZ-POUCEL is a consultant in Mexico. He has been an adviser to two energy secretaries and a lecturer and researcher in several academic institutions. He worked as a principal economist in the Industry and Energy Department at the World Bank and as an infrastructure economist at the Inter-American Development Bank. He has extensive experience in structural reforms, energy and water pricing, public enterprise restructuring, optimization studies, and loan operations. He directed the World Bank ESMAP assistance for the reforms of the energy sector in Poland and the power sectors of Colombia and Bolivia.

JOSE D. LOGARTA, JR. (VIKING) has worked in the Philippine energy sector since 1980. His experience ranges from power plant operations to regulation, policy, and research work in the executive and legislative branches of government. He was a technical and economic adviser to the energy and environment committee of the Philippine Senate, where he was involved in crafting power sector reforms. He works for the Energy and Clean Air Project funded by USAID.

JOSEPH ANDREW MCALLISTER is the director of operations for the San Diego regional energy office in San Diego, California. He worked for 9 years with NRECA International in the electric sectors of countries in Central and South America, Southeast Asia, and Africa, and as an energy efficiency analyst at Lawrence Berkeley National Laboratory. He has published on various energy topics in academic, trade and popular journals.

AHMED OUNALLI is a consultant on energy with the African Development Bank. His experience in the energy field started in 1974 as an energy economist with the Tunisian Electricity and Gas Utility. He worked with the Ministry of National Economy, Energy Management Agency, and as a regional advisor within the Regional Energy Development Programme at the Economic and Social Commission for Asia and Pacific. He was also a consultant with the African Development Bank, the European Economic Commission, the World Bank, and other United Nations organizations.

MICHAEL J. SHIEL joined the Irish Electricity Supply Board (ESB) in 1947 to work on the country's first rural electrification program. He was one of the first registered rural electrification consumers in the west of the country and his house was often used as a "showhouse" to persuade reluctant farmers of the benefits of electricity. He became commercial/distribution director in 1971. His book, *The Quiet Revolution: The Electrification of Rural Ireland,* draws on his personal experiences in implementing the rural electrification program.

VORAVATE TUNTIVATE is an energy specialist and survey research scientist with extensive experience designing and implementing projects focused on household energy policy, rural energy, and rural electrification. He has designed several short and long term monitoring and evaluation studies that have provided critical data in understanding the success and challenges of ongoing and completed projects and programs.

DANIEL B. WADDLE is the vice president of operations at NRECA International. He oversees projects in Latin America, Africa, and Central and South Asia that focus on expansion of electricity access through establishment and strengthening of rural electric utilities. His technical expertise includes specialization in biomass power conversion, electric power system design and analysis, and agricultural processing systems. His professional experience includes design and implementation of national rural electrification programs; geographic information system design and analysis; electric distribution utility performance benchmarking and management consulting; and solar, biomass, and hydroelectric project design for rural electrification applications.

PAUL WOLMAN is a historian, writer, and editor. He coedited the volume *Teacher Unions and Education Policy: Retrenchment or Reform?* and coauthored *No Child Left Behind? The Funding Gap in ESEA and Other Federal Education Programs.* During the 1990s, he was editor of *The ESMAP Connection,* a newsletter on energy alternatives for developing countries, and participated in the development papers and briefings on energy topics for the World Bank, including rural electrification, solar power, energy efficiency, and energy trade and regulation.

XIANGJUN YAO works in the Department of International Cooperation at the Chinese Ministry of Agriculture. She worked at the Institute of Energy and Environment Protection of the Chinese Academy of Agricultural Engineering as a rural renewable energy specialist for more than 20 years. She chaired projects on national rural renewable energy policy research, rural energy development planning at regional and national levels, and the design of large-medium biogas plants in intensive livestock farms. She also participated in international cooperative projects sponsored by the United Nations Development Programme, the World Bank, the African Development Bank, the Food and Agriculture Organization, and the European Union.

The Challenge of
Rural Electrification

CHAPTER 1

The Challenge of Rural Electrification

Douglas F. Barnes

*M*ORE THAN 1.6 BILLION PEOPLE in the world are without electricity. Most of these people are in rural areas of the developing world, where the pace of electrification remains painfully slow. Why is this so? Providing electricity to remote, rural people is often easier said than done. Well-publicized problems plaguing some programs have led to wariness about rural electrification among energy policymakers. Some highly subsidized programs, for example, have drained the resources of many state power companies, with highly damaging effects on their overall performance and quality of service. The result is widespread brown-outs and blackouts for their existing customers and a reluctance of the power companies to reach out and provide electricity service to the poor.

Rural electrification programs can undoubtedly face major obstacles (World Bank 1975, 1996). Low population densities in rural areas result in high capital and operating costs for electricity companies (Denton 1979; Fluitman 1983). Consumers are often poor and their electricity consumption low. Politicians interfere with the orderly planning and running of programs, insisting on favored constituents being connected first and preventing the disconnection of people not paying their bills. Local communities and individual farmers may cause difficulties over rights of way for the construction and maintenance of electricity lines.

Yet despite these problems, many countries have been quietly and successfully providing electricity to their rural populations. In Thailand, well over 90% of rural people have a supply. In Costa Rica, cooperatives and the government electricity utility provide electricity to more than 95% of the rural population. Again, in Tunisia, more than 95% of rural households already have a supply. Thus, there are many good examples of successful programs to counterbalance those that have experienced problems. This book focuses on the characteristics of successful rural electrification programs by examining the accomplishments and difficulties that have been overcome.

Rural Areas Still Lag Far Behind in Access to Electricity

Worldwide energy availability issues are under increasing scrutiny, and access to electricity services is a special concern. One reason for this scrutiny is the commitment of international development agencies to promote the Millennium Development Goals for the purpose of halving poverty by the year 2015 (United Nations 2003, Sachs 2005). Neither energy nor electricity is stated as a goal under the Millennium Development Goals, but electricity actually provides the foundations for most of them (Modi et al. 2006). Without access to modern energy services, it is generally agreed that the achievement of these goals would be difficult, if not impossible.

The growth in the number of people who have gained access to electricity over the past few decades has been quite remarkable. Today more than 1 billion more people have electricity compared to 25 years ago. But as impressive as this accomplishment is, population growth over the period has meant that big gaps in access to electricity remain. About 1.6 billion people—around a quarter of the world's population—lack access to electricity (International Energy Agency 2002). Moreover, under today's energy policies and investment trends in energy infrastructure, projections show that as many as 1.4 billion people will still lack access to electricity in 2030. In sub-Saharan Africa only 8% of the rural population has access to electricity, compared with 52% of the urban population. A similar disparity exists in South Asia, where only a little more than 30% of the rural population has access, compared with approximately 70% of the urban population. Indeed, four out of five people without access to electricity live in rural areas of the developing world, mainly in South Asia and sub-Saharan Africa.

Although higher income and mainly urban households now have access to modern energy, the world's poorest households do not (Table 1-1). With the exception of towns and cities in Africa, most urban areas in developing countries now provide electricity to their residents, although the reliability of this supply is sometimes intermittent. Thus, the problems of electricity access are now far greater in rural than in urban areas. Although urban population growth rates will continue to exceed those in rural areas, this actually means that the rural populations must become more productive and efficient at satisfying ever-increasing demands for food and other farm products.

In many African and South Asian countries, the rate of the number of people gaining access to electricity is even lower than rural population growth. In Africa, 9 out of 10 rural people do not have access to electricity or appliances. In South Asia, which has a large number of poor people, more than 800 million people do not have electricity. These dramatic figures have recently become a central issue in the debates over how to achieve improvement in education, reduction of diseases, and overall quality of life for rural people in developing countries.

The conclusion is that even though progress has been made, there still is a long way to go to raise the world's poorest populations above the poverty line. Without access to modern energy services—including electricity—it would be virtually impossible to meet the challenge of achieving the Millennium Development Goals or more generally to reduce poverty in the developing world. Having

Table 1-1. *Electricity Access in Developing Countries, 2005*

Country or Region	Population without electricity (millions)	Percent of population with electricity	Percent of urban population with electricity	Percent of rural population with electricity
South Asia	706	51.8	69.7	44.7
Africa south of the Sahara Desert	547	25.9	58.3	8.0
North Africa and the Middle East	48	85.8	91.5	77.5
East Asia	224	88.5	94.9	84.0
China*	8.5	99.4	100.0	98.9
Latin America	45	90.0	98.0	65.6
Developing countries	1,569	68.3	85.2	56.4

* For China, figures are for 2002.

Source: IEA 2006, WDI 2006, and author's calculation.

said this, there has been some controversy over the effect of rural electrification on development in the past (Barnes 1988), and it is still true that electricity is a necessary but not sufficient condition for development. Thus, the next section deals with the role of electricity in promoting both social and economic development.

Why Worry about Rural Electrification? A Review of Important Issues

Countries are often faced with a dilemma concerning the provision of electricity. Over the long term, the benefits of providing electricity to poor rural households can be quite high, as evidenced by the well known positive relationship between electricity consumption and gross domestic product. This correlation is mirrored by the relationship between a country's rate of electrification and the percent of households that are above the poverty line of two dollars per day (Figure 1-1). This figure illustrates that the rate of electrification is related to the percentage of a country's population that is above the poverty line. Their rates of electrification are higher than what would be expected given their level of development, but despite this relationship, the initial cost of developing the infrastructure is high and unaffordable for poor people. The benefits must be evaluated and compared to the costs involved in providing electric service. Building extensive central grid distribution systems with miles of medium- and low-voltage lines is expensive to light a few light bulbs in the rural areas that have low densities of consumers.[1]

The social and economic benefits of rural electrification have been researched over the past 30 years. One notable review was conducted in the early 1980s covering several countries (USAID 1981; Butler et al. 1980; Goddard et al. 1981;

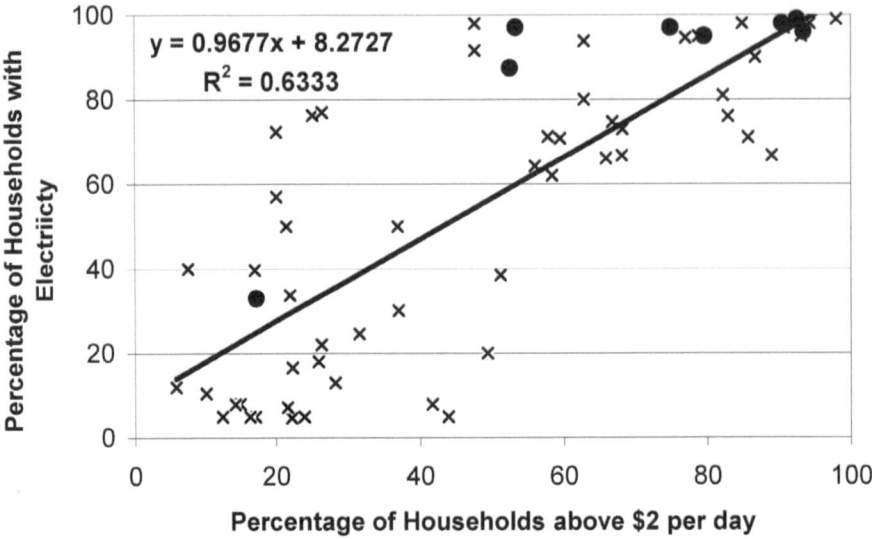

Figure 1-1. The Relationship between the Percentage of Electrification and the Poverty Rate for Developing Countries

Note: The eight developing countries in this report are indicated by solid circles. Data for rural electrification rates are for 2002, and for the % households above poverty, they range from 2000 to 2004.

Source: World Bank 2002. Development Data Group, World Development Indicators database. Tables prepared on electrification rate for business renewal strategy. Energy and Water Department, World Bank, Washington, DC.

Madigan et al. 1976; Mandel et al. 1980). Intuitively, one can easily understand that in households with electricity, people are better able to undertake activities that require higher levels of lighting, such as reading and studying (Barnes et al. 2003; Samanta and Sundaram 1983). They can also listen to the radio or watch television, and attend to more household chores (World Bank 2004a, b). In contrast, the kerosene lantern or candles in the household without electricity emit a dull light inadequate for reading or close work (Nieuwenhout et al. 1998; Van der Plas and de Graaff 1988). In such households with no electricity, the family may retire early after a fairly unproductive evening.

Such accounts may seem to overstate the actual value of rural electrification, but they are typical of the expected benefits for rural areas anticipated by both politicians and those still living in rural communities without electricity. In this section, we review the evidence of the social and economic effects of rural electrification, but not the benefits versus the costs, as this is a completely different area of research (for a review, see Webb and Pearce 1985; Barnes and Halpern 2000; World Bank 2002b). In addition, there is discussion of equity issues and subsidies for rural electrification.

Importance of Social Effect and Household Benefits

The arguments for rural electrification often have centered on the transformative effect that it can have for rural households. At the micro level, the effect of rural electrification on a household can be substantial. At the macro level, the arguments for rural electrification have revolved around the productive work that can be done in rural areas with electricity.[2]

In rural households that adopt electricity, lighting is the first choice by households as they begin to use electricity. Virtually 100% of households with electricity use it for lighting, as electricity allows activities to continue through the evenings. Cooking is not changed significantly in most households with electricity, except in Latin America, where electricity is used for cooking in some urban areas. In general, rural households prefer the traditional wood, coal, and charcoal stoves to the more expensive electric stoves or even heating plates or coils. However, there is some emerging evidence that in households with electricity, women are spending less time in fuel collection and meal preparation, even though they do not change their cooking fuels (World Bank 2002a, 2004a). The apparent reason is that with lighting in the evening, women can prepare the main meal just before it is eaten rather than preparing some dishes during the day and then reheating them at night.

Women and children are prime beneficiaries of rural electrification. Lights and appliances had a significant effect on household work in the early stages of rural electrification programs in the United States, as appliance use reduces the drudgery of household chores. A study in India found that electric appliances helped decrease the amount of toil and thus increased the time available for family and leisure activities (World Bank 2004a). At a minimum, all households used electricity for lighting. The other major uses were space cooling (fans) and watching television. This report also established that in general women from homes with electricity were better able to balance paid work, household chores, and leisure than women from homes without electricity. Similarly in Bangladesh (Barkat et al. 2002), women in households with electricity spent less time on household chores. Other studies have found that lighting alone made a dramatic difference in one's ability to do household chores at night and to read for education and leisure (Lay and Hood 1976; Khandker 1996; Gordon 1997; Filmer and Pritchett 1998; Kulkarni and Barnes 2004). However, the socioeconomic background of the household often determined the trade-offs they made with their time and the extent to which they could enjoy the advantages of electrification.

To conclude, an overriding impression from some of the recent reviews is that rural electrification has a significant social effect. For instance, the positive benefits include increased appliance use and more reading—especially for children. The findings of the effect of electricity on migration are somewhat mixed (Herrin 1979), with no conclusive evidence. Education and electrification definitely appear to be mutually reinforcing programs (Saunders et al. 1978; Velez et al. 1983; Khandker et al. 1994; World Bank 1999; Kulkarni and Barnes 2004; Arcia 2000). In this section, the social benefits of electricity have been reviewed without reference to the division of the benefits between social classes. In other words, the benefits of rural electrification may well be distributed unevenly among the rural population. The next section examines the equity of the effect of rural electrification.

Importance of the Economic Impact

Due to the importance of economics for rural life, a brief review of the economic effect of rural electrification is in order.[3] This section concentrates on the effect of electricity on agriculture and on the growth of rural businesses. However, to a great extent the economic effect of electricity depends on government policies directed toward either household or productive uses. In some countries and among some donor agencies, the overemphasis on the economic benefits of rural electrification has meant a lack of proper perspective. This emphasis does not mean to deny the importance of electricity for economic development, but policies supporting both social and economic effects seem to lead to favorable program results.

For instance, the rural electrification policy in India since the early 1960s has focused on the promotion of electric pump sets, which has had a large effect on agricultural productivity (Das 1990; Bose 1994; Barnes 1988). This effect is in part due to the fact that India has in place an aggressive agricultural development program, including the dissemination of hybrid seeds, fertilizers, and other agricultural inputs, along with a policy to subsidize electricity for water pumping. Credit programs also have helped agricultural development in India, whereas in some other countries there are no similar programs to complement rural electrification.[4]

India's effort to improve rural development through electrification has been relatively successful, but it is not unique (Barnes et al. 2003). For instance, the growth of electric pump sets in Bangladesh is much lower than those experienced in India, but it has been higher than was expected at the beginning of the program (Barkat et al. 2002). However, no similar effect has been measured in other countries. For instance, one survey in a relatively rich rice-growing region in Indonesia found that the rate of growth of pump sets was low because most irrigation was accomplished through gravity-fed sources (Brodman 1982; U.S. Census Bureau 1980). This finding is similar to those in the rice growing regions in India (World Bank 2002a; Barnes et al. 2003). Also, the price of diesel fuel in Indonesia is heavily subsidized, making it less attractive for farmers with pumps to switch to electricity. Thus, the productive effect of rural electrification can be substantial, but the effect depends on factors such as government policy and complementary programs.

Businesses in rural areas of developing countries include home businesses, small commercial shops, grain mills, sawmills, coffee and tea processing, as well as brick kilns (for a review, see Cabraal and Barnes 2006). The effect of rural electrification on small businesses is determined by the nature of the local community, the complementary programs, and the ability of rural entrepreneurs. Although electricity is an important and often essential input that helps in the development of small rural industries, the other complementary conditions include access to good rural markets and adequate credit. Perhaps because these complementary conditions are not present in all rural areas, the anticipated growth of industries in rural areas provided with electricity is somewhat slow (Zomers 2001). However, areas without electricity have an even worse record of business development. For instance, in a recent study in the Philippines, small home businesses were found to be more active in areas with electricity, contributing to family incomes (World Bank 2002b). The majority of these businesses are small general stores for food and other necessities.

Finally, an overemphasis on rural productivity can divert attention away from the household benefits. As indicated earlier, there are substantial social benefits of rural electrification, which accrue mainly in the evening hours, when small businesses and commercial establishments are not operating. Thus, the same investments can serve two complementary purposes at the same time.

The conclusion is that electrification is an important condition for the development of rural businesses and that under the right circumstances, it has resulted in significant economic growth. However, it is unrealistic to expect that it will produce an explosion of industry and commerce in a short time, especially in the absence of other development programs. Concerted effort is needed to coordinate rural electrification with other relevant programs. Without such complementary programs, the full socioeconomic effect of electrification probably will not be realized, and the required substantial capital investments may not be fully exploited. Electrification projects properly coordinated with such programs or implemented under the right regional conditions will increase productivity and improve the quality of rural life.

Distribution of Benefits and the Equity Controversy

The Achilles heel of many rural development projects is that their social and economic benefits are unequally distributed. To ensure the participation of the poor in development programs, the United States Agency for International Development (1981) many years ago adopted a policy to help the "poorest of the poor." The World Bank has as its main goal to eradicate poverty. Questions have been raised over whether the "poorest of the poor" can benefit from large, capital-intensive projects such as rural electrification. Many critics of rural electrification claim that the expensive electricity distribution systems will serve only the wealthiest families and thus reinforce existing inequities and distribution of wealth. It is well documented that the adoption of electricity by rural households is highly dependent on income level. Thus, the rich will be able to partake of all of the benefits of electrification, such as the use of modern appliances and lighting. In contrast, the rural poor may not be able to afford electricity; in fact, they may not even be permitted to have access to electricity because their houses are of substandard quality.[5] The implication is that considerable money allocated to rural electrification systems would be better spent on projects that more directly improve the lives of the rural poor.

Although equity is an important part of the work on rural electrification, many studies on rural electrification fail to deal with it in a meaningful way. Associations between income and electricity often are reported as evidence that electrification results in an improvement in rural household incomes, when the causal relationship could be the reverse. Households with higher incomes may be those that chose to have electricity. On the other hand, poor regions are proclaimed not to benefit from rural electrification without a proper examination of the long-term benefits and disadvantages of electrification. It is declared that poor households do not directly benefit from electricity use, but unfortunately no attempt is made to measure indirect benefits, such as employment creation or the effect on women and children.

The empirical evidence that does exist suggests that the direct effect of rural electrification for rural households, especially over the short term, may worsen rural inequality. The poor are not totally excluded, but in just about all countries, the poor adopt electricity at a lower rate than do more wealthy households. For instance, in 1980 in the Philippines, it was estimated that households below the poverty level could not afford electricity (Mandel et al. 1980). However, a more recent Philippines survey found that whereas households above the poverty line adopt electricity at a much higher rate, nevertheless, many poor households that are below the poverty level do adopt electricity (World Bank 2002b). In higher income countries, such as Costa Rica or Colombia, the adoption rate of electricity also is much higher, meaning that the negative equity effect is much less pronounced. For instance, in Costa Rica, which is now at the end of its program, almost all households that are within the reach of electricity lines adopt electricity, and this figure includes those households considered below the poverty line (see Chapter 2).

Thus, on the question of household equity, the news is both good and bad for rural electrification. The bad news is that in countries with extremely low incomes or poor records of income distribution, the poor will not be able to afford electricity at first. In fact, those wealthy households that can afford electricity will be able to purchase more appliances, thus potentially widening the gap between the rich and the poor. This lack of affordability by the poor can be partially addressed through the use of appropriate subsidy policies. The good news is that for those households that adopt electricity, their overall quality of life is enhanced compared to nonelectrified households, and to some extent the gap between the middle and wealthy households is narrowed. Also, women and children as a group benefit more from rural electrification than men, which is somewhat uncharacteristic of those rural development programs not directed specifically toward women.

This discussion of equity naturally leads to an assessment of subsidy issues because it generally is accepted that without some kind of initial program subsidies, it is unlikely that rural electrification will be able to reach the world's poorest populations.

Making Rural Electrification More Equitable: Important Subsidy Issues

Most rural electrification programs in the world have involved some kind of subsidy. The issue of subsidy justification generally is not addressed in most of the country studies in this book, but, the type of subsidies and means of financing rural electrification is covered in great detail within the chapters. However, it is necessary to touch on subsidy justification in this chapter because of its importance to encouraging success or sometimes dictating the failure of some programs.

Despite laudable objectives in many poorly managed programs, subsidies have often failed to meet their stated objectives of making services more affordable to the poorest families or households. All too often, subsidies have become the grist of politics and have been provided to those already with access to modern services. It is no coincidence that in developing countries the populations with access to energy services are the middle- and upper-income households. Even well-intended subsidy programs can have problems (Barnes and Halpern 2000; Komives et al. 2005). Subsidies have often been implicit, such as default or non-payment of electricity bills. They also have been untargeted, such as a subsidy for energy used by all. Another characteristic is that they have been indiscriminate, such as a subsidy for a quantity that is well above that needed by poor or rural populations. Finally, most subsidies become *complex,* or difficult to administer to targeted groups, and overly restrictive with respect to end use or technology, depriving users of choice.

The effective programs in this study can generally be considered as being based on good subsidy policies. Of course, some of the countries have performed better than others, but generally all have achieved a measure of success in relation to subsidy policies. What is the reason or justification for this statement? According to most subsidy theory, several criteria need to be reviewed to evaluate whether a subsidy is justified or not. We can call them the three Es—efficiency, equity, and effectiveness (World Bank 2002b).

Efficiency refers to maximizing the social (or economic) benefits under the assumption that even the best energy project has an opportunity cost. That is, is this an efficient investment for society? For this question, one must calculate an economic rate of return. For most projects, the anticipated rate of return is positive. Projections are developed, costs are estimated, and the benefits to users are calculated. We do not quantify the economic rate of return in the case studies involved in this study because we are examining only workings of different rural electrification programs, and not projects. However, most of the projects within the countries have had to pass a rate of return test to determine if the investment is good for society in general.

Equity refers to the efficacy of the subsidy. In other words, do the subsidies actually reach poor people who do not have electric service? Rural electrification is a process of providing new connections to households that have never had electricity. In the early stages of the program, the project generally does not reach the poorest households in society. However, this fact means that if a country already has 70% of households with electricity, then a rural electrification program is well targeted to reach the poor because it is providing access to the poorest 30% of society. It would be inequitable to leave those remote areas with no electricity and without access to the benefits enjoyed by the rest of society.

Effectiveness refers to the fact that justified subsidies have to be in a program that works because otherwise they are by definition poorly targeted. Nothing could be worse than pouring subsidies into a program that does not work properly. Many rural electrification programs have suffered problems. Sometimes distribution companies would build the electricity lines and then would find one excuse after another to keep consumers from using electricity. India is an example of a country that has invested heavily in rural electrification, but in some states the program has not created the proper incentives for the electricity distribution companies—called state electricity boards—to serve a high percentage of rural households. There are many reasons for not providing service to rural consumers even after lines are built, but the main one is that electricity prices for consumers in rural areas often are set low, and companies actually have a disincentive to serve them.

The main emphasis of this study is on identifying the characteristics of effective rural electrification programs and insights that can be gained from the types of subsidies used in such programs. The case studies were selected based on the criterion that distribution companies within countries had reached a high percentage of their rural populations and provided high-quality service to consumers. The rationale for examining these best practices of rural electrification was that the problem programs garnered most of the attention of development practitioners and the best programs gained little fanfare.

So how did it happen that rural electrification was subsidized without running into significant political problems? The country studies make it clear that there would not have been any significant rural electrification without a political decision to take some kind of initiative to make electric supply more inclusive. Similar arguments have been put forward by political theorists who are developing democracy-enhancing approaches to political and social equality (Jechoutek 2005). The important point here is that it may be mainly the enabling environment that counts. The ability to make choices freely depends on the capacity to exercise equal rights as equally respected citizens. Having theoretical access to electricity may be of little use to villagers under the thumb of local elites who will keep them excluded from being able to improve their lives. States without a determination to mitigate overall inequality probably will not be successful in establishing equal access for rural electrification either. Thus, successful programs are more likely to emerge in countries with a longer and more complex view of development.

What Are the Challenges?

Expanding the coverage of electricity service and improving its quality poses formidable challenges. Some challenges are unique, but many are inherent to the rural environment. These are challenges with which industrial as well as developing countries grapple, and they must be addressed by any national rural electrification strategy or program.

Rural areas are characterized by low population density with a significant number of households that are poor. This density results in low levels of household demand for electricity, which generally is concentrated at evening peak times. The low population densities mean that electricity distribution costs must be spread over relatively few people, resulting in high costs for each unit of electricity consumed. Demand normally matures slowly (over two to three years and even longer) as consumers wire their houses, invest in appliances, and make the switch from other fuels for lighting and cooking. As the demand grows, the cost per customer for rural electrification declines. Unfortunately, this progression is difficult to predict, making returns to investment in grid extension to poor rural people uncertain.

Thus, grid expansion costs are typically high in rural areas because loads to be served tend to be small and widely dispersed. The cost of rural electrification can, however, be minimized if design standards are modified appropriately, and the choice of technology is based on both financial and potential socioeconomic benefits to a community or region.

Operating and maintaining systems in rural settings poses additional difficulties. For large centralized utilities, retaining and supervising a cadre of technical staff is more costly and problematic in a rural setting. Larger distances make supervision difficult and expensive, resulting in low-quality maintenance, high levels of corruption, and high rates of absenteeism.

Most rural electrification programs involve some form of subsidy to encourage rural consumers to adopt electricity. This subsidy has caused two types of problems. The first is that because governments are providing subsidies to rural electrification, politicians feel that they have a right to intervene in the operation of the distribution company to get electricity to their constituents. After connection, they also intervene on behalf of their constituents to restore service that has been cut off due to lack of bill payment. This interference often makes the cost per consumer even higher and causes financial stress for the company providing the service. The second problem is that subsidies that are poorly designed can lead the distribution company away from a primary emphasis on serving consumers. Instead, they may maximize the amount of subsidy they can extract from the government with rural service as a secondary goal. Once such a consumer orientation is lost, the quality of service is sure to suffer.

Main power companies often have institutional difficulty meeting special demands of rural distribution (Zomers 2001). For integrated power companies, the rural consumer makes up such a small part of their business that the power companies often do not pay attention to the numerous possible ways to minimize costs of service to them. The result is that rural electrification becomes a tolerated loss-maker for the company, and ways are found to cut corners in terms of customer service. For instance, rural consumers more often than not are the first to be cut off when there are problems with power supply in developing countries.

Local community-level problems often provide an obstacle to rural electrification. For instance, the poles and lines cut across the rural countryside, and sometimes local elites object to having lines on their property or to the compensation methods that have been developed to pay for the rights of way. Thus, ways have to be developed to involve communities in the process of rural electrification.

Thus, the way to successful rural electrification is paved with problems to be solved. The chapters in this book illustrate how each country has devised solutions to these problems. Some countries have been more successful than others in meeting the challenges to rural electrification.

The Approach of This Study

The main goals of this study are to illustrate how a variety of countries have successfully addressed the problems inherent in having successful rural electrification and to draw lessons from these programs for countries that are just now beginning to tackle the challenge. By examining the many ways in which programs have succeeded, other countries at the beginning or in the middle of this long journey can benefit from this body of experience. These diverse experiences should make it possible for others to follow in their footsteps. Thus, in the conclusion of this study, we will describe the practices that should be emulated and those that should be avoided.

This study is the most complete compilation of rural electrification case studies ever put in one volume, and it is hoped that this body of work can provide both encouragement and guidance to those countries sincerely interested in and committed to providing electricity to their rural populations. Creating a viable market for rural services that includes appropriate and adequate incentives for private investors is a complex process. At the heart of the challenge is the fact that large investment capital is required, combined with significant benefits for countries but poor financial returns for electricity distribution companies. This challenge is even more daunting in that the achievement of the Millennium Development Goals is undoubtedly impossible without adequate availability of rural infrastructure.

Research Methods and Issues Examined in Country Studies

The method chosen to uncover and evaluate the best practices was to examine a particular set of issues for countries with successful programs. This method not only provided a description of how the programs achieved their goals within countries but also provided a way to compare across the countries. Thus, all case studies were conducted based on a similar set of issues or questions.

The authors of each individual country study conducted field visits, collected data, and conducted interviews with officials who are influential in rural electrification policy decisions. Each case study provides a history of the program, including the important decisions that were made during the course of rural electrification, as well as identifying and explaining all relevant issues affecting the rural electrification program for each country.

One significant issue involves how countries make it attractive for poor rural consumers to connect to the grid. Therefore, the studies include the policies for financing the initial connection charges to increase consumer access to electricity. It is also important to understand the pricing of electricity and how the distribution companies cover the high costs involved in rural electrification programs. The studies were expected to elaborate on any and all charges in monthly bills such as monthly fees for meter rent, along with the usual charges for electricity itself. On the investment side, there have been several ways to finance rural electrification to make electricity service more affordable by rural people. As a consequence, the source of the subsidies, including subsidized loans, cross-subsidies, and others have been examined in the chapters.

As indicated, many rural electrification projects have problems because of over-expansion to regions with little electricity demand. The reasons for this can be pressures to serve political constituents or just poor planning. Thus, it was important for each case study to examine the regional or village level of "objectivity" in fixing priorities for investments in rural electrification projects. This prioritization might be characterized by the degree of autonomy that the utility has in order to make appropriate decisions on distribution planning. The chapters provide details on the ways in which criteria were used for establishing priorities and goals, including how decisions are made on data collection and investment planning. Finally, another issue involves understanding the degree of coordination of the rural electrification program with other rural development or infrastructure initiatives.

The country studies in this report also examined how companies dealt with the typically lower levels of electricity demand that are found in rural areas. Attention was paid to how customers were billed and how the distribution company related to them. For instance, are there any programs to inform consumers about their service, about ways to use electricity efficiently, and about ways to use electricity productively? Many of the well-run programs have active monitoring and evaluation of their customers' use of electricity.

The regulatory framework is also important for rural electrification (Reiche et al. 2006; Brown et al. 2006). The framework usually includes issues such as how rural distribution companies purchase power from the main grid company and, of course, how prices are set so that the companies can remain financially viable.

Thus, the method used to determine the best practices in rural electrification are both historical and qualitative in scope. The reviews stand on their own, but they also provide valuable lessons for how to accomplish rural electrification in a world with more than 1.6 billion people without electricity.

Successful Approaches in the Country Studies

This book focuses on rural electrification programs that have been successful. The main criteria for selecting countries for this study was that the countries demonstrated a significant level of growth of rural electrification during the past 20 years in ways that are financially sustainable. The eight developing countries that were selected for the study are listed in Figure 1-2, and the time frame is for 1980 and 2000. Because they are so advanced, the United States and Ireland are not included in this figure. The lines in this figure represent the level of real per-person income as represented by gross domestic product adjusted for inflation and purchasing power[6] and the percentage of people in rural areas with electricity from 1980 and 2000. As can be seen in the figure, all countries had real economic growth during the period, and this growth correlates with growing access to electricity by people in the rural areas. The countries with the highest levels of growth and high rates of rural electrification are Tunisia and China. The countries started at quite different levels of electrification, and this difference contributes to the diversity of experience that can be explored through the case studies. For instance, Bangladesh—a poor country—started at close to 10% of its population with electricity in 1980 and progressed to a little less than 30%. Both Thailand

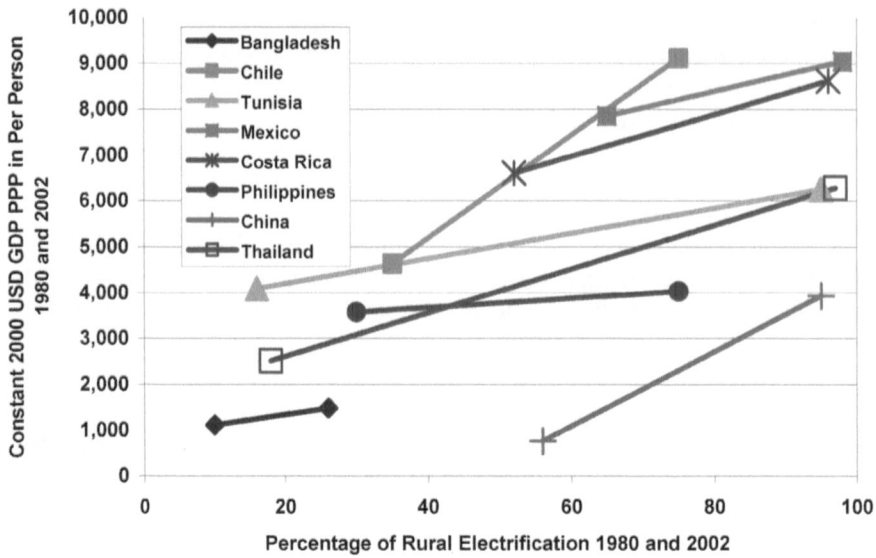

Figure 1-2. *Country Case Study Levels of Rural Electrification Development, 1980 and 2000*

Note: Due to currency fluctuation, the data for the Philippines is from 1985 and 2000.

Source: World Bank statistics and data tables 2006.

and Tunisia had low levels of economic development in 1980 and progressed to quite high levels by the year 2000. They managed to provide electricity for almost all of their rural populations without electricity, even though they started at close to 20% in 1980. Some countries, such as Mexico, Costa Rica, and Chile, started at higher levels of development and made significant progress during the 20-year time frame.

The countries with successful rural electrification programs do not seem to follow one institutional model, indicating that rural electrification does not seem to be constrained by different ways of providing electricity to rural people. In this study, we have divided the case studies into three categories. The first category involves a model of rural electrification exemplified by the rural electric coopera- tive model that is derived from the experience of the United States (Ross 1972). Examples of developing countries following this model include Costa Rica, the Philippines, and Bangladesh. There is also is a chapter on how the rural electric cooperative model was developed in the United States. The cooperative system of rural electrification has a great deal of merit. The people being served by the cooperative are the owners of the distribution company. However, the supervis- ing agencies overseeing such cooperatives had the foresight to recognize that such companies must be run according to business principles.

A common path to rural electrification is through public companies. The chap- ters on successful public companies, the second category, include Thailand, Mexico, and Tunisia. Ireland, a developed country, has a model public rural electrification program. Recently, many development specialists have tended to ignore the success of well-run public companies by insisting that companies must become private to be run efficiently. It is true that privatization of public companies is a worthy goal because many public companies are inefficient and driven by political agendas (van der Fehr and Millan 2003). However, many public-sector electricity companies were created because the private companies that they replaced were in the business of making profits rather than extending service to unprofitable markets made up mainly of the rural poor. So the case studies illustrate that there has been an interesting swing from the private sector serving mainly urban areas, to public companies overextending themselves, and now back to the private sector again. One of the most interesting conclusions of this study is that the institutional form is not as important as the adherence to strict business principles in operating rural electricity distribution companies.

In the third category, there are two chapters that involve private or decentralized electrification companies. Chile has had private-sector and cooperative electricity distribution companies for more than 20 years, and they have a unique subsidy program to encourage these companies to serve people in rural areas. Although definitely not classified as a private program, China's electricity distribution program developed in a decentralized manner. Local companies, mostly at the county level, were supported and nurtured by the central government to become full-fledged and often independent companies. These two chapters illustrate clearly that even with privatization or decentralization, most rural electrification programs involve some form of subsidies, and without them progress is likely to be limited. Even for private companies, a great deal of attention is necessary to get the right balance of providing subsidies to encourage expansion, while not creating a dependence on them. Most of the successful countries that make up the main chapters of this book also started off at low levels of rural electrification and had the patience and commitment to see the program through many difficult periods.

The question rightfully can be asked whether programs that have not been successful could also hold lessons for rural electrification programs. There are plenty of examples of programs with problems or that have not yet been developed. India is a good case of a program that has had significant problems because of an unsuccessful national development model of central public planning that was applied to rural electrification through large public state utilities. The financial commitment was and still is substantial in India, but the program does not satisfy most of the principles derived from comparison of the often innovative and well-managed programs detailed in this research. This may change with the current gradual dismantling of the "license raj" in India that began in the early 1990s and the resulting improvement in development pace has the possibility of translating into a better program for rural electrification. Many African countries provide similar examples, where a failed overall national development model shows equally disappointing results for promoting rural electrification. Currently, the development of infrastructure in Africa is an international issue that will not go away, and many of the lessons learned from the case studies are applicable to these countries with low levels of rural electrification. For this book, it was felt that the main lessons are in successful programs and that such lessons can then be applied to countries either with problems or just starting down the path of developing their rural infrastructure.

The Importance of Rural Electrification

Well-planned, carefully targeted, and effectively implemented rural electrification programs provide enormous benefits to rural people. Indeed, once an area has reached a certain level of development, further progress in raising standards of living to socially and politically acceptable levels will depend on the availability of electricity. As restructuring of national power utilities gathers momentum around the developing world, it is essential that this idea is borne in mind and that the appropriate institutional frameworks and incentives are created to ensure that rural electrification takes place.

The fact that rural electrification programs have been implemented relatively smoothly and efficiently and have enhanced access to energy for a significant number of people in rural areas is a purely instrumental fact that should lead us to the bigger questions. Rural electrification, and for that matter also rural development programs as a whole, are one part of a major process of social, economic, political, and cultural transformation. In Amartya Sen's terms (2000), successful development can be observed when marginalized individuals and groups gain the freedom to make choices in all aspects of life and are able to improve their capabilities and "functioning" to exercise any freedoms obtained. Both individual freedom and social equity have to be advanced to arrive at a "successful" transformation and modernization of society.

Thus, successful rural electrification cannot be divorced from the country context. The recent work on rural development in Peru indicates that the combination of infrastructure services is much more effective than single interventions (World Bank 1999b), and yet most investments in this field involve single interventions. In addition, the evidence from this study indicates that the method of rural electrification has to fit in with national ways of achieving development. Cooperatives may be the answer in one country, whereas national public grid extension is the solution in others. Whatever the solution, the freedoms made possible by rural electrification and other infrastructure are often underestimated because of the concentration on single intervention models that lack coordination with other development programs.

The overall main message from this study is positive. There are major opportunities for increasing the pace and widening the scope of rural electrification. If these opportunities are grasped, it will enable large numbers of new consumers to enjoy the benefits of an electricity supply at acceptable costs and will avoid burdening national governments and power utilities with unsustainable subsidies. Although there is no "one way" to accomplish rural electrification, there is an underlying set of principles that need to be followed to have successful programs. These principles are exemplified in the following chapters.

Notes

1. Costs could be reduced to about US$400–500 per consumer by using solar home systems instead of grid electrification in areas that are remote from the grid and where the main use of electricity is lighting.

2. An interesting quote on the anticipated effect of rural electrification in the United States during the 1920s, a time when there was little rural electrification, anticipates some of the effects in developing countries today: "The effect of cheap (central grid) power . . . is likely to add somewhat to the movement of factories away from the great cities and to change the map of our industrial power-using centers by adding to the number of small- and medium-sized manufacturing cities and towns. . . . the massing of population in urban centers, which is partly a result of the urban locations of factories, will also be affected by cheap power. Central grid power will operate almost certainly as a force making for decentralization. With the factories as they move to the country will naturally move also the workers who formerly clustered at the mill gates in the city, and with this spreading out of population will disappear some of the ills that come from having too many families per acre of ground" (Bradford 1925).

3. For a review of rural electrification and economic development, a report on the Philippines has examined the effect of household electricity use on small enterprises (World Bank 2002b). Some work in Bangladesh also has examined the role of electricity in both agriculture and small industries (Barkat et al. 2002).

4. In India during the 1960s, a series of droughts caused a significant reduction in agricultural production, resulting in a significant shortage of food supplies, reaching almost crisis proportions. The response of the government was the introduction of intensive agriculture by means of irrigation, so that land would become more productive and could be farmed for more than one season. As part of the promotion of intensive agriculture, during the 1970s the use of electric pump sets was an important part of the strategy, along with other forms of irrigation and inputs. The growth in agricultural production during this time came to be known as the "green revolution."

5. This fact is actually more of a problem with the rural electrification program because there are practical ways to provide electricity to households in homes that are built of substandard materials.

6. The use of purchasing power parity (PPP) is commonly used as an adjustment to exchange rates to reflect the real purchasing of consumers in a country. Thus, for gross domestic product, PPP is used in Figure 1-2 as a proxy for income that has been adjusted for both inflation and the local purchasing power of people in the country.

CHAPTER 2

The Cooperative Experience in Costa Rica

Gerald Foley

C OSTA RICA IS ONE OF rural electrification's unique success stories. By 1995, the country had succeeded in bringing a reliable and sustainable electricity supply to 93% of its total population, and today the figure is over 97% (World Bank 2006). In urban areas, coverage has achieved 100%, and rural coverage has increased to 87%. Over the previous 40 years, electricity coverage had increased 63%, at a time when population growth and social changes had resulted in a more than fourfold increase in the number of households.

The important factors that enabled Costa Rica's rural electrification program to develop and flourish were not merely technical. A variety of social, political, and economic factors created a particularly favorable environment within which to launch the country's rural electrification efforts. Early and full coverage of urban areas provided a secure technical and financial foundation from which to extend the benefits of electrification into rural areas. By the mid-1960s, rural electrification had become an important social and developmental issue forcing its way onto the political agenda.

Within the context of a government genuinely committed and willing to invest in rural development, combined with the support of an experienced and effective supply utility and a strong egalitarian tradition, Costa Rica's rural electric cooperatives were established and have thrived. Today, the country is in the enviable position of being able to choose from among a number of well-working organizational models with which to complete and continue expanding its rural electrification service. Although the exact conditions that led to Costa Rica's success may or may not be replicable elsewhere, valuable lessons from its experiences in success can be applied in many developing countries seeking to electrify their rural areas.

Background on Costa Rica

A variety of social, political, and economic factors created an environment that made Costa Rica particularly suitable for rural electrification at the time of the country's initial efforts in the 1960s and 1970s. Many of these factors, indeed, can be traced back to the early colonial experience and subsequent developments that resulted in a stable, relatively prosperous country of well-educated, small-property owners.

Costa Rica was part of the Spanish colonial empire for almost 300 years. The lifestyle of the majority of its colonists, unlike those in other Latin American colonies, was relatively modest. Most of the colonists cultivated their own small farms (*fincas*), and there were few of the large ranches commonly found in other Latin American colonies. Today, Costa Rica remains a country of independent-minded property owners. Many rural residents own their farms and houses, the quality of which is high in comparison to that of much of the developing world.

Development of Lucrative Exports

In the decades following Costa Rica's independence from Spain in 1821, coffee, grown in the valleys and on the mountain slopes of the Meseta Central, became the country's main crop, much of it exported to Europe. With the rise in the coffee trade came an influx of Europeans, many with progressive educational and political ideas that were readily adopted.

In the 1870s, a decision was made to build a railway from San José to the Atlantic port of Limón to facilitate the export of coffee. The task proved more formidable than expected, taking 19 years to build and costing the lives of many construction workers. When the railway company turned to the government for financial assistance, a deal was negotiated whereby the government gave the company 800,000 hectares along the railway lines to Limón for the cultivation of bananas for export to defray railway construction costs.

The banana trade expanded rapidly, and, by the 1890s, Costa Rica was the world's leading exporter of the crop. In 1899, members of the railway company board founded the United Fruit Company, which went on to play a significant role in the social, economic, and political life not only of Costa Rica but also of Central America and the Caribbean as a whole.

Political Stability

One of the most notable aspects of Costa Rica's history since its independence has been the relative absence of political violence, with the exception of the 1948 civil war that followed a disputed presidential election. The constitution of 1949, drawn up by the interim government after that war ended, remains in force.

The 1949 constitution included a variety of reforms and civil rights guarantees, including the establishment of an independent electoral tribunal charged with overseeing free and independent elections every four years. The interim government also took the radical step of abolishing the country's army to demonstrate its neutrality and to free resources for investment in health, education, and rural development.

Costa Rica's system of government aims to provide a variety of checks against corruption and the build-up of vested political interests. Similar to the governments of France and the United States, Costa Rica's system is separated into executive, legislative, and judiciary branches. The president is limited to one four-year term of office. The two main political parties tend to alternate power from one election to the next; however, their policies are broadly similar, meaning that the country is politically stable. This stability, in turn, ensures the uninterrupted continuation of long-term development programs, such as building the national electricity supply system and rural electrification.

Geography

Costa Rica is a small country located in the isthmus of Central America between the Pacific and Atlantic oceans. Despite its tiny size—its total land area is only 51,000 km²—the country boasts a wealth of topographic, ecological, and climatic diversity. A chain of mountain ranges forms the geological spine of the country, running from the northwestern border with Nicaragua to the southeastern border with Panama. Numerous rivers and streams flowing from these mountains to the Atlantic and Pacific coasts provide a wealth of opportunities for small-scale hydropower development. Two intermontane valleys found in the geographic center of the country are referred to as the Meseta Central. Though these valleys cover only about 3,200 km², or 6% of Costa Rica's total land area, they make up the country's most fertile, densely populated, and economically important area. Rainfall is plentiful, averaging about 2,400 mm annually.

In Costa Rica, wide, low-lying plains that extend inland in the northern part of the country border the Caribbean coast. Tropical jungles, swamplands, and extensive mangrove forests cover much of this area. On the Pacific side, the land slopes steeply upward from a much narrower coastal plain to the central mountain ranges. The large, relatively flat Guanacaste Peninsula lies northward, and the smaller Osa Peninsula lies to the south.

Population Distribution and Infrastructure

Costa Rica's population of about 3.3 million is divided almost equally between rural and urban areas. With an average family size of 4.2, the total number of households is about 800,000. The annual population growth rate in 1990 was estimated at 2.25%. The Meseta Central, which accounts for about 70% of Costa Rica's total population, is home to San José, the country's capital and by far its largest city, with a population approaching 1 million. The bulk of the remaining population is scattered along the Pacific coastal plain and the lower mountain slopes. The Atlantic coast, by contrast, is sparsely populated. The province of Limón, which covers the whole of the Atlantic plain, for example, has a total population of only 260,000.

Table 2-1. *Electricity Coverage by Province in Costa Rica, January 1995*

Province	Total Population	No. of Residences	% Rural	% Rural Electrified
San José	1,198,283	281,435	20	97.5
Alajuela	589,059	134,365	54	90.6
Cartago	371,091	80,822	44	98.7
Heredia	264,740	66,588	36	96.9
Guanacaste	261,611	48,989	73	82.4
Puntarenas	368,208	64,948	77	79.7
Limón	248,218	59,432	69	89.2

Source: Ministerio de Industria, Energía, y Minas 1986.

Costa Rica has a well-developed infrastructure of roads, water supply, and communications networks. The country's national and regional roads are paved and maintained to a generally high standard. Ninety-two percent of households have a piped water supply. Access to public telephone service is 92%; television coverage is nationwide, with 30 cable stations; and there are about 130 radio stations. Health and education are national priorities, as reflected in statistics comparable to those of wealthier industrialized countries. Infant mortality, for example, is 13 per 1,000 live births; life expectancy is 72 years for men and 77.5 years for women; and the total literacy rate is 93% for both men and women. Today more than 90% of the rural population has electricity (Table 2-1). Both rural cooperatives and the national public utility serve rural consumers.

The country's seven provinces are divided into 82 cantons and 436 districts. Larger towns have municipal councils that are elected every four years at the time of national elections. Costa Rica is a prosperous nation by the standards of developing countries. During the 1990s, gross domestic product (GDP) per capita rose from $1,809 in 1991 to more than $4,000 today, positioning the country toward the upper end of the World Bank table of middle-income countries.

Evolution of a Stable Climate for Rural Electrification

On August 9, 1884, in the presence of Costa Rica's president, members of government, and a large crowd, a new electric lighting system consisting of 25 carbon-arc lamps was switched on in the capital city of San José to great excitement and jubilation. This event meant that San José was the third city in the world, after New York and London, to receive public electric lighting. The contract for the supply was signed between the San José municipality and a private company, Compañía Eléctrica de Costa Rica, with power provided by a small hydroelectric plant constructed close to the city's center.

Despite early technical and financial problems, the public lighting system was quickly expanded, and a new five-year contract was signed by the municipality and Compañía Eléctrica in 1887. Under this contract, lighting was provided from nightfall until dawn, except on moonlit nights. Other towns in the Meseta Central, including Cartago and Heredia, were quick to follow San José in obtaining their own public lighting systems.

Improvement and expansion of these services followed rapidly as the Costa Rican supply companies adopted the technical improvements being made in the United States and Europe. Industry began to discover multiple ways of using electric power, and businesses and wealthier households began to rely increasingly on electric lighting. At this stage, all power was generated by small hydroelectric systems.

Safeguarding the Spirit of Public Service

Recognizing the importance of these new developments and concerned that supply companies might monopolize access to the water supply to the public's detriment, the Costa Rican government passed a law in 1910. It declared that the right to exploit energy obtainable from the country's rivers rested inalienably with the state. Any concession for their exploitation would be subject to permission by the state and would be for a limited time under conditions the state determined.

These safeguards, however, did nothing to slow the pace of development. A variety of Costa Rican and foreign companies were granted concessions to exploit hydroelectric resources. Because the companies offered few guarantees of quality and there was no control over the tariffs they charged, consumer protection became an issue.

Establishment of a Regulatory Framework

As a result of rising public concern and congressional discussion, Costa Rica established the regulatory agency, Servicio Nacional de Electricidad (SNE), in July 1928. SNE was empowered to develop electrical power, grant others concessions to do so, and set the tariffs at which electricity could be sold to the public.

Despite the creation of SNE, public concern over performance of the private power utilities mounted. By the 1940s, the Compañía Nacional de Fuerza y Luz (CNFL), owned mainly by the American and Foreign Power Company, Inc., had become the country's largest electricity supplier. There was also considerable public dissatisfaction with the performance of supply companies in Cartago, Puntarenas, Turrialba, and Limón.

Reform and Build-Up of the National Supply System

After the civil war, the interim government of José Figueres passed a law in 1949 establishing the Instituto Costariccense de Electricidad (ICE) as the national electric power utility. The law did not elaborate on ICE's mandate, but it was extremely firm with regard to the institute's political independence. ICE was to have a board of seven directors nominated by the president. Perhaps even more important, their tenure was eight years, and appointments could only be made in the first and third years of an administration. Thus, no government could pack the board with its own nominees, thereby acquiring political power over the institute.

One of ICE's principal initial tasks was to develop a national electricity supply strategy. In 1952, in response to the rapidly deteriorating supply position in Cartago and a number of other towns, ICE took control of their supply companies with a compensation of ¢7.5 million. That same year, ICE began construction of the 30-MW La Garita hydroelectric station, the country's first major power station, which was opened in 1958. Meanwhile, the 12-megawatt La Colima diesel plant was built to supply the rapidly rising electricity demand in the Meseta Central. Subsequent developments were rapid. Construction of the Río Macho (174 MW) and Cachí (111 MW) hydroelectric stations began in the 1950s, and they were completed in 1963 and 1966, respectively.

In 1962, ICE took control of 95% of the shareholding in CNFL, which was then owned by the U.S. Electric Bond and Share Company. Again, the reason was the poor quality of the supply and the rising dissatisfaction of both connected customers and those who could not get a supply. That same year, ICE acquired the CNFL telephone system, and, in 1968, it took full control of CNFL.

Development of the National Grid

In addition to building up the generating capacity of the national supply system, ICE also developed the national grid, referred to as the Sistema Nacional Interconnectado (SNI). SNI's operating voltages are 230 kV and 132 kV. There is a 230-kilovolt backbone line running along the length of the country, mainly on the Pacific side but looping inland through the Meseta Central. This line has interconnections to Nicaragua and Honduras to the north and Panama to the south. The total lengths of the 230-kV and 132-kV lines are 667 km and 670 km, respectively. Spur lines at 132 kV run to Limón and the Guanacaste Peninsula.

Initiating Rural Electrification: The Search for Financial Assistance

Whereas ICE was predominantly concerned with building up the country's supply system and developing the national grid, distribution remained the concern of the municipal and private electricity companies. Public pressure for better performance by the private companies and for a much faster pace of rural electrification led a responsive Costa Rican government to seek ways to initiate rural electrification programs.

ICE lacked financial resources to assist such programs, and the World Bank and Inter-American Development Bank (IDB) expressed interest only in promoting urban electrification (Goddard et al. 1981). However, the United States, through its U.S. Agency for International Development (USAID), proved more receptive. It particularly favored the rural electric cooperative as a model for assistance because of the cooperative's successful history in the United States, where the concept originated, and because Costa Rica had a history of establishing successful agricultural and other cooperatives. In addition, the Costa Rican government was highly supportive, and a legal framework had been established for initiating and running cooperatives. Thus, Costa Rica was a natural choice for U.S.-supported rural electrification.

Overview of the Power Sector

Electricity generation in Costa Rica has traditionally been by hydropower, which has always been readily available. In fact, the country is reputed to have the world's highest hydropower potential per unit of land area (Ministerio de Industria, Energía, y Minas 1986). Total theoretical capacity is estimated at 25,000 MW, and some 75 sites with an economic potential of about 9,000 MW have been identified. The country's high rainfall, long rainy season, and steep-sided mountain valleys through which its numerous rivers flow keep the financial and environmental costs of hydroelectric development relatively low.

ICE, the country's national electric power utility, is widely regarded as an effective modern utility. At the end of 1995, ICE's total capacity was 1,075 MW, of which 729 MW were hydroelectric, 285 MW were thermal, and 60 MW were geothermal. There was one isolated diesel power plant of 1.3 MW.

The Arenal-Corobicí hydroelectric complex, a principal component of the ICE system located in the country's northwestern region, came on stream in 1982. The complex has a capacity of 331 MW and provides most of Costa Rica's hydropower during the dry season.

Other major plants in the ICE system include the 144-MW gas turbine plant located on the Atlantic coast near Limón, the 38-MW San Antonio gas turbine plant, the 30-MW Sandillal hydroelectric plant, and a number of smaller thermal and hydroelectric plants.

Electricity Distribution

ICE is also Costa Rica's largest distribution company. In 1995, it supplied 351,000 consumers directly, and its wholly owned subsidiary, CNFL, which supplies San José, had 327,000 customers. Together, ICE and CNFL serve 80% of the country's 851,000 consumers.

Two municipal companies—Junta Administrativa de Servicios Eléctricos de Cartago (JASEC) and Empresa de Servicios Públicos de Heredia (ESPH)—serve Cartago and Heredia, respectively. In 1995, JASEC had a total of 50,000 consumers and Heredia had 36,000. JASEC's hydroelectric capacity was 24 MW, and ESPH's was 23 MW; both companies received the remainder of their supply from ICE.

Three of the country's four cooperatives serve the Meseta Central, and one serves the Guanacaste area. None has its own generating capacity, though one buys electricity from a privately run hydroelectric plant of 3.2 MW in its supply area. Otherwise, the cooperatives obtain all of their bulk electricity supplies from ICE. The total number of consumers served by the cooperatives in 1995 was 87,000, slightly more than 10% of the national total, or 20% of those living in the rural areas.

Construction and installation standards used in the distribution system are based mainly on U.S. codes of practice. Connections to domestic consumers are single-phase 120 or 240 volts; those for light industrial use are 3-phase 240 volts. The supply frequency is 60 cycles per second.

Electricity Consumption and a Changing Economy

Since the mid-1960s, electricity consumption has grown uninterruptedly, though at different rates. From 1970 to 1978, a time when the country's economic growth averaged 6–7% annually, total electricity consumption grew at an average annual rate of 9%, reaching a little more than 2,000 GWh in 1980.

Falling coffee prices, rising import bills, and increased interest charges on foreign loans were, however, beginning to weaken the country's economic position. The regional political situation, particularly the war in Nicaragua, added considerably to Costa Rica's economic difficulties, and a major economic crisis occurred in the early 1980s. The gross national product fell by 4.6% in 1981 and a further 9% in 1982, with inflation reaching 90%. The average GDP per capita fell from $2,030 in 1980 to $950 in 1983; foreign debt service accounted for more than half of export earnings, and industry was working at about half capacity.

In response, Costa Rica's government implemented a number of corrective measures, including a freeing of exchange rates and a reduction in subsidies. A standby credit arrangement was made with the International Monetary Fund, and considerable external assistance, mainly by the United States, was provided. As a result, the economy was stabilized in 1983. By the late 1980s, growth had resumed at about 5% per year, and in 1995, electricity consumption was 4,340 GWh.

Surprisingly, from 1965 to 1995, average monthly household consumption of electricity fell by 8% at the national level and by 18% in San José. One likely factor contributing to this decrease is that early and full electrification in the relatively prosperous San José brought about a high level of saturation of appliances such as cookers, water heaters, and refrigerators, and, by the early 1970s, there was little room for growth in their ownership. Improvements in appliance efficiency and the effects of an intensive energy conservation campaign by ICE would therefore have tended to reduce average household consumption. The reported growth in the use of liquefied petroleum gas (LPG) for cooking is also likely to have been a significant factor.

In other provinces, household consumption, starting from a much lower base in 1965, grew gradually over the same 30-year period. In Alajuela, Cartago, and Heredia, where the bulk of consumers are located in the Meseta Central, 1995 consumption levels approached that of San José. In the more remote, poorer provinces of Guanacaste, Puntarenas, and Limón, however, household consumption remained at considerably lower levels.

Recent Patterns of Domestic Energy Use

All of Costa Rica's urban households are electrified. Electricity is universally used for lighting and ironing, and most households have a television. Most middle-income families have a refrigerator, electric water heater, fan, and other electric appliances, but domestic air-conditioning is virtually unknown. Electricity is also the major urban cooking fuel. Poor families and people living alone tend to use a single hot plate for cooking and heating water, but four-plate cookers with an oven are used by middle- and upper-income families. LPG cookers are also common.

Rural families with an electricity supply use it for lighting, television, ironing, refrigeration, and other appliances. It is also commonly used for cooking in areas where fuelwood is scarce. In addition to domestic applications, electricity is widely used for pumping water and powering hand tools. On larger farms, it is also used for grain milling, refrigeration, sawing, welding, and general workshop applications. On dairy farms, electricity has an important use in milking machines, coolers, refrigerators, and boiling water for cleaning and sterilization.

Where readily available, firewood is the preferred cooking fuel (Goddard et al. 1981). In coffee-growing areas, for example, shade trees, which protect the coffee crop from sunlight during the growing season, are cut back when it is time for the pods to ripen. This waste wood, as well as trimmings from the coffee plants, is a major source of free firewood.

Better-off families use an enameled iron stove with several cooking plates, an oven, and an integral water heater. A simpler and cheaper cast-iron stove with an oven is also commonly used. Some of the poorest rural families still use a mud-and-brick stove, which they build themselves, but cooking on an open fire is now virtually unknown.

Tariffs

In adjudicating on tariff revisions, the SNE endeavors to balance the statutory obligations of ICE and the distribution companies with the needs of consumers. Each time ICE or the distribution companies wish to increase their tariffs, they must apply to SNE. The agency then analyzes their submissions (which must include detailed sales projections and cost analyses to demonstrate the need for the increase), accepts or modifies their applications, and publishes the verdict in the official newspaper. Although the process is relatively bureaucratic, it appears to work effectively.

The Cooperative Experience

In the more than 40 years since their establishment, Costa Rica's four rural electric cooperatives have proven to be stable, effective, and financially viable organizations. Although they supply only about 20% of the country's rural residents, they increasingly appear as attractive models for decentralized electricity supply.

An Egalitarian Tradition

Costa Ricans had no problem understanding or accepting the basic idea of the rural electric cooperative when it was first proposed in the 1960s. The concept fit well with the country's egalitarian attitude and the value it placed on the cooperative movement. The principles of *cooperativismo*, for example, were taught in all the schools. Thus, much of the same spirit that infused the cooperative movement in the United States in the 1930s has also characterized the establishment and development of the Costa Rican cooperatives.

One strongly held egalitarian principle is that everyone in the service area of a cooperative is entitled to a supply. Through the traditional cooperative approach of rural electrification, which was used in the United States, a distribution network is evenly spread over the area to be supplied to ensure that access is more or less uniformly available to all potential consumers. Although this approach to rural electrification can be more expensive than a conventional linear approach, the cooperative views itself primarily as a business enterprise in which full cost recovery is axiomatic.

Organizational Structure

Ultimate authority in the cooperative organization rests with the general assembly, which meets annually. Delegates representing communities in the service area are elected to the general assembly every two years. The general assembly elects an administrative council (effectively a board of directors) to oversee the cooperative's management. Responsibility for the day-to-day running of the cooperative rests with the general manager, who is appointed by this council. The organization is run on a nonprofit basis, with any surplus placed in reserves or used to reduce tariffs.

In practice, control rests with the general manager, as, indeed, it does with the chief executive in public shareholder companies. Nevertheless, members, through their votes at annual meetings of the general assembly and their appointment of administrative council members, provide a measure of defense against gross mismanagement.

Each cooperative also has an education committee, whose primary duty is to inform and educate members, thereby enabling them to participate actively and knowledgeably in the cooperative's business. The cooperative also provides its members with technical information on new and improved ways of using electricity for domestic, farm, and commercial use.

Selection and Funding

In 1963 and 1964, consultants from USAID and the international division of the National Rural Electric Cooperative Association (NRECA) visited Costa Rica to determine the feasibility of establishing rural electric cooperatives (Benjamin 1964). Based on an area's potential to meet specific goals, primarily in food growing, and the extent to which other development projects were under way or being planned, three areas were selected: San Marcos (south of San José on the edge of the Meseta Central), San Carlos (northwest of San José in Alajuela Province), and Guanacaste (north Pacific coast).

The program was funded by a US$3.3 million loan from USAID, together with a contribution equivalent to $800,000 in local currency from the National Bank of Costa Rica. Design and construction of the program, carried out during 1965–1969, were coordinated and implemented by ICE, with technical and administrative training provided by NRECA to initial core staff.

Each of the three cooperatives was expected to contribute a minimum of US$118,000 to the start-up funds, and arrangements for the establishment of local savings committees were put in place with the Department of Cooperatives of the National Bank of Costa Rica. The cooperatives assumed responsibility for the administration, maintenance, and expansion of the distribution systems under their control. They were also legally liable for repayment of the USAID and other loans that had funded the construction and establishment of the systems. This meant that each cooperative began operations with a foreign debt of approximately US$1 million. A fourth cooperative, funded by the Arkansas Electric Cooperatives, Inc., was established in 1971. The following sections describe the development of cooperatives that were formed during this period.

COOPESANTOS—*Cooperativa de Electrificación Rural de los Santos*

COOPESANTOS serves a 600-km² area south of San José, which is almost completely surrounded by mountains. San Marcos de Tarrazú, the main town of the area and the site of the cooperative's headquarters, lies about 30 km from San José (Muñoz and Gonzalez 1995). The total population of the service area, according to the 1963 census, was 43,000, representing 5,300 households. The majority of people were small farmers who owned their homes and land. About 30% of landholdings were smaller than 0.7 ha (1.7 acres). Coffee was the main cash crop of the area, followed by dairy products, sugar cane, maize, pigs, beans, poultry, and vegetables. There were various small agroindustrial enterprises. The average annual family income in 1967 was estimated at US$820. There was a relatively poor-quality, but year-round, network of roads. Houses were situated along these roads, generally within 300 m of each other.

History of Electricity Supply. Beginning in the early 1900s, various attempts were made to install electricity supply systems in the area's small towns. In San Marcos, a small hydroelectric system was installed in 1912, providing lighting for 15 houses through a 600-MW distribution system. In 1920, a hydraulically driven coffee-processing plant began providing lighting to the surrounding houses. During the 1930s, a 20-kW generator installed at another coffee-processing plant provided a limited public supply until 1958, when it finally broke down.

The history of electricity supply was similar in several other towns in the area. In Santa María de Dota, for example, a small and not particularly successful hydroelectric station was installed in 1916 by a man described as "more idealistic than practical" (COOPESANTOS 1994). A more successful attempt was made around 1920, when the owner of a hydraulically driven sawmill installed a dynamo that provided electric lighting for the sawmill; however, his attempts to extend the system failed. Nevertheless, interest in setting up a supply system continued, and, in 1929, a small gasoline-powered plant was built. This plant provided lighting for the town plaza and some of the surrounding commercial premises; however, after a few months, financial losses led to the system's shutdown. A limited supply was eventually provided from a coffee-processing plant owned by the National Bank.

In the 1950s, a new feeder road connecting the San Marcos service area to the Meseta Central was built, which greatly improved access to agricultural markets.

Coffee production expanded, and demand for electricity grew. In 1961, three cantons set up a pro-electrification committee to petition ICE to electrify the area. ICE, however, lacked funds for rural electrification at that time because of prior investment commitments. Thus, the committee turned its attention to the possibility of building its own small hydroelectric station.

Beginnings of the Cooperative. A pro-electrification committee member, together with a director of ICE, attended a USAID-sponsored meeting on rural electric cooperatives in Uruguay. As a result, when NRECA examined the potential for establishing rural electric cooperatives in Costa Rica in 1963 and 1964, the San Marcos area was included in the feasibility study, and a cooperative for the area was proposed.

When detailed studies were carried out, the areas to receive electricity were expanded beyond the three initial cantons, and ICE embarked on the technical design of the distribution system with assistance from NRECA. The Department of Cooperatives of the National Bank of Costa Rica undertook the initial educational and organizational work required for establishment of the cooperative and also devised savings schemes so that future members could contribute to establishment costs via their initial connection fees. Local savings committees were set up in each community to be electrified, with their efforts coordinated by a central committee.

Administration, Initial Investment, and Financing. Early 1965 marked the inaugural meeting of COOPESANTOS and the appointment of its first manager. This manager was nominated by the National Bank of Costa Rica, which also nominated five of the nine members of the administrative council. As the financial guarantor of the cooperative, the bank retained the right to nominate the manager and members of the administrative council until 1973, when this right was transferred to the cooperative's general assembly.

The loan contract, which was signed with USAID in 1966, was for 40 years, with a 10-year grace period at an annual interest rate of 1% and a 2.5% rate during the amortization period. The National Bank acted as intermediary at no additional charge to the cooperative. Table 2-2 shows the amounts contributed by initial investors.

Initial construction of the distribution system was financed by the USAID loan. At the completion of this first phase, with some financial resources remaining, the

Table 2-2. *Initial Investment in COOPESANTOS in Costa Rica, by Contributor*

Contributor	Amount (US$)
USAID (loan)	1,200,000
National Bank	75,757
ICE (construction)	114,788
Cooperative members	169,464
Total initial investment	1,560,009

cooperative decided that new consumers would be required to contribute 30% of their connection costs.

By 1974, with initial loan resources exhausted, the cooperative decided that all new consumers would be required to pay the full cost of connection. To ease their burden, the cooperative set up a credit arrangement with the National Bank of Costa Rica, whereby new consumers were allowed to borrow the cost of their connection from the bank at the rate applied to agricultural credits, with a four-year repayment period and the cooperative acting as guarantor of the loan.

Growth of the Cooperative. COOPESANTOS was an immediate success when public supply began April 25, 1969. At that time, the number of consumers was 2,231, representing 42% coverage of the service area. By 1993, the number of households had increased to 19,768, representing 98% coverage of the service area, whose population had expanded almost fourfold since the cooperative was formed. The number of employees had risen from 17 to 62 over the same period, a rate of increase considerably less than that of consumers served. Electricity consumption had grown from 3,077 MWh to 40,874 MWh, representing a relatively small growth in monthly domestic consumption, from an initial 115 kWh to 172 kWh.

By 1995, the service area was almost completely electrified, except for some isolated pockets. In keeping with the public service ethos of the cooperative, ways were actively sought to connect the remaining, far from commercially viable, groups of potential consumers.

COOPESANTOS has been operating effectively and enthusiastically and is fiercely protective of its membership's interests. Its management has been dedicated to running the cooperative as an efficient, fully commercial organization and has actively sought ways to improve its technical and financial performance. For example, it decided that, given its financial position, the cooperative could provide its own insurance service for vehicles and personnel, and thus arranged with its bankers to do so. COOPESANTOS has also partnered with the country's other cooperatives to develop the San Lorenzo hydroelectric station to obtain a cheaper electricity supply than that provided by ICE.

The operational statistics for 1995 show that the turnover for the year was ¢516 million, with a net surplus of ¢26.6 million. Net assets were ¢1.127 billion. There were 21,788 consumers, of whom 19,462 were residential, 1,609 were commercial, and 38 were industrial. There were 71 staff, of whom 28 were administrative, 30 were maintenance, 8 were meter readers, and 5 were transformer repair workers. The length of the medium-voltage-distribution network was 1,200 km, and system losses were a little less than 10%. Average monthly residential consumption was 125 kWh, and the average tariff was ¢14 (US$.78 in 1995) per kWh.

COOPELESCA—*Cooperativa de Electrificación Rural de San Carlos*

COOPELESCA is located in the San Carlos region of Alajuela Province in north central Costa Rica, a tropical area lying at the inner edge of the Atlantic Plain. With a service area of 600 km², the cooperative is headquartered in Ciudad Quesada, a town about 70 km northwest of San José.

Cooperative Setting. According to the 1963 census, the population of the San Carlos area was 27,955, representing 4,155 households. At that time, San Carlos was one of the most prosperous rural areas in the country, with an average income twice that of Guanacaste, for example. The 1963–1964 feasibility study carried out by NRECA found that most families in San Carlos owned their own land and houses and that the average annual income was US$1,485.

COOPELESCA is described as having been established in "nearly a boom town setting" (Goddard et al. 1981). The timing of its creation paralleled other development investments by the government in roads and agricultural credit programs. The service area was experiencing a high rate of immigration. Meat, milk, and timber production were increasing, as were outputs of sugar, coffee, and rice.

Before COOPELESCA was established, the only public electricity supply was in the towns of Ciudad Quesada, Florencia, and Venecia, where there were a number of privately run, small diesel or hydroelectric plants. In addition, private commercial enterprises had numerous generators in use; one cooperative survey identified 114 motor generators and 24 hydroelectric plants.

Formally established in March 1965, COOPELESCA began supplying electricity in May 1969. The initial connection fee was ₡500 (US$58), which included basic interior wiring with two or three sockets and light outlets. Through an arrangement with local banks, cooperative members could borrow this amount over a five-year period at an annual interest rate of 8%. When the cooperative became operational, its loan obligation to USAID was a little more than US$1 million.

A 1976 NRECA evaluation of COOPELESCA concluded that "this is a generally sound cooperative" (Lay 1976a). It had met its entire debt obligation up to that time and foresaw no problems in continuing to do so. It was also in the process of obtaining an additional US$1 million loan from the IDB.

The evaluation found that the number of consumers was 4,207, of whom 3,072 were residential. The estimated population of the service area at the time was 75,500, meaning that the cooperative was serving about 25% of the families in the area. These figures were almost identical with projections made when the cooperative was being planned. Residential consumption, which had been 51 kWh per month in 1970, had grown to 85 kWh per month by 1976, almost exactly as predicted.

A relatively small survey carried out at that time showed that all families with an electricity connection used electric lighting, and 96.5% had electric irons. The proportion with refrigerators was 27.5% and 24.6% had televisions. The proportion using their supply for cooking was, however, relatively small, and only 18.7% had electric stoves.

Before the formation of COOPELESCA, 521 of Alajuela Province's 15,180 farms had an electricity supply; of these, 354 were supplied from privately owned hydroelectric or motor-driven generators. The 1976 NRECA study found that though the number of farms with an electricity supply had scarcely increased, a substantial number of farmers had shifted from the use of their own generators to the cooperative supply.

A 1971 USAID-funded study conducted by a team of sociologists from the University of Florida had observed that larger farmers, especially dairy farmers, many of whom had run their own diesel plants before the establishment of COOPELESCA, found the cooperative supply more reliable, convenient, and cheaper (Saunders et al. 1978). Those who had switched to the cooperative usually sold their generators to farmers outside the electrified area. In some instances, farmers continued using their own generators but intended to switch to the cooperative supply once their generators became too expensive to maintain or had to be replaced.

The 1971 study found that small farmers did not use electricity for farm tasks. Similarly, the 1976 NRECA review showed that the availability of an electricity supply had had little effect on small farms. Thus, the widely held view that rural electrification stimulated "productive" activities on small farms did not appear to be borne out in practice for the earliest stages of rural electrification (Wasserman and Davenport 1983).

The service area's more prosperous businesses and households appeared to have already provided themselves with an electricity supply before the arrival of the cooperative. But the availability of a centrally provided supply greatly extended the use of electricity and brought improvements in convenience and reliability.

The 1976 evaluation noted considerable increases in public lighting, as well as the use of electricity in businesses, schools, and health facilities, all of which clearly benefited from the greater reliability of the supply.

Growth of the Cooperative. COOPELESCA has continued to expand and prosper in the ensuing two decades. In 1995, its service area was 4,875 km², and the total number of consumers was 34,000, of whom 29,000 were residential. The number of staff was 128. The average consumption per residential consumer was 189 kWh per month, and the average tariff was ¢12.5. The total length of medium voltage distribution line was 1,820 km, and the degree of electrification of the service area was 87%.

The cooperative continues to expand its network in the relatively underdeveloped northern part of the country bordering Nicaragua. Its development plan for 1996 envisaged an investment of ¢122 million in new distribution lines. Together with the upgrading of lines and other improvements, its investment was planned at ¢358 million, financed mainly from its own operating surplus. The cooperative has also carried out a feasibility study regarding construction of a 6-MW hydroelectric plant of its own and is awaiting authorization from the SNE to proceed.

Total turnover of the cooperative in 1995 was ¢1.421 billion, and the net operating surplus was ¢143 million. The accumulated capital at the end of 1995 was ¢4.25 billion, and debts amounted to ¢353 million, a little more than 8% of the capital. In short, at the end of 1995, COOPELESCA was a prosperous and conservatively administered organization.

COOPEGUANACASTE—*Cooperativa de Electrificación Rural de Guanacaste*

COOPEGUANACASTE is located in Guanacaste Province. This north Pacific coastal area covers the bulk of the Nicoya Peninsula, a unique region of about 800 km² that differs geographically and economically from the rest of Costa Rica. The cooperative's headquarters is located in the town of Santa Cruz. According to the 1963 census, the population of the original service area was 26,497, representing 3,361 households.

Cooperative Setting. Of the three areas that USAID selected for cooperative development, Guanacaste was the poorest, with the least developed infrastructure; during the rainy season, for example, some areas were virtually inaccessible. Much of the area was occupied by large farms (*latifundios*) dedicated to cattle ranching, which accounted for an estimated 75% of the area's earnings. However, there were also a substantial number of small farms of less than 10 ha. The area's average annual family income of US$655 was heavily skewed; it is estimated that the income of 85% of farm families was only US$250 per year. There was little industry; the NRECA 1963–1964 feasibility study identified only three small industrial consumers likely to connect to the new service.

At the time of the feasibility study, ICE operated a 245-kW diesel generator in Santa Cruz that supplied electricity to 430 customers. (ICE was happy to hand over the distribution system to the new cooperative.) A sugar plant with a generation capacity of 1.1 MW, using bagasse as fuel, supplied electricity to local households six months of the year. In addition, an estimated 85 privately owned gasoline and diesel generators were operating in the service area when the cooperative was established.

Early History. COOPEGUANACASTE was incorporated as a cooperative in January 1965, and the first electricity supply was provided in December 1968. Because of the low population density and the extent of the service area, much of the area was not covered by the distribution system initially. The cooperative charged a connection fee of ¢400 (about US$12), which could be paid off in small monthly installments as part of each monthly bill. It was also possible for consumers classified as poor to be connected at a concessionary rate of ¢100.

When the cooperative began operating, the initial number of members was about 2,900, of whom 2,300 were residential. In 1969, the first full year of operation, sales totaled 3,500 MWh. By the end of 1974, sales had increased threefold, and the number of consumers had increased to 5,000, of which 4,200 were residential.

Subsequent Reviews. A 1976 NRECA evaluation of the cooperative found that the total membership in August of that year (eight years after establishment) was 6,200 (Lay 1976b). The population of the area, according to the 1973 census, was 93,239, with 15,285 households; the cooperative thus supplied about one-third of families living in the service area. At that stage, the cooperative had 543 km of distribution line. Some 70 small towns in the area, compared with just three before the cooperative was set up, had public lighting.

This was a considerably better performance than had been projected by the 1963–1964 feasibility study, which estimated that the number of consumers would reach 4,054 in the seventh year, with a consumption of 720 kWh per year. In fact, the number of consumers in the sixth year had reached 5,816, with an annual consumption of 1,544 kWh, more than twice the amount predicted. Sales to all

Table 2-3. *Analysis of Monthly Consumption by Type of Service in Costa Rica, 1975*

Type of Service	No. of Users	Total Monthly Consumption (kWh)	% of Total	Average Monthly Consumption (kWh)
Residential	4,984	469,140	23.7	94
Commercial	832	279,526	14.1	335
Industrial	46	240,765	12.1	5,234
Public lighting	66	80,848	4.1	1,225
Cooperative	1	8,443	0.4	8,443

classes of consumers were three times the amount predicted. Table 2-3 shows the average monthly consumption by each type of service used in 1975.

Out of the service area's 9,000 farms, only 340 had electricity; of these, 189 had their own supply before establishment of the cooperative. It was only in dairy, pig, and poultry farming that electricity was used for farm purposes. Milking machines, electric fences, refrigerated storage, heating for hens and piglets, and pumped water for animals were among the uses; on some of the larger farms, water pumps were used for rice irrigation. On the smaller farms, the availability of an electricity supply had had little effect.

The study found, however, significant welfare differences between precooperative conditions and those in 1976. The number of educational institutions with lighting had risen from 7 to 48. Various night classes had been established where previously there had been none. The number of health centers had risen from 7 to 23, and a new hospital had been established. The number of towns with public lighting had increased from 3 to 70 (Lay 1976b).

The study stated that reliable central-station electricity had had an "almost immeasurable" economic effect on the rural service area. The total number of businesses, many of them small family concerns, had increased from 15 to 86. A limited survey showed that all had electric lighting and many had freezers, refrigerators, jukeboxes, and other items that use electrical equipment. The number of industrial plants had increased from 10 to 15, and the number of motels and hotels had risen from 0 to 11. All of these changes could not, of course, be attributed solely to rural electrification, but there was no doubt that COOPEGUANACASTE had brought substantial and appreciated benefits to its members.

A 1978 NRECA review found the cooperative in a financially healthy state, with growth in membership and consumption levels running well ahead of forecasts (Lay and Hood 1976). The review, however, observed that the administrative council had become politicized and highly partisan. Few board members were reelected for a second term; consequently, there was virtually no build-up of management experience or continuity. There were also management weaknesses and problems in labor relations.

Management Crisis and Recovery. These management concerns proved entirely justified. Declining service and obvious mismanagement led to a membership revolt at the cooperative's Santa Cruz headquarters. The fact that members living in Santa Cruz had a clear perspective on the management abuses seems to have been decisive. Members in other towns were uninvolved in the revolt, even though they were equally affected by the poor service (Girer 1986).

In addition, the cooperative defaulted on its loan payments. A special management audit was conducted by NRECA, and external management assistance was drafted into the cooperative in 1983. Following a series of management changes, tariff increases, and reinforcement of financial controls, the cooperative had fully recovered by the mid-1980s.

Growth of the Cooperative. By the end of 1995, COOPEGUANACASTE had 28,748 consumers, of whom 25,125 were residential. Its turnover was ¢1.411 billion. The average residential tariff was ¢11.6 per kWh, and average consumption was 166 kWh per month.

The low population density of much of the service area, however, means that there are substantial numbers of potential consumers for whom it is extremely expensive to provide a service connection. The cooperative has therefore carried out a pilot project using photovoltaic (PV) solar home systems. A total of 30 domestic systems and one water-pumping system to provide drinking water for cattle have been installed.

Three payment methods are available to consumers. The first allows consumers to purchase equipment from the cooperative over a two-year period, while paying 30% interest on the outstanding balance and subsequently owning the whole installation. Using the second method, consumers rent the full installation at ¢2,500 per month. The third method allows consumers to rent only the PV panel for ¢1,000 per month. Under both the second and third methods, the cooperative is responsible for battery replacement. Under all three payment methods, the cooperative provides maintenance advice.

COOPEALFARO—*Cooperativa de Electrificación Rural Alfaro Ruiz*

COOPEALFARO is the smallest of the country's rural electric cooperatives, with a service area of about 250 km². Its headquarters is located in the town of Zarcero, which lies about 40 km northeast of San José.

The cooperative was set up in 1971 in response to local pressure for an improved electricity supply. At that time, customers were dissatisfied with the performance of the two private power producers that were supplying the central sections of a number of small towns in the area. ICE informed the residents who petitioned for a supply that it would take many years before funds were available to provide one.

In response, residents in and around the existing supply area requested assistance from the Arkansas Electric Cooperatives, Inc., which agreed to finance the project.

Line construction and house wiring were carried out during 1973–1974, and the cooperative became operational in 1974, with 57 km of distribution lines and 950 consumers. In January 1975, a loan agreement was signed for US$121,350, with a 30-year term, a 5-year grace period, and an interest rate of 5%. An additional loan of US$56,761 was signed later that same year, with a 10-year term, no grace period, and an interest rate of 9%.

Like the other three cooperatives, COOPEALFARO has clearly answered local needs. By the end of 1982, it had doubled the number of consumers to 2,000, of whom 1,600 were residential. At that time, the average monthly consumption was 144 kWh. By the end of 1995, the number of consumers had more than doubled again; out of these 4,008 consumers, 3,059 were residential. The average residential consumption had risen to 191 kWh per month, the average tariff was ¢11.9 per kWh, and turnover was ¢171 million. Bordered on three sides by ICE service areas and by COOPELESCA on a fourth side, COOPEALFARO had little room left for expanding its services at the end of 1995.

Important Factors Supporting Success of the Rural Cooperatives

All four of Costa Rica's rural electric cooperatives have survived and prospered. After more than 30 years, it is worth looking at some of the important aspects and common features of the country's cooperative experience. The Costa Rican experience highlights the need for meticulous preparation and continued institutional support when setting up new organizational structures for rural electrification. Even then, unexpected problems can arise.

Tariff Structure and Taxes

Despite the existence of a body of cooperative law and the extensive preparations that went into forming the first three rural electric cooperatives, they nonetheless faced unforeseen difficulties when they began to operate. ICE had no tariff category for bulk sales to rural electric cooperatives. The country's major industries, such as the national oil company, were the closest analogue as purchasers of large quantities of bulk power. Because, under the Costa Rican tariff structure, large users paid considerably more than domestic consumers, this meant that the cooperatives were paying three times more than the municipalities for their bulk electricity supplies, with obvious effects on their tariffs. Vigorous lobbying by the cooperatives resulted in a subsequent change in the law when, in 1973, a special electricity tariff category for bulk sales to rural electric cooperatives was established.

It also became evident that the tariff structure designed for the cooperatives contained a basic pricing error. The fixed charge for the first 30 kWh had been set extremely low (¢6), which at the time amounted to US$0.70 per month. Faced with the high charge for consumption above this level, many consumers restricted themselves to 30 kWh per month, and the cash flows of the cooperatives suffered accordingly. By the year 2000, the cooperatives set their minimum charges for 30 kWh per month at about US$2.50–2.75.

Another early problem was that the cooperatives were not exempt from being taxed on their materials and equipment supplies, as were other electricity suppliers. This anomaly was also resolved by the legislative changes of 1973.

Loan Repayment

Repayment of the USAID loan was another cause of early disputes. At the time the original loan was made, the exchange rate was ¢6.84 to US$1.00. No thought was given to the possibility of the exchange rate changing substantially. Even with the relatively slow rate of depreciation in the 1970s, the increased debt repayments became increasingly onerous for the cooperatives. Again, vigorous lobbying by the cooperatives resulted in reverting to the original exchange rate for loan repayments.

Training

Cooperative staff in the predominantly rural service areas lacked the administrative and technical skills required to meet the demands placed on them. In response, NRECA, with financing from USAID, provided a variety of training courses and opportunities for staff, including visits to rural electric cooperatives in the United States and twinning arrangements with specific U.S. cooperatives. ICE and the National Bank of Costa Rica also provided early support (Lay et al. 1984).

Training has continued in all the cooperatives using internal resources, as well as resources of ICE and other local organizations. NRECA continued to maintain a small office in San José until recently; it provided a course in stringing distribution lines in 1995 and another in line inspection and supervision in 1996. The costs of both courses were paid for by the cooperatives.

Cost Recovery: A Fundamental Principle

From the beginning, the country's cooperative managements and memberships never questioned the need for full cost recovery. Apart from the initial setup period, which was clearly foreseen at the planning stage, the cooperatives have been obliged to cover their operating costs, as well as their debt repayments and interest (NRECA 1980). The SNE has overseen their performance in this respect. This practice, combined with skilled, honest, and dedicated management, largely accounts for the healthy financial condition of Costa Rica's cooperatives.

At the village or local level, the cost recovery policy is operative when extension of the service is being considered, often in response to a petition from a group of families living in the area. First, the cooperative designs the scheme and works out its cost. Then, based on the number of consumers prepared to take a supply, it predicts the annual cash return to determine whether the project is financially viable. If the required investment is greater than the cash return, the project goes ahead only if the local community is willing to pay the difference. In 1995, for example, when COOPELESCA extended service to 655 families in some 20 small communities, these families paid roughly 50% of the construction costs of the extensions.

In the case of individual consumers, the standard connection fee entitles them to a connection, provided they are within a certain minimum distance of the distribution line. All consumers beyond this point must pay the full additional cost of their connections. However, if families can demonstrate that their income falls below a certain minimal level, the cooperative will grant them a concessionary connection fee. The cooperatives view this policy as part of their public service. Relatively stringent qualification criteria and the social pressures of the rural areas are major deterrents to abuse of the system.

Although rural people appreciate the value of and are willing to pay for an electricity supply, they do occasionally attempt to bend the system to their own advantage. When, for example, a supply is being brought to an area, some consumers may try to avoid being among those who contribute initially, hoping to connect into the system later for the standard connection fee. COOPESANTOS has overcome this tendency with the following rule: Those deferring their own connections in a newly electrified area still must pay their full share of the community costs for the first five years. Rarely are people willing to wait five years for a supply to save money.

In summary, cost recovery provides the cooperatives a freedom of action that enables them to

- carry out maintenance work, which is so often neglected in electricity supply systems in developing countries;
- extend their networks;
- improve their services; and
- use a limited amount of funds to fulfill agreed-on social objectives.

If the cooperatives depended on subsidies to cover their operating costs, they would risk a cutback during times of economic stringency. They would be tempted to skimp on maintenance, and funds for expanding and improving the supply would be lacking.

Low-Cost Capital

Avoiding dependence on operating subsidies does not mean, however, that rural electric cooperatives should not obtain their investment capital at a concessionary rate. When the cost of capital is high, the tariffs needed for the supply company to break even in areas where there are small numbers of thinly dispersed consumers may be impossibly high. Similarly, it may be unrealistic to expect the local community to contribute the funds needed to bridge the gap when the investment capital is charged at full commercial rates.

If, however, the supply organization can obtain lower cost capital, then electrifying such an area may become financially viable. Using such concessionary capital does not breach the principle of cost recovery. The supply company is still obliged to repay its debt, as well as cover its energy and maintenance costs; but the overall scope for rural electrification is widened.

By providing low-cost capital while maintaining the principle of full cost recovery, Costa Rica was able to ensure that rural electrification took its place as part of the balanced development of the rural areas in a financially responsible and sustainable way. The original USAID loan for the cooperatives was at highly concessionary rates. Without the availability of this low-cost capital, the degree of rural electrification that has occurred in Costa Rica would not have been possible. The significant achievement of the country is that the funds were used in a highly responsible manner. Rather than undermining the long-term viability of the supply companies, the concessionary funding was used to provide them a firm and self-financing consumer base.

Framework for Effective Management Autonomy

In the Costa Rican system, the management autonomy of the cooperatives—their ability to make decisions based on technical and financial grounds without political interference—is set within a clearly defined framework of enforceable rules and responsibilities. The cooperative structure itself provides some degree of management responsibility, as the COOPEGUANACASTE experience shows; however, the degree of supervision that membership can realistically exercise is limited.

The SNE plays a decisive role. Its insistence on professionally audited accounts, maintenance of proper administrative procedures, and inspections and management audits largely prevent mismanagement. The SNE is also critical in balancing the needs of cooperatives and consumers. In its approval of tariff proposals put forward by the cooperative management, for example, the SNE's role is to weigh the cooperative's need for working cash against consumers' need for low-cost electricity, thereby ensuring management incentives to run the cooperatives competently and efficiently within the terms of their mandates.

Opportunities for Renewable Energy Initiatives

Contrary to much earlier thinking, the best opportunities for renewable energy sources, such as small hydroelectric plants and wind, often arise, not as alternatives to the grid, but when there is an existing grid system into which they can feed. The 1990 legislation permitting sales to the SNI has opened the way for Costa Rica's four cooperatives to consider a range of renewable energy initiatives.

CONELECTRICA, a consortium of the country's four rural electric cooperatives, has been formed to build the San Lorenzo hydroelectric station in the San Carlos area, about 160 km from San José. Financial analysis shows that the plant, which will feed into the grid, will have the capacity to provide the cooperatives significantly cheaper electricity than ICE currently can.

The station's capacity is 16 MW, and cost is estimated at US$20.7 million. Capital is being provided in the form of US$16.6 million in bank loans, with the remaining US$4.1 million coming from the cooperatives' own funds. Cooperative financing is being shared in direct proportion to membership numbers: COOPELESCA, 45%; COOPEGUANACASTE, 33%; COOPESANTOS, 18%; and COOPEALFARO, 6%.

The cooperatives are also interested in small, locally controlled renewable energy projects that would not be of interest to ICE at a national level. These projects include the PV pilot project being run by COOPEGUANACASTE and potential wind-energy projects in the Cordillera de Guanacaste and Nicoya Peninsula areas.

Another project being considered by COOPESANTOS at the time of this study is the installation of a 15-kW Pelton turbine into a high-pressure water supply system as part of the pressure reduction arrangement. The cooperative has obtained the necessary quotation for off-the-shelf equipment from a U.S. company. With an estimated cost of US$15,000 and an annual savings of US$6,196, the project is a highly attractive investment. Because the cooperatives are local and relatively small in scale, they will likely discover that many projects that could not practicably be implemented by ICE may now become interesting.

Social Dimension: Strengths and Weaknesses

The cooperative approach is unique in that its consumers are its members and have a voice in its affairs. In return, through its education committee, the cooperative is positioned to feed information back to its members regarding the cooperative and broader issues of electricity use.

Cooperative membership also brings a number of social benefits. For example, a mutual fund has been set up to benefit the families of deceased members, and a scholarship fund has been established to assist needy families in the secondary education of eligible children. Life and accidental death insurance are available to members and their families. The cooperatives also provide discounts on domestic electrical appliances, and there is a range of employee benefits.

Overriding all of this is the cooperative spirit, which fosters a public-service management ethic. Rather than striving to maximize profits without regard to the social cost, the cooperative's management objective is to ensure that everyone in the service area is provided with a reliable electricity supply at the lowest possible price.

While all of this is true, the practical involvement of the membership in the running of the cooperative tends to be slight. Even in the case of COOPESANTOS, where there appear to be strenuous management efforts to promote membership involvement, most consumers simply want a reliable electricity supply as cheaply as they can obtain it. In its 1995 annual report, COOPESANTOS complained about the continuing difficulty in getting members to take an active interest in the cooperative.

A study in the mid-1980s had reported that in practice, the cooperatives are run by a small group of employed personnel, with oversight from directors and outside interested parties, such as ICE, SNE, and NRECA. With centralized control and little communication to members, it is not surprising that participation is minimal. The goal of democratic institution building has not substantially been met by the cooperatives (Girer 1986).

The 1976 NRECA review had noted the need to stimulate member interest, and an even earlier study on cooperative electricity distribution stated that "the electric cooperative serves principally, if not exclusively, simply as a means of distributing electricity. From the standpoint of the users interviewed, it really does not matter whether electricity is supplied by a public utility or a rural electric cooperative" (Saunders et al. 1978, *169*).

Nonetheless, the cooperatives have played an important role in bringing rural electrification to their service areas considerably sooner than would otherwise have happened. As discussed earlier, the cooperatives formed an attractive focus for external funding and technical assistance from USAID and NRECA. In addition, they provided and continue to provide a mechanism whereby local cash contributions, which people are clearly willing to make, can be channeled into rural electrification projects. As locally based organizations, the cooperatives are also well placed to resolve disputes that can arise over rights of way, compensation for damage during construction, and other local issues.

The National Rural Electrification Program

The cooperatives are the best known and most widely reviewed component of Costa Rica's rural electrification program. However, ICE and the municipal companies have contributed greatly to the process, and private suppliers played an important initial role. As early as 1953, ICE became involved in rural electrification when it took over the Saxe electricity supply company. The company had been serving the town of Cartago and the ports of Limón and Puntarenas, as well as their surrounding areas, and 70% of its customers were rural. Saxe was universally viewed, however, as providing unacceptably poor service.

ICE obtained a US$2.7 million loan from the IDB with which to rehabilitate the Saxe networks. However, the work proceeded slowly, and in 1964, discontented citizens of Cartago decided to set up their own municipal supply company, known as the JASEC. ICE continued with the remainder of its program, which was completed in 1971, when 18,700 new consumers had been connected.

During this same period, ICE acted as the implementing agency in the design and construction of the country's three USAID-funded cooperatives. Subsequently, ICE obtained a loan of US$3.8 million from the IDB in support of a US$6.8 million project involving the generation, transmission, and distribution works in Guanacaste and Limón; part of the loan was then loaned to COOPE-GUANACASTE for expansion of its system.

Launch of the National Rural Electrification Program

In 1975, the government of Costa Rica launched the first phase of its national rural electrification program, part of a broader national development program, with ICE acting as the implementing agency. The program aimed to assist in rural community development through rational and productive use of electricity, support to farmers, and the slowing of emigration to urban areas.

The program was successfully implemented between 1976 and 1979, when some 1,600 km of medium-voltage distribution lines were built, extending service to the principal towns of most of the cantons in the country. During this phase, 18,000 new consumers were connected.

Rural Electrification Program Combined with Complimentary Investments

The second phase of the national rural electrification program was planned as part of a balanced investment program for rural development. It took an integrated approach that combined ICE's activities with those of the municipal companies and cooperatives. ICE itself limited its rural electrification activities to 10% of its total investment. The program's goal was to increase the national level of electrification, which stood at 65% in 1977, to 75% in 1983, providing a base from which 90% could be reached by the turn of the century.

The program envisaged the construction of 2,639 km of medium-voltage lines, a 30-km extension of 132-kV transmission lines, and the building of three 75-kW isolated diesel plants. In this way, the supply system would be extended to more remote parts of areas covered in the first phase. The target was 20,000 new consumers, with an additional 10,000 connections by 1992. The cost of the program was US$38.4 million, of which US$19.8 million was in foreign exchange.

A socioeconomic study of the area was conducted, which found that the average monthly family income was ¢1,010 (US$117), about half that considered to mark the poverty line. Even so, these families expressed interest in obtaining an electricity supply; moreover, a commercial demand for electricity clearly existed. In various places, for example, small private suppliers were providing a limited evening service for nearby houses. A typical monthly charge was ¢6 (US$0.70) per light bulb.

Field studies were carried out in all areas where line extensions were being considered. It was assumed that 80% of potential consumers would take a connection immediately, with the percentage rising to 90% over the following 10 years. Initial consumption was assumed to be 74 kWh per month and expected to double in 10 years. Both sets of assumptions were derived from experience in comparable electrified areas then being served by ICE or the cooperatives.

Detailed cost estimates were prepared for the construction of the distribution system in each area. The number of consumers per kilometer of electricity line varied between 3 and 21, with an average of 5. Connection costs, including meter installation, were ¢750 per consumer. The cost breakdown for the overall project was 88% for the distribution system, 8% for the transmission system extension, 2% for the connections, and 2% for the diesel generators.

In the economic analysis, the opportunity costs of the electricity supplied were taken as zero. The small increase in the consumption of electricity generated from existing hydroelectric stations was judged not to involve any significant costs at a national level. The economic benefits were calculated by multiplying the projected sales revenue by a factor of 2.7 to allow for consumer surplus. The calculated economic rate of return for the project was 15.4%.

The financial analysis, however, revealed that, at the going annual interest rate of 7.5%, the project would produce a financial deficit for the rural electrification cooperatives of about US$1 million per year over the first 10 years of operation, involving an unacceptable degree of cross-subsidization from existing consumers. The alternative—a concessionary loan from the IDB's Special Operations Fund at 2% interest with a grace period of 8.5 years—showed an accumulated financial surplus for the distribution companies of US$9 million during the first 10 years. A go-ahead for the project, therefore, depended on the loan being available on these terms.

In fact, ICE obtained this loan, but the economic crisis that hit the country in 1980 led the utility to postpone the project until the mid-1980s. The program was then implemented, and ICE, together with the municipal companies and the cooperatives, has continued to expand the rural electrification network.

Shifting the Balance between Private and Public Suppliers

In the early years, private suppliers played an important role in building up electricity demand in advance of the availability of a public supply, though service was often limited and of poor quality. By 1972, nearly two-thirds of villages with a population between 501 and 1,000 had private power, but quality and coverage varied greatly. Often it was no more than an unreliable evening supply available only to people living in the immediate vicinity of the supplier. Nevertheless, it ensured that, in the absence of a public service, many people had electricity.

ICE, together with the cooperatives and municipal companies, expanded their distribution systems, and private suppliers were gradually replaced by the public service. By 1972, more than 76% of communities in the Meseta Central obtained their supply from public companies. Yet public power had reached only 7.7% of consumers in the Atlantic region, 3.7% in the north Pacific (Guanacaste), and 0% in the northern plains, reflecting a strong regional bias toward the Meseta Central.

Conclusion

Costa Rica has accomplished what still remains a distant aspiration for a majority of the world's developing countries. Through the combined efforts of its national supply utility, rural electric cooperatives, municipal companies, and private suppliers, the country has succeeded in bringing electricity to over 97% of its citizens. Although the exact combination of social, political, and economic factors that led to Costa Rica's success in rural electrification is unique, this success story has significant lessons for programs in many other developing countries.

Costa Rica has a number of distribution models that appear to be working well. With decentralization on the political agenda in Costa Rica today, it is likely that changes will be made in the way rural electrification services are provided and managed in the future. In addition to the cooperative model, within ICE rural electrification is already separated from its generation and transmission businesses. The utility has shown that it is fully capable of effectively implementing and operating rural electrification programs—an observation that cannot be made about many utilities in developing countries. Municipal companies provide a model in which provision of rural electrification services is under the control of the democratically elected representatives of the people served.

Thus, the rural electrification program in Costa Rica provides a useful example that multiple approaches to electricity distribution can be successful even in the same country. In fact, the public and cooperative companies have sometimes acted like they are competing for "bragging rights" as to who can provide the best service. Of course, this would not have been possible without the political and institutional support for rural electrification within the country. However, the clear winners are the rural people in Costa Rica, who now have access to high-quality electricity services.

CHAPTER 3

Power and Politics in the Philippines

Gerald Foley and Jose D. Logarta, Jr.

T HE PHILIPPINES'S RURAL ELECTRIFICATION PROGRAM began
more than 30 years ago with the establishment of two pilot cooperatives,
and its achievements have been impressive. Despite its problems and setbacks,
the program has succeeded in bringing electricity to millions of households.
The 1970s saw rapid expansion, with more than 100 rural electrification cooperatives established and 1 million rural families provided an electricity supply.
Although expansion continued during the early 1980s, underlying organizational and financial problems gradually surfaced, bringing many rural electrification cooperatives (RECs) and the overall rural electrification effort to
virtual bankruptcy.

A vigorously implemented series of institutional and financial reforms, which
began in the late 1980s, have now largely restored the health of the program.
By 2004, electrification had reached all of the country's municipalities, and 80%
of the rural population was connected to the supply. Much work remains to be
done, however. Connecting the remaining 20% of the rural population, who are
generally poorer and live in more remote areas, will be an even greater challenge
(World Bank 1992). Nevertheless, progress continues, and government and public support for the program remains strong.

The Physical Challenges to Rural Electrification

The physical layout of the Republic of the Philippines presents many problems
for programs of rural electrification. The country is a large, widely dispersed
archipelago located in the South China Sea. The total land area of 300,000 km^2
consists of more than 7,000 moist tropical islands, 2,800 of which are inhabited.
Most of these islands are extremely small; only 154 have areas larger than 8 km^2.

The responsibility of covering these mostly rural areas with electricity today is delegated to 107 rural electric cooperatives.

The challenge of providing electricity to the people of the Philippines is great not only because the country is divided into many islands, but also because much of the land is mountainous. Luzon, in the north, is the country's largest and wealthiest island, spanning 141,000 km² (nearly half the area of the country). Mindanao, in the south, is the second largest island, with a total area of 102,000 km². Seven large islands clustered between Luzon and Mindanao, known as the Visayas, cover 56,000 km²; of these islands, Cebu is the largest. The long, narrow island of Palawan lies to the west of the archipelago. Hundreds of smaller islands lie in between and in the offshore waters of these major islands. Much of the interior of Luzon and Mindanao is covered with steep, heavily wooded mountains, though less than half of the original lush forest remains because of logging and conversion of upland areas for agricultural use. Most of the smaller islands have a flat coastal strip rising into a core or spine of steep, wooded mountains. Because of the mountainous terrain, only 40% of the country's land is cultivable.

Further complicating electricity coverage, the country is periodically hit by typhoons, which often damage rural electrification networks, particularly in the eastern islands. The country has 18 active volcanoes. When Mount Pinatubo erupted in 1991, it caused enormous physical and economic damage. The Mayon volcano near the southern tip of Luzon has erupted 50 times in the past 400 years. Various regions are susceptible to earthquakes, landslides, and floods.

Of the country's total population of about 68 million, 38 million live on Luzon and 14 million on Mindanao. Today about 60% of the population is rural, and about 80% of these rural people have access to and use electricity. Some 10 million people live in Manila, the country's capital, which is located on Luzon. Most of the residents of Manila are served by a private electricity distribution company. Davao, Mindanao's largest city, has almost 1 million people. Cebu City, an important center for business, tourism, and university life, has 650,000 people.

Rural electrification is often complementary to other rural development programs. The government-run national network of health-care facilities, supplemented by private facilities, varies considerably. Education is a national priority, as reflected in 100% enrollment in primary school. Because state schools often lack adequate facilities, wealthier families tend to send their children to private schools. However, compared with other developing countries, the literacy rate in the Philippines is fairly high.

Historical Overview of Rural Electrification

In 1960, the government of the Philippines announced total electrification of the country as a national policy objective. Over the previous 70 years, private companies had controlled the electricity supply, with the government's role restricted to regulation of installations. Belief in the primary role of the private supplier, still strong in 1960, was reflected in the early functions of the Electrification Administration (EA). Established by the government to implement its new policy, the EA granted franchises and encouraged suppliers to set up local distribution systems in rural market towns. These either relied on the National Power Corporation (NPC) for their bulk electricity supplies or produced their own, usually from diesel generators.

Prelude to the Rural Electrification Program

Between 1962 and 1969, the EA provided loans and technical assistance for the establishment of 217 small supply systems, each with less than 500-kW capacity (Cabrera 1992). In general, these local networks were poorly constructed and maintained. Supply quality was low, with large voltage fluctuations and frequent breakdowns. Financial discipline was weak, and the EA frequently could not collect payments on the loans it extended. Many franchise holders abused their positions and diverted borrowed government funds to their own private uses. Substantial numbers of the systems failed or fell into disuse.

In 1970, only about 18% of the population had electricity; more than 70% of consumers were limited to six urban centers. In rural areas, where 75–80% of the population resided, only 8–19% received any service, most of which was unreliable. Moreover, most of these connected residents were the rural elite living within 200–300 m of local market centers (Denton 1979).

When the Philippines became independent from the United States in 1946, its economy was in ruins. Manila was reputed to have been the second most war-damaged city after Warsaw. The country's first president died suddenly after two years in office, and his successor was killed in a plane crash. Ferdinand Marcos, elected president in 1965, came to power at a time when the country was suffering from high levels of crime and widespread corruption in public life. The Marcos administration, aware of the weaknesses in the private franchise system, was determined to accelerate national electrification. In particular, it viewed rural electrification as a means of reducing rural unrest by demonstrating that the government cared about rural development.

In 1966, the government sponsored a rural electrification feasibility study carried out by a team of U.S. consultants funded by the U.S. Agency for International Development (USAID). The feasibility study determined that rural areas had considerable potential for establishing financially viable electricity supply systems. Among other findings, the report recommended that the Philippine government launch a nationwide rural electrification program based on the U.S. rural electric cooperative (REC) model that had been developed in the 1930s.

Recognizing the great differences between rural areas in the Philippines of the 1960s and rural America of the 1930s, the study proposed that two pilot cooperatives be established. These would test the feasibility of the cooperative model under Philippine conditions and make any needed adaptations based on practical field experience (NRECA 1967). An approach for assistance was made to the U.S. government, which declared its willingness to assist in establishing the two pilot cooperatives. After considerable debate, two locations were chosen: Northern Mindanao in the province of Misamis Oriental and Negros Island in the Visayas. The U.S. and Philippine governments agreed to fund these two pilot projects in 1968. Consultants from the National Rural Electric Cooperative Association (NRECA) provided assistance in setting up the cooperatives. Preparatory survey and design work was quickly undertaken, public opinion was mobilized in both pilot project areas, and the two cooperatives were established.

Why the Cooperative Model?

Widely scattered islands and poorly developed roads and communications systems meant a range of difficult problems for implementing and managing rural electrification in the Philippines. From the outset, it was obvious that effective management of many small distribution systems would require local control, no matter what type of organizational solution was adopted.

Based on the U.S. model of the 1930s, the REC provided a well-proven mechanism for devolving management to the local level and was persuasively advocated in the initial USAID-funded study. Run properly and openly, the REC could give local people a stake and a supervisory role in the system. Instead of relying on a remote government, they would see what was happening firsthand and become involved.

The model was not without its risks, as cautionary precedents of the 1950s had clearly shown. For example, the Agricultural Credit and Cooperative Financing Administration, set up to help establish marketing cooperatives, registered 300,000 members but failed to manage its finances. Its replacement, the Agricultural Credit Administration, also failed. Loans were seldom repaid, and borrowed funds were misused. Attempts to establish rural water-user associations also failed because water sharing tended to be based on proximity to the source or local political influence. These experiences demonstrated that an effective REC system required a delicate organizational balance between centralization and devolution. The country's subsequent history of rural electrification has shown how difficult it was to achieve and maintain the necessary balance.

Full-Scale National Rural Electrification

Not wishing to wait, the Philippine government decided to proceed with large-scale rural electrification before either pilot cooperative had become operational. In 1969, it passed the National Electrification Administration Act, which proclaimed total electrification of the Philippines on an area coverage basis, primarily through the use of RECs, as a national policy objective. It replaced the Electrification Administration with the National Electrification Administration (NEA) as the implementing agency. Thirty-six RECs, each covering a franchise area with a population of about 100,000, were selected for the first phase of the program. The U.S. government expressed a willingness to provide funds and technical assistance through USAID.

In 1972, the Philippine government formally launched the program when it issued basic policies for the electric power sector. Its presidential decree reiterated the goal of total electrification on an area coverage basis, effected primarily through the setting up of island grids with central generation facilities and cooperatives for the distribution of power. This decree also authorized the NPC to build transmission lines and set up generating facilities in Luzon, Mindanao, and on other major islands. In areas not covered by the NPC grids, cooperatives, private utilities, and local governments were entitled to own and operate isolated grids and generation facilities.

In a parallel presidential letter of instruction, the NEA was directed to proceed with the establishment of the 36 RECs. It was authorized to convert or integrate existing private systems that had exorbitant rates and poor service, those delinquent in their payments to NEA or NPC, those connected to NPC or the Manila Electric Company that did not have their own generating plants, as well as those within the areas of the proposed 36 cooperatives. In effect, responsibility for rural electrification now rested solely with the NEA.

Although implementation of the program quickly got under way, a skeptical public needed convincing. Eager to build public confidence in the seriousness of the program, the government sought early visible results to ensure that this happened. The now functioning pilot cooperative on the island of Mindanao provided a practical demonstration of what could be accomplished and soon became a national showcase. The program gathered momentum and, by the end of 1973, 46 RECs had been established. In the process, the NEA acquired a reputation as a competent and effective organization.

The potential problem of corruption within the cooperatives was recognized and addressed at the outset. The public service ethic and need for dedication and honesty were emphasized in all training provided to staff and board members. Where appropriate, these messages were reinforced by incorporating a religious element and involving local religious leaders.

NEA representatives frequently visited new cooperatives, inspecting their books and listening to local people's views. In this way, any operational irregularities quickly became evident. When problems did emerge, the NEA responded decisively. In one high-profile case, a general manager was discovered to have placed cooperative money in his own account and was dismissed, despite his protestations. In another case, a board that had been elected at a politically packed meeting was dissolved. Similar well-publicized incidents created a widespread feeling that wrongdoing would likely be detected and punished and that the standards of cooperative staff and boards would improve as a result.

Progress of the 1970s and Early 1980s

The government strongly backed the rural electrification program, and international banks and donor agencies provided considerable support. The NEA functioned as the overall coordinating body, channeling government and donor funds into the program.

When RECs were set up in areas that lacked a supply company, the NEA constructed a new distribution network, and the cooperatives thus received a totally new system. In other cases, new RECs took over existing networks of small, private suppliers, a high proportion of which experienced financial difficulties after the oil price increases of 1973. On the major islands, existing regional grids supplied the

RECs with bulk power. On smaller islands and areas not yet covered by a regional grid, small power stations were built to meet the new RECs' needs.

Little cash was involved in the loans provided to the new RECs. The NEA provided the distribution network and other facilities. The RECs assumed responsibility for paying back the investment costs, which obligated them to generate funds to cover their operating expenses and repayment of loans. There was no shortage of government or external donor funds for initial capital investments, and RECs were rapidly established. By 1980, 112 had been created and were supplying more than 1.4 million consumers (USAID 1980).

By the early 1980s, RECs had been established in a high proportion of potentially viable areas, and the rate of creating new ones inevitably declined. Existing RECs, however, continued to expand their coverage. By 1985, the total number of RECs had increased to 119, and the number of consumers had nearly doubled to 2.7 million. Total external donor funding for REC formation and expansion up to that time amounted to more than US$400 million.

For any developing country, this would have been a remarkable achievement, let alone for the Philippines, with its widely dispersed national territory and major logistic difficulties. The World Bank's 1985 performance audit (World Bank 1985) of its 1978 rural electrification loan to the NEA (US$60 million) commented:

> The Philippines rural electrification program was characterized by full central and local governments' backing, active and frequently enthusiastic support by the local population, strong leadership by NEA, and competent guidance in the early critical years by USAID-financed experts familiar with the U.S. co-op system. While rural electrification in the Philippines could, in principle, have been based on several alternative institutional set-ups, it followed the U.S. co-op model because of USAID start-up funding, and this arrangement has served the country well.

Evaluation in the Early 1980s

In 1981, NRECA conducted a USAID-funded impact study of selected cooperatives, providing a valuable snapshot of the state of the program at that time (NRECA 1981). Of the seven cooperatives reviewed, three were on Luzon, two on Mindanao, and two in the Visayas. They provided a representative sample of the RECs then existing, and 770 consumers were interviewed.

The study found that all seven projects were succeeding in providing increasing levels of electricity service at competitive, affordable prices. Rapid progress in the numbers of consumers and sales of electricity were being recorded, with an average increase of 50% from 1978 to 1980. On average, each cooperative had 18,000 members.

The review reported that the cooperatives appeared to have sound organizational structures with comprehensive training programs in place. Managers and boards of directors, with few exceptions, appeared highly motivated and dedicated to the program's objectives. Electricity tariffs were found to be easily competitive with kerosene for lighting and diesel for motor power. Sales were approximately

40% for domestic use and 60% for industrial and commercial uses. Regarding enhancement of economic activity, the study found that electricity use:

- extended operating hours of private and public enterprises;
- broadened the types of services that private and public enterprises provided;
- maintained increased levels of manufacturing and agricultural enterprise production and facilitated the formation of new industries;
- reduced costs, particularly of small businesses, through the use of electric equipment; and
- increased enterprises' efficiency and attractiveness as investments, which, in turn, expanded operations and contributed to area employment and income generation.

The study also hinted at problems to come. For example, it noted that current financial performance was lacking in a number of cases, mainly as a result of poor collection rates for customer billings. In some cases, local politics was interfering with the healthy management of the cooperatives. On the whole, however, the study found the cooperatives to be competently run organizations providing a high degree of user satisfaction.

Build-Up to Crisis

Despite the favorable external impressions created by the program, serious problems had been building, which erupted into a major crisis of confidence in the mid-1980s. Unknown to outside observers and the donor community, many of the program's safeguards were being ignored or bypassed. The strict criteria that characterized the early REC program were being widely disregarded. Franchises were granted in areas that had little hope of sustaining a viable electricity supply. Engineering and financial judgments were being overruled in many cases, and distribution system lines were being extended in response to local political pressures rather than on the basis of technical and financial merits. Cooperatives were frequently viewed as vehicles for exercising political influence in job appointments and providing economically unjustifiable electricity supplies to isolated areas.

REC board membership was viewed as an important step on the political ladder. There was a tendency to subdivide potential franchise areas as a way of increasing opportunities for the exercise of patronage by local politicians. Prominent local community members viewed REC management committee membership as a route to political power, favors, or even outright theft. One commentator reported, "In many cases, coffers were cleaned out to meet political and personal ends, leaving the cooperatives deep in debt" (Cabrera 1992).

Oil price increases of the 1970s were a major distraction that created havoc in the national finances of many developing countries, including the Philippines. By the end of the 1970s, when some energy experts were predicting oil prices of US$100 per barrel, the development of alternative sources of energy was a major item on the agenda of the Philippine government. Funding for renewable energy projects was freely available from international donor agencies, and the NEA became involved in promoting a variety of renewable energy projects, some quite ill-conceived, in collaboration with the RECs, which badly diffused available managerial and financial resources.

Even more serious, tariffs charged by the RECs failed to keep pace with the increased costs of their bulk electricity supplies and the rate of inflation; as a result, their revenue flows began to fall short of their requirements. Payments to the NPC for bulk power supplies and loan repayments to the NEA fell behind. Line clearing and system maintenance were cut back, and many networks began to fall into serious disrepair. A survey conducted on the distribution systems of half the RECs during 1986–1989 found that 35% were in an unsatisfactory state and 30% showed no signs of any maintenance (World Bank 1989).

At the same time, because the government emphasized extending coverage of rural electrification, loans for system expansion remained readily available. Consequently, many RECs badly overextended themselves. Inevitably, the quality of service, supervision, maintenance, and financial performance declined. Slackening of financial discipline spread throughout the rural electrification system, and increasing numbers of consumers defaulted on their bills. Pilferage of electricity through meter tampering and other means also became a major problem.

Reckoning of the Mid-1980s

By the early 1980s, economic mismanagement and large-scale corruption had begun to take their toll on the country. Martial law was used to suppress criticism and, despite various populist measures, the Marcos regime increasingly shed its pretensions of democracy. Although martial law was formally lifted in 1981, Marcos retained vast authoritarian powers. Rapidly accumulating foreign debt sent the country's economy almost completely out of control in the mid-1980s.

Deteriorating international opinion of the country's performance meant that foreign grants and loans were increasingly difficult to obtain. With the drying up of new funds, underlying weaknesses of the NEA and many RECs began to surface. The increased openness that accompanied the end of the Marcos era revealed that the rural electrification program was in serious difficulty.

During 1981–1986, the NEA had maintained a small positive net income, but subsequently its financial performance drastically declined. Deprived of its cash flow from loan repayments, it was forced to operate at an increasing deficit. The average efficiency of collecting REC payments dropped to 36% during 1987–1989, and the NEA became effectively bankrupt. In 1990, the national average bill collection efficiency among RECs was just 52%.

A review of the performance of the NEA and the RECs carried out by the Philippine government and the World Bank revealed that most RECs were facing serious operational and financial problems. A high degree of political interference leading to a decline in the professional standards was one of the major problems noted in the report (World Bank 1989), which stated

> The politicization of NEA created an environment that enabled the RECs to become politicized. Although NEA has sound rules governing the conduct and remuneration of REC boards and individual directors (National

Electrification Administration 1981), those rules are flaunted more often than not. Available data suggest a high correlation between interference by board members in the day-to-day activities of the RECs and poor management of those institutions. While the RECs are, in principle, accountable to their consumer members, most RECs whose weak performance may be attributed to poor management show little effort to develop membership involvement. In effect, these boards (and, in turn, their RECs) are accountable primarily to the political interests that sponsored their elections.

The World Bank's 1989 assessment also indicated that only 22 out of the 119 RECs then existing were well managed and commercially viable. Commercial viability was felt to be within reach for another 24 RECs, provided that managerial and financial adjustments were made. The remaining 71 RECs were considered either in need of substantial remedial action or beyond rescue. The review concluded that the problems were too pervasive to be addressed by simple solutions. Rather, it called for the government to implement an integrated program to revitalize the sector through (1) restructuring the NEA; (2) making institutional reforms, including financial restructuring of all RECs responsible for distributing electricity to smaller areas; and (3) a thorough refocusing of operational practices and investment priorities.

This outcome was a major disappointment for everyone who had been involved in the early optimistic days of the program. However, many positive lessons were gleaned from the experience up to that time. Despite the various pressures on them, some RECs had maintained their technical and administrative standards, showing that they could function effectively, even under adverse circumstances. Moreover, the vast majority of RECs were still functioning, managerially and technically, and delivering an electricity supply to almost 3 million rural families.

Restoration of the Program

The problems of the Philippine rural electrification program had been evident long before the World Bank presented them, and the government recognized the need for action. In 1988, the government signed an agreement with USAID to assist in restoring the program (USAID 1988). This rural electrification project had a budget of US$53.5 million, of which US$40 million was a grant to the Philippine government. The project's major objectives were institutional development of the NEA and RECs and system loss reduction.

The main implementing organization was NRECA, with subcontracts provided to a number of companies, including a Philippine consulting firm based in Manila. This firm was given the task of preparing system planning reports (SPRs) for 100 RECs. Each SPR included a survey of the existing system; a detailed listing of any rehabilitation work required; and a long-term development plan outline for the system, thus providing a framework for the RECs' own work plans and investment programs. The project also provided for computers and training of REC staff, development of an investment planning model, and training in accounting for the NEA and RECs.

During the same period, the World Bank and the Philippine government formulated the Philippines Rural Electrification Revitalization Project, which was approved in 1992. The total project cost was US$118.5 million, of which the World Bank share was US$91.3 million in the form of a loan to the Philippine government. An additional US$9 million was provided from the USAID project, with the remainder contributed by the NEA and RECs. The objective of the project, which covered 54 RECs, was to provide for system rehabilitation and reinforcement, connection of new consumers within reach of existing lines, and economically justifiable system expansion.

Parallel with this project, the Japanese Overseas Economic Cooperation Fund provided approximately US$100 million to fund a similar project aimed at an additional 44 RECs. It too aimed at supporting the rehabilitation of networks, economically justifiable expansion, and the provision of support facilities.

The Philippine government and the NEA also agreed in 1991 to implement a comprehensive restructuring of the rural electrification program. This restructuring included converting P5.1 billion to equity, assuming responsibility for all of the NEA's foreign loan obligations of P6.3 billion as they came due, and writing off numerous doubtful or uncollectable loans. (The exchange rate in 1996 was about 26.20 pesos per U.S. dollar.)

These loans included P1.4 billion in loans to 25 remote island cooperatives with limited prospects for financial viability, P1.9 billion in loans for uninstalled operational alternative energy equipment, and a further P1.8 billion for nonoperational alternative energy equipment and other doubtful assets and loans on NEA's books. The measures also included stricter accountability of RECs and development of a rural electrification master plan in collaboration with NRECA.

The World Bank commented at the time (1991):

> Although this price tag is high in nominal terms, the fiscal implications are negligible; the government has been aware ... that NEA had no revenues corresponding to these liabilities and therefore no hope of meeting them. Moreover, NEA and the RECs acquired these obligations as a result of intemperate decisions by the previous government. No other entity can appropriately take responsibility for these liabilities.

A new statement of operating policy was agreed on by the NEA's board in November 1991. The NEA's relationship to the RECs was redefined as an "interested lender," which divested the NEA of the social, renewable energy, and other peripheral responsibilities it had acquired during the 1970s and early 1980s. Its responsibility was now to provide loans and technical services to RECs and to supervise and monitor their performance.

A major reorganization and depoliticization of the RECs was carried out. Almost half of their general managers were replaced. In some cases, the authority of boards of directors was reduced to advisory status or abolished. In a number of cases, adjacent RECs were merged to form more viable entities. A new tariff formula designed to restore the financial viability of RECs was introduced in 1990.

The problem of electricity pilferage, which had long vexed even the most conscientious RECs, was also resolved. A quirk in Philippine law had made the standards of evidence required for convictions so severe that obtaining them, except in the most flagrant cases, was virtually impossible. Although the RECs had demanded a change in the law, its passage through Congress was delayed many times. The necessary legislation finally came into effect in 1995, making it easier for RECs to deal with this issue in the future.

The Sector Revitalization Project, as originally envisaged, had an ambitious three-year implementation period (1992–1995). There were, however, worries about possible problems and bottlenecks in procurement and distribution of materials. These concerns have been borne out in practice, and in early 1998, the World Bank listed the program as among those needing management improvements. Nevertheless, progress continues in restoring the functioning of the NEA and individual RECs, as well as extending the scope of rural electrification. In 1997, the distribution system on Taganak-Turtle Island was energized, finally completing the task of bringing an electricity supply to the country's 1,607 municipalities. By April 1998, 24,871 *barangays* had a supply, representing electrification coverage of 69%. A total of 4.5 million rural households were connected, giving a rural coverage rate of 60%.

The financial health of the program had also improved by 1998, with a collection efficiency of 94% in amortization payments to NEA. The RECs' own collection efficiency was 93%. Total investment in the rural electrification sector over the period 1992–1997 was P4.95 billion, about US$200 million. The Philippine cooperative system had clearly recovered from the state of near collapse in which it found itself a decade earlier.

Description of the Rural Electric Cooperatives

RECs are organized as nonstock, nonprofit membership corporations whose goal is to distribute electric power in designated areas, where they act as regulated monopolies. All households and enterprises within the REC franchise area are eligible to become members. The households and businesses pay P5 to join, and these membership fees represent the REC's paid-in capital. Each REC has a membership services department, whose role is to involve members in cooperative activities.

Members elect a board of directors for two- and four-year periods. The board plays a policymaking and supervisory role, and its members are also trustees, with legal responsibility for protecting cooperative assets. The board recommends a candidate to the NEA for appointment as general manager of the REC; this individual becomes responsible for managing the REC in accordance with board policies.

The board has no authority over day-to-day management of the cooperative. Board members do not receive a salary but are reimbursed for expenses incurred in attending meetings and carrying out functions. They are prohibited from becoming involved, either directly or indirectly, in materials purchase and procurement or adjudication of tenders.

Well-functioning RECs are characterized by their independent, egalitarian, and public-service attitude, as well as their deep and genuine belief in the value of rural electrification. Extending the distribution system to supply all potential consumers within the franchise area is an ever-present management preoccupation.

RECs act as distributors of bulk power, which is purchased from the NPC. Where the NPC grid is within reach, the REC is connected into it, generally through a 69-kV line. The NPC is responsible for the construction, operation, and maintenance of this line, whereas the RECs assume responsibility for the 69-kV substations.

In the past, isolated RECs were responsible for generating their own supplies from plants that they operated under NEA supervision. These isolated plants were usually diesel generators, but small hydroelectric plants were used in some cases. Many RECs, however, lacked the necessary depth of technical expertise, and generation responsibilities were transferred to the NPC in 1988.

This distribution system is essentially that of the U.S. model developed during the 1930s. It is particularly appropriate in areas where lines are long and loads are small because initial construction costs are relatively low. The system can easily be upgraded to two- or three-phase systems as demand grows. Use of the area coverage system, which was referred to in the presidential decree that set up the NEA, is liberally interpreted in the construction of the REC distribution systems. These systems usually rely on 13.2-kV, three-phase backbones constructed along major roads to link main towns and villages with lateral two-phase or single-phase spurs running to serve areas of demand on either side.[1]

To lower billing and metering costs, the concept of the Barangay Power Association (BAPA) was developed and applied by many RECs to reduce billing and metering costs. A BAPA consists of a cluster of 30 or more consumers who draw their supplies from a common meter, which registers their overall consumption. These consumers, who also have their own individual meters, pay their bills to the BAPA; in turn, the BAPA pays the overall bill to the REC. The BAPA receives a quantity discount on the electricity supplied through it, thus reducing billing and collection costs. In addition, the BAPA acts to deter pilferage.

Development of Institutional Support for Rural Electrification

Institutional issues have been dominant factors in the evolution and progress of the Philippines rural electrification program. The NEA, as the prime promoter of rural electrification and coordinator and supervisor of the RECs, is the program's key institution. When the NEA has operated as a tightly focused, well-managed entity, rural electrification has prospered. Conversely, when the NEA became diffuse and inefficient in its functioning, the country's overall program suffered. The performance of many RECs deteriorated along with the program.

The RECs are the operating agencies of the electricity distribution systems in their franchise areas. Although their scope for independent action is heavily constrained by the NEA, the country's rural electrification history shows that the quality and vigor of individual REC managements can have a significant effect on their performance. This section attempts to learn from the remarkable achievements and major obstacles faced by these institutions by taking a detailed look at their managerial structures and organizational functions.

Historical Role of the National Electrification Administration

From its beginning, the story of rural electrification in the Philippines has revolved around the NEA. As the core agency responsible for implementing the national rural electrification program, the NEA selects the franchise areas of all distribution companies, including the RECs; allocates funding; checks and approves REC investment and expansion plans; monitors REC performance; coordinates materials procurement and distribution; and supports and supervises the country's 119 RECs. The NEA is a government-owned stock corporation headed by a chairman and a four-member board of administrators. Responsibility for running the organization rests with the administrator, who acts as chief executive officer and ex-officio member of the board.

Over the past 30 years, the NEA's role has undergone significant changes. From the time of its establishment in 1969, it immediately faced major institutional problems. Low salaries made it difficult to attract and retain well-qualified staff. Many of the initial 150 staff members were political appointees who expected to draw a salary but not work. The weakness of the NEA was so obvious that staff seconded from the Development Bank of the Philippines, assisted by external advisers, were used to help establish the country's two pilot cooperatives.

The government realized that if rural electrification were to succeed, NEA's capacities across a wide spectrum of activities would need to be upgraded. It would have to become capable of analyzing and setting overall policy, providing education and training for its staff and those of the RECs, raising and disbursing program funding, arranging procurement and distribution of equipment and supplies, and supervising and monitoring individual REC performance. It would also need to set the overall tone and standards of competence and integrity that would govern the rural electrification program as a whole.

Because the NEA lacked the necessary legal powers and resources to carry out such a range of tasks, the government decided to establish it as a corporation, which would greatly increase its freedom in day-to-day activities. It would be able to pay adequate salaries, hire and fire staff without following civil service procedures, and borrow and lend money. Although it would still function as an arm of the government, it would be distanced from immediate political pressures.

Establishment as a New Corporation. In August 1973, the NEA was established as a corporation under Presidential Decree 269, a major policy document that set out the aims of the national rural electrification program and the NEA's role in implementing it. Placed under the direct supervision of the office of the president, the NEA was granted extensive powers to establish and oversee electric cooperatives, make loans to them, acquire physical property and franchise rights of existing suppliers, and borrow funds to implement national electrification. Salaries were also freed from some of the restrictions applied to other civil servants.

In its early years, the NEA's major function was to establish RECs and set up distribution systems in areas that lacked an electricity supply. This pioneering work required a high degree of dynamism and aggressive promotion to achieve the impressive results obtained, which are clearly reflected in the program's first decade of work. However, the institution was subjected to frequent changes in roles and responsibilities, which diffused its efforts, stretched its resources, and blurred lines of responsibility.

Progressive widening of the NEA's scope of activities began in 1975, when it was transferred to the Office of Public Works. There, it soon became involved in a self-help community project in collaboration with the Ministry of Education and Culture to extend the benefits of rural electrification to schools. In 1977, in collaboration with the Ministry of Human Settlements, it became involved in a program that provided rural water supplies. In 1979, it was authorized to invest in and grant loans for the development of power generation companies, including dendrothermal and minihydroelectric power plants.

In 1980, the NEA created the Directorate for Promotion of Industries, which promoted handicrafts, small manufacturing, and service industries. It also implemented a number of alternative energy projects: developing improved techniques for charcoal-making, promoting charcoal-fueled gasifiers for motor power, setting up a dendrothermal power station program, and developing small hydroelectric resources.

Diffusion of Efforts Results in Problems. Much of the NEA's work in alternative energy, carried out in collaboration with the RECs, imposed heavy financial and technical burdens on all the institutions concerned. For example, the small hydroelectric program intended to provide RECs with their own sources of bulk power was carried out under heavy external donor pressure. Immediate results were expected, but inadequate technical planning and poor site and hydrological investigations hampered the program (Sathaye 1987).

Seventy-five small hydroelectric sites were identified, and equipment was ordered and delivered for many of these. But the NEA's growing financial worries and the deteriorating state of the national economy in the early 1980s meant that funds were not available to cover local construction costs, and, in many cases, even delivered equipment could not be installed.

Having limped through the 1980s, the program sought to salvage what it could in the early 1990s, turning over equipment and working installations to the NPC and encouraging private-sector companies to take over the establishment of remaining viable sites with uninstalled equipment. In 1993, the NEA signed a financing agreement with the Development Bank of the Philippines, under which money was made available for completing technically and economically viable projects.

The NEA's dendrothermal project, also carried out in collaboration with RECs, was similarly pressured by external donors to produce ambitious results. Donor agencies envisaged the construction of 17 plants powered by wood chips produced from special plantations of fast-growing trees run by tree farmer associations.

The program ended in total failure. Most of the plantations did not produce the anticipated wood yields. Some of the planned wood-harvesting methods proved impractical and the design of the power plants was seriously flawed. Three of the plants were abandoned at the planning stage, five were suspended after the equipment was delivered, and three others were suspended during construction. Six were completed, but only three became operational. By the early 1990s, all were closed down.

Redefinition for the Future. With the demise of the Marcos government in 1986 when Corazon Aquino came to power, NEA once again was moved to a different part of the government. It was moved to the Department of Energy, Environment and Natural Resources. In 1992 the Department of Energy was established as the overall coordinating body for the NEA, NPC, and the Philippines National Oil Company.

By the late 1980s, virtually all of the country had been covered with franchises, and establishing new RECs was now a rare event. The RECs' major task was now to increase electricity coverage. Even in the more prosperous RECs of the mid-1980s, this coverage meant that often no more than 50–60% of the rural population had electricity in their homes; in the smaller RECs, the figure was more like 20–30%.

The crisis of the mid-1980s had precipitated a major review and redefinition of the NEA's role. The NEA shed many of the diffuse responsibilities that it had acquired in the 1970s and 1980s. Its new role with RECs was limited to lending and technical assistance. Even in relation to the RECs, its role was changing. In order to liberalize and democratize REC operations, NEA's role in supervising and providing assistance to RECs that had achieved a rating of A and A+ was reduced. Rather than having to rely on NEA for all of their financing, RECs were given authority to obtain short-term loans from established financial institutions. They also were given the right to procure materials for Philippine-funded projects, and the NEA remained responsible for donor-funded projects.

Perhaps the most fundamental change involved a 1996 decision that allows REC members to determine their cooperatives' future direction and status. A REC would be allowed to convert from a nonstock to a stock cooperative, which would allow the entry of equity investors into rural electricity distribution. Whenever the first conversion to a stock company occurs, the NEA's role as lender to the REC will be to protect the government's loan.

NEA's Role in Cooperative Development and Oversight

The role of NEA in the early stages of the rural electrification program was the development of new electricity distribution companies to cover all areas in the rural Philippines. However, its role also went beyond just the development of the cooperatives, as it also involved oversight to ensure that the cooperatives were operated and managed in a way that would ensure continued service to rural consumers. This section reviews the role of the NEA in managing the process of rural electrification as it evolved in the Philippines.

Development of New Cooperatives. The NEA's first essential task in establishing a REC in a new area was to acquire the necessary economic, demographic, and topographical information to delineate the franchise area of a potential cooperative. A provincial electric cooperative team was established in each province to assist in this effort. Team members included NEA representatives, along with local delegates from the government's community development, education, and cooperative administration; office of the provincial governor; and league of municipal mayors. They also included representatives of local civic and religious organizations. During their 10 days of training in Manila, provincial electric cooperative team teams were imbued with a strong public service ethic (Denton 1979).

Provincial electric cooperative team members were responsible for collecting information required for selecting franchise areas, conducting public information campaigns, assisting in setting up the REC, and helping register it and applying for its start-up loan. A major advantage of this approach was that the NEA could devolve the difficult task of choosing the franchise areas for the initial cooperatives to the local level, at the same time retaining the final right of approval.

NEA usually recommended that the franchise area selected should cover 5–10 municipalities, with a total population of 100,000–200,000. Other criteria included the level of economic activity and population density to ensure a reasonable level of demand and that costs of the initial network were minimal. To avoid disputes over territorial coverage or ownership of facilities, new cooperatives were established only in areas that did not have an existing supply system.

Once the area for the proposed franchise had been selected and approved, a district electrification committee was set up. Its eight members included representatives of education, business, barangay councils, religious organizations, youth and civic groups, government, and farm groups. To minimize local political manipulation, formation of the district electrification committee was carried out under close supervision of the NEA in accordance with a standard procedure. NEA organized two workshops for the district electrification committees on the workings of RECs.

The district electrification committees were expected to recruit cooperative members and begin the process of educating them about their responsibilities. The district electrification committee acted as a liaison between the NEA and the local community and dealt with the often-difficult problem of securing rights of way for the construction of distribution lines. Each district electrification committee elected its own chair from among its members, and the committee became the REC's initial board.

Involvement of local communities through the district electrification committee and the new cooperatives was critical to building up the rural electrification program. People could see firsthand that this was not a government handout scheme as local communities were expected to do much of the work themselves. Local involvement mobilized local resources and minimized local criticism and cynicism because the RECs had to struggle with slow contractors, unreliable shippers and suppliers, and natural disasters.

The NEA's formal establishment of the REC followed a regular pattern. All relevant officials and individuals—from the provincial leadership to the cooperative membership and local media representatives—were kept involved and informed. Registration of the cooperative and the swearing in of board members involved considerable ceremony. These formalities were followed by the first board meeting, during which officers were elected, with NEA representatives in attendance. In all, this process involved 15 days of intensive activity.

Subsequently, training courses necessary for effectively running the REC were provided for the cooperative management and administrative and technical staff. Courses covered cooperative accounts, warehouse management, meter reading, billing and collection, line work, pole preservation and testing, and industrial safety.

When the initial network design for the REC was prepared, the market town (*población*) was the first to be provided with electricity service. Lines were often strung together along the backbone line. As the system developed, distribution lines were extended laterally to smaller groups of consumers. This simple design provided the REC with a basis for expansion. Provided that reasonable care was taken to ensure that cost and revenue requirements were met, the planning requirements were relatively minor. System expansion was governed primarily by availability of funds. However, when the pace of rural electrification began to accelerate, initial feasibility studies became outdated, and in some cases, problems began to surface with system expansion.

Rates of Return for New Investments. As part of the sector revitalization process, NEA's Corporate Planning Division established a system under which RECs prepare medium-term investment plans. Under these five-year plans, RECs are required to identify their expansion programs, which are then analyzed by the NEA and ranked in order of their economic return. Projects that pass the technical and economic screening process are included in the RECs' approved five-year development plans and current year's work plans. Those projects that fail to meet the criteria are returned to the RECs for revision (NEA 1994).

Regional investment planning workshops, organized by the NEA and attended by REC managers, are held annually to discuss and agree on allocation of available funds for the following year. In addition to meeting the NEA's technical and economic criteria, all REC investments are evaluated in light of available funds.

Economic evaluation is the main criterion for final selection of projects. The National Economic Development Authority sets the economic internal rate of return for evaluating development projects. The NEA applies this rate to the RECs' expansion projects. Typically, the economic rate of return required for approval of such projects is 15%. Projects with lower rates of return are implemented only if the government provides the NEA with the needed subsidy.

Procedures for Setting Electricity Prices. One positive feature of the cooperative system of electricity supply is the relatively clear way they set prices for their customers. The Department of Energy Act brought all electricity pricing, including bulk supply tariffs of the NPC, under the mantle of the Energy Regulatory Board. Before the NPC became responsible for all supply sources, much of the power was generated by the RECs themselves, and tariffs varied widely among them. Now these tariff differences are considerably smaller.

Until 1990, the NEA interpreted its mandate to approve REC tariffs as both establishing the tariffs and regulating the results. Its approval procedures were cumbersome, stringent, and slow, which added to the cooperatives' financial problems, rendering even the most disciplined ones unable to keep pace with inflation and rising costs. When a redefined NEA allowed the RECs to start setting their own tariff rates, the 30–50% increases that resulted caused only minor political opposition (World Bank 1994).

The 1992 tariff formula required RECs to cover their operating costs and NPC amortization payments and energy bills. The formula also provided for self-financing of future investments. Relatively simple, the formula had no variation for quantity consumed and little difference between consumer categories.

The tariff rates of RECs hinge on a number of factors. A major consideration is the cost of bulk power supplied by the NPC. This cost includes a fuel supplement linked to the price of the generating fuel, which varies depending on the price level and fuel use, as well as the fuel mix between regional grids. In 1996, bulk electricity charges to the RECs were in the 0.069–0.085 U.S. cents (P1.8–2.4 range) per kilowatt-hour range.

After the NPC became responsible for all supply sources, RECs could still obtain their supplies from isolated diesel or other generating plants, but they were charged as though they were connected into one of the regional grids. Because the costs of isolated generation (typically using 1–5-MW diesel plants) greatly exceed those of the large power stations in the regional grids, such RECs receive a subsidy. The amounts are relatively small at a national scale, involving less than 1% of the NPC's overall operation costs; however, they are extremely important for the RECs, particularly in cases where charging the full local generation costs would involve tariff increases of up to 25–30 U.S. cents.

Two other factors that enter into the tariff decisions are the level of system losses and the operating expenses. The NEA caps losses that the RECs are allowed to pass on to their consumers through the tariff system. In 1997, this amount was lowered from 22 to 20%, was reduced to 18% in 1998, and fell to 14% in 2000. Any excess losses have to be absorbed by the RECs through economies elsewhere. Subject to NEA scrutiny and approval, operating expenses are intended to cover normal REC operating and maintenance costs. Taken together, bulk supply costs, systems loss allowance, and operating expenses make up the total operating cost per kilowatt-hour for the REC.

Added to the total operating cost is the RECs' amortization cost, which depends on the level and status of their outstanding debt. Generally, NEA investment loans to the RECs for system development have not been written off. The main exception to this rule is investments in failed renewable energy projects and other such problems. A further allowance of 5% of total operating costs is added to provide for reinvestment in the system. This amount is insufficient for normal expansion and upgrading requirements that must be covered by loans from the NEA, and it is felt to be unduly restrictive by some RECs. This situation is especially the case of the RECs in low-density service territories, where investment requirements are higher than others.

The total permissible tariff charged by the RECs varies considerably. Operating expenses alone can account for more than 30% of the total cost in smaller or poorer RECs, where average consumption levels are low. In cases where fixed expenses can be spread over higher overall consumptions, the expenses component may drop to 15% or less. In 1996, the tariffs in operation ranged from just a little more than US$0.12–0.20 per kWh (P3 per kWh to almost P6 per kWh). Even at the lower end of the range, these were relatively high by international standards, so the highest rates are well above international norms.

Continual Evaluation of Cooperative Management. Another effective feature of the cooperative systems in the Philippines is the continual evaluation of the performance of the RECs. Each year, the NEA reviews and formally rates the RECs. Scores are awarded in six categories: amortization payments, system losses, collection efficiency, payment to the NPC, nonpower costs, and demerit points. Although the NEA uses categorization as a monitoring tool, its main purpose is to motivate the RECs to improve their performance. This improvement is reinforced by linking certain staff benefits to these categories.

Under the amortization payments category, for example, RECs that are up to date are awarded 25 points, whereas those that are more than three quarterly payments in arrears score 0. Allowance is made for RECs that have maintained their payments, have obtained a moratorium, or have been restructured. A restructured REC that is more than three quarterly payments in arrears is penalized 10 points. An REC ahead in its payments, by contrast, receives a 5-point bonus.

RECs with system losses of 11% or less receive 25 points. As the system loss increases, the score decreases to 0 for losses of 26% and higher. RECs with collection efficiencies of 95% or higher score 20 points; those with 75% or lower efficiencies score 0. RECs that are up to date with NPC payments earn 10 points, and those that are three quarterly payments in arrears earn 0; those with a restructured loan are penalized 10 points.

The NEA sets the allowable budget level for RECs each year. RECs that stay within the agreed-on budget score 15 points. Those exceeding their budgets are downgraded, receiving a score of 0 for an excess of 15% or higher. Single-point penalties are also given for every P50,000 of unliquidated cash advances to officers and employees, for failing to complete budgets on time, and for not providing 5% of revenues for reinvestment. Scores are added, and RECs are categorized (Table 3-1).

Table 3-1. *REC Rating System in the Philippines*

Score	Category	Rating
90 and above	A+	Outstanding
75–89	A	Very satisfactory
65–74	B	Satisfactory
55–64	C	Fair
30–54	D	Poor
29 and below	E	—

Source: Philippines' Rural Electrification Administration.

The cooperative rating system is designed to encourage competition among the cooperatives. The idea is that the cooperatives will learn from each other's experiences. The cooperatives that are rated fairly low can improve their standing over time. Also, it allows the NEA to determine which of the RECs require serious consideration for a change in their managers.

Rural Electric Cooperative Experiences: Responding to Problems

Understanding the context within which the Philippines's major types of RECs were established, the obstacles they overcame, and the problems that remain provides important lessons in rural electrification. This section considers these examples: the country's two initial pilot cooperatives, a larger grid-connected cooperative, and two representative island cooperatives.

The First Pilot Cooperatives: MORESCO and VRESCO

Initial planning for the Philippines rural electrification program began in the mid-1960s. Before a decision was made on the exact form the program should take, however, the national government decided to establish two pilot RECs: Misamis Oriental Rural Electric Cooperative (MORESCO) on the island of Mindanao and Victorias Rural Electric Service Cooperative (VRESCO) in the Visayas.

Misamis Oriental Rural Electric Cooperative. MORESCO, the first pilot cooperative, was established on a peninsula in northern Mindanao in 1970. The franchise area is located along a costal strip of agricultural land that rises into a range of rugged mountains. MORESCO is bordered by two small but reasonably prosperous urban cities: Cagayan de Oro to the east and Iligan to the west. In the late 1960s, the franchise area was predominantly rural. A single all-weather road, served by seasonal feeder roads, ran its length. Economic activity was limited to low-productivity farming. Coconut was the main cash crop, rice and maize were grown as food crops, and fish farming was also important. The total population was 108,000, literacy was about 75%, and 85–90% of primary-age children attended school. Income levels were relatively uniform and not greatly above subsistence.

In early 1967, a team of consultants from NRECA conducted a detailed study to determine the feasibility of setting up an REC in the area. During fieldwork, public meetings were held in each of the eight municipalities. At these meetings, rural electrification committees were formed to represent each municipality. Representation was widespread, including businesses, churches, government, education, and women's groups.

Local leadership was an important factor in this initial development stage. At the time, rural electrification initiatives had little credibility, and a major barrier of local distrust had to be overcome. Fortunately, a prominent senator and former vice president of the country, who lived locally, became an enthusiastic supporter of the cooperative. Local politicians gave the project their full support and refrained from engaging in partisan politics, which would have undermined the effort.

Once the local committees had been elected, they were given the task of informing people about the cooperative and helping to promote and organize its membership. By the beginning of 1968, 5,543 families out of the total 18,000 families in the franchise area had expressed an interest in joining the cooperative.

NRECA consultants carried out a financial analysis on the basis of the probable initial cooperative membership and its likely growth rate. The conclusion was that adequate working capital and reserve funds would be available for the next 10 years. Thus, a decision was made to go ahead. The cooperative was formally constituted, and construction started in 1970. Despite some construction delays and cost overruns, the network was energized in September 1971.

MORESCO rapidly became a national showpiece. National and foreign dignitaries were led though the countryside to see the system and the households benefiting from it, and senior NEA staff were sent there for training. The sight of electricity meters attached to the walls of one-room houses of thatch and matting was a revelation to many people. It showed skeptical planners at the national level that even people in poor circumstances wanted electric power and were prepared to pay for it. It also meant that people in other areas, as they learned of the project, began to demand, with increasing insistence, that they too be provided with an electricity supply.

Over the past 20 years, MORESCO has survived and prospered moderately. Now named MORESCO I, the cooperative had, by the end of 1995, energized 81% of the barangays in its franchise area. It had a coverage rate of 62% (23,592 consumers out of a potential 38,181), with residential consumers accounting for 93% of the total. There were 921 commercial and 20 industrial consumers. Average household consumption was 48 kWh per month, still well short of the 60 kWh predicted for 1975. Nonetheless, MORESCO was functioning well and obtained an NEA rating of A+ in 1995. That same year, its collection rate was 98%, and system loss was only 1%.

Victorias Rural Electric Service Cooperative. VRESCO, the country's second pilot cooperative, was established along the west coast of Negros Island in the Visayas. The franchise area, located about 50 km north of Bacolod, the provincial capital of Victorias, consists of a relatively large coastal plain that stretches back 20 km to the mountains.

At the time it was set up, VRESCO had a population of 200,000 in three municipalities. The area was devoted mainly to sugar production on large estates, and most of this sugar was exported to the United States at subsidized prices. Although profits were large, incomes for migrant sugar workers were minimal (Denton 1979). Huge income disparities separated wealthy, often absentee, landlords from poor laborers and tenant farmers. Unlike other rural areas in the Philippines, the island of Negros had few middle-class landowners. Thus, the environment was extremely difficult for testing the rural cooperative concept.

Two of the municipalities had private electricity suppliers, but service was unreliable, expensive, and confined to the market center. These deficiencies led sugar planters to set up their own cooperative in 1965. Power was provided by the local sugar-milling factory but proved extremely limited in capacity and subject to interruption in accordance with its operating needs. As a result, the planters' cooperative could provide an electricity supply to only 53 of its members.

Although VRESCO's technical problems were relatively minor, the new cooperative faced serious social challenges. The existing electricity supply systems catered only to the rich, ignoring impoverished worker families. Another pervasive problem was the existing culture's evasion of payment for electricity and other public services. Friends and relatives of powerful figures could evade payment, illegal connections were common, and meter readings were distorted (Denton 1979).

It took considerable courage for VRESCO's management to tackle these social issues. However, the necessary decisions were made, and a firm policy of disconnection for nonpayment was adopted. By 1976, the problem had virtually disappeared. The distribution system was steadily extended to low-income households, and by mid-1976, the cooperative had 11,600 members.

Despite its difficult beginnings, VRESCO survived. At the end of 1995, the proportion of electrified barangays in the service area was 82%. However, the proportion of households with a supply was only 41%. Residential consumers accounted for 89% of the total. Average consumption was 52 kWh per month, still significantly below that projected for the fifth year of operation. Average annual consumption of the cooperative's 215 industrial consumers was 147,000 kWh. In 1995, the NEA gave VRESCO a rating of B. Although its performance was creditable, the cooperative still faced ongoing difficulties in providing poor rural workers an electricity supply.

The Turbulent Performance of BATELEC 1

BATELEC 1 is in a picturesque region with good beaches and areas for diving, and has seen a substantial build-up in tourist facilities in recent years. A significant amount of industrial development has also occurred, as evidenced by expansion of a steel rolling mill. The electric cooperative serves 12 municipalities in Batangas Province. Inland, where one finds isolated farms and farming villages, the population gradually decreases as the mountain slopes become steeper. The bulk of the population lives along the coast, where there is a string of small towns, fishing villages, and holiday resorts.

BATELEC 1 was among the first RECs established in the Philippines. Following a feasibility study, the NEA granted approval for establishing the cooperative in May 1972. After initial mobilization of the local community, BATELEC 1 was formally established and registered in June 1974. Construction of the REC's headquarters and the distribution network began the next year. The main backbone line followed the coast, connecting the existing settlements, with spur lines extended inland and upward along the lower mountain slopes.

The first town to be energized was Calaca, site of BATELEC 1's headquarters, which received power in late 1975. Shortly thereafter, the REC assumed control of the existing private supply system in the town of Lemery; a year later, it took over the private supply in the town of San Luis. After its headquarters opened in May 1975, the cooperative expanded rapidly, obtaining a second loan from the

NEA in 1977. The first board election in which local candidates, as opposed to nominated members of the provincial electric cooperative team, could seek office was held in 1979. The cooperative continued to expand rapidly, receiving a third NEA loan in 1981 and a fourth in 1984, the same year it received an award for having the most new house connections. By 1987, it had energized 89% of the barangays and 85% of the households in its franchise area. Two years later, all of its barangays and 92% of its 59,100 households had been connected.

The success of the rural expansion of the cooperative started to unravel in the 1980s, when bill collection efficiency dropped to only 75%. This deficiency, combined with problems of management and the board of directors, caused the cooperative's rating to drop to D. In 1987, BATELEC 1 showed an operating loss of more than US$14,000 (P330,000). The NEA responded to the problems by assuming direct management of the cooperative. Its board was dismissed in 1989, and a new acting general manager was appointed. The REC's loan of P8 million was restructured, and a major tariff increase of about 50% was imposed in 1990. The new management achieved a major improvement in financial and managerial performance, and in 1990 BATELEC 1 received the country's award for most improved electric cooperative.

Progress continued in 1991, when the acting general manager won an NEA award for outstanding service. That same year, the cooperative settled its loan repayments and power account with the NPC and reduced its system losses to 10%, for which it received the award for best in low system loss. In acknowledgment of these achievements, its NEA rating was upgraded to A.

Rapid population growth in the franchise area is reflected in the increasing number of electricity connections. In 1987, the number of connections was 50,010, representing 84% of the potential total of 59,104. In 1995, the number of connections had reached 67,890, or 91% of the potential total of 74,369. By September 1997, the total number of connections had increased to 77,896, barangay electrification had reached 98%, and the total value of sales was more than US$14 million (P290 million).

Breakdown by consumer for 1996 shows that 64,604 were residential (about 88% of the total), 3,224 were commercial, and 13 industrial. Annual consumption by residential consumers was 798 kWh, just 66 kWh per month. Such a low level of consumption indicates that the main electricity uses were lighting, television, and small appliances, with little cooking or water heating. Commercial consumers had an average consumption of 6,220 kWh per year. Annual revenue per domestic consumer was more than US$100 (P2,828).

Electricity consumption has grown rapidly and continuously since the late 1970s. As Table 3-2 shows, overall consumption rose almost fivefold between 1979 and 1996. The peak demand on the system up to September 1997 was 27.49 MW.

Residential use accounted for almost 47% of total consumption in 1979 and grew to almost 60% by 1996. Commercial use grew from about 9% to 23% over the same period, reflecting the increased economic activity in the area. Industrial use fell from about 37% of the total to under 12%, but this loss does not reflect the full picture. A number of large industrial users in the area, such as steel mills, obtain

Table 3-2. *Electricity Consumption Growth by End-Use Category in the Philippines, 1979–1996*

Category	Annual Consumption (thousand MWh)				
	1979	1985	1990	1994	1996
Residential	8,648	14,177	23,838	37,337	51,163
Commercial	1,741	4,348	7,992	12,649	20,052
Industrial	6,905	6,331	12,507	12,744	11,157
Public buildings	337	392	698	1,099	2,630
Other	848	1,485	1,866	2,968	1,470
Total	18,479	26,733	46,901	66,797	86,472

Source: Philippines' Rural Electrification Administration.

their supply of high voltage directly from the NPC. These consumers sign a waiver agreement with the cooperative and pay a royalty on their consumption.

In 1996, total system losses had increased to 21.8%, a highly unsatisfactory figure in comparison with the 10% achieved in 1991. In large part, these losses resulted from rapid growth in demand in the franchise area, which caused severe overloading in parts of the network. Subsequently, three substations (two 10-MVA and one 5-MVA) have been installed, greatly relieving the overloading at critical points. BATELEC 1 has also purchased a mobile 10-MVA substation, which will be used for temporary reinforcement or backup as needed. A variety of other measures were taken, including line clearing and general rehabilitation of the distribution system. Defective meters were replaced, and antipilferage action was stepped up. As a result, losses were reduced to 12.3% by July 1997.

BATELEC 1 makes significant efforts to benefit its members and local community in the franchise area. Outreach includes a quarterly newsletter, news broadcasts on a local radio station, and member meetings at various barangays and schools. The cooperative holds a monthly raffle, provides a 2% discount to consumers who pay their bills within 10 days, and makes scholarships available to eligible children of member families. Through an arrangement with a private health-care company, employees are entitled to annual medical exams, free outpatient consultations, hospitalization, medicine, and surgery.

The cooperative also participates in a number of membership-oriented livelihood programs. Under a wood pole production program, for example, some 100,000 seedlings have been distributed to farmers since 1994. Under a pig-breeding program, about 1,000 pigs have been distributed to member families. A pilot quail-raising program has also been launched.

The experience of BATELEC 1 provides insight in how necessary it is to have institutional support for rural electric cooperatives in time of crisis and the continuing challenges that it faces in providing electricity service. BATELEC 1 now appears to be a well-functioning organization. The initial task of electrifying the franchise area has been almost completed, but population growth and immigration ensure an ongoing demand for new connections. Commercial and industrial development of the area, which is well within Manila's economic reach, are likely to continue, leading to increased demands on the distribution system and the need for ongoing capital investment in the system.

The Island Cooperatives of BANELCO and FICELCO

Some of the country's remote, relatively poor islands demonstrate how cooperatives can operate reasonably successfully under conditions that usually would be considered on the margins of viability for rural electrification. The following accounts of two such island RECs show what can be achieved when committed, disciplined management works in collaboration with local communities.

BANELCO Service Area. The Bantayan Island Electric Cooperative (BANELCO), formally incorporated in 1978, has served this small island since 1983. Located near Cebu in the Visayan Sea and covering an area of 90 km², Bantayan comprises three municipalities and 40 barangays, with a population of about 90,000.

Bantayan's poor topsoil makes agriculture difficult, except for coconuts, which flourish near the island's shoreline. Farm holdings average about 1.5 ha. The main crops are maize, coconut, cassava, and other fruits. Known for its egg production, the island has a number of large battery farms that sell to other areas of the country. Rich fishing grounds surrounding the island provide a living for about 60% of the population. Even so, insufficient production of maize and inadequate amounts of water needed for rice cultivation make Bantayan dependent on imported food for subsistence. The island has little industrial development but a reasonably active service sector. Incomes are low and fluctuate seasonally.

BANELCO operates in a difficult physical and economic environment. Typhoons regularly sweep through the area, causing major damage to the distribution system and to consumers' property. The majority of families are poor, and their consumption of electricity is low. Nonetheless, by careful control of its costs and a tariff that allows it to cover them, this cooperative has managed to cover its operating costs. The NEA's 1995 rating of BANELCO's overall performance was A+, reflecting the cooperative's disciplined and innovative management.

The cooperative increased the number of connections per employee from 60 in 1985 to 186 in 1990. It has adopted a rigorous policy of ensuring that system losses are kept to an absolute minimum and, wherever possible, borne by the consumer. Meters for its industrial consumers are positioned before, rather than after, the transformer. Where there is a long service drop to a domestic consumer, the meter is positioned on the distribution system pole rather than the dwelling. As a result, system loss was only 7% in 1995. BANELCO has also adopted an aggressive antipilferage policy. Meters are often positioned well away from dwellings in highly visible locations to deter meter tampering. Bill collection efficiency is 100%.

By the end of 1995, all of the barangays on Bantayan had an electricity supply. Out of the total 17,544 potential consumers, 10,317 were connected, representing a 59% coverage rate. Of connected consumers, 9,185 were residential, 726 commercial, and 47 industrial.

In 1995, BANELCO's electricity tariff was US$0.14 (P3.65) per kWh, subject to a minimum consumption of 10 kWh per month. This minimum monthly tariff was close to the low-end range of expenditure on kerosene, showing that, even for the most modest level of consumption, electricity, because of its superior lighting quality, was competitive.

BANELCO charges a connection fee based on distance from the nearest distribution system line. In 1995, this fee was charged at P20 per meter of service drop, plus labor charges. Taking house wiring costs into account, the minimum installation cost for a household was about US$57 (P1,500) that year. About half the families connected availed themselves of the cooperative's two-year loan of US$38 (P1,000), which was offered at an interest rate of 1% per month.

Larger industrial and commercial consumers requiring a three-phase supply pay for the necessary wiring, meter, and labor costs. Those unable or unwilling to cover the cost of the transformer may rent it from BANELCO.

As a result of financial difficulties during the mid-1980s, BANELCO's rating fell to C in 1985. A rigorous program of reform, however, transformed its finances, and, by 1992, had moved the cooperative into a surplus in its operations. This positive change resulted from tariff increases, a loss-reduction campaign, 50% growth in the number of consumers, and a modest increase in average consumption.

FICELCO Service Area. The First Catanduanes Electric Cooperative (FICELCO) serves Catanduanes, a small, sparsely populated island province southeast of Luzon. The island is prone to typhoons, which regularly cause major damage to FICELCO's distribution network. Catanduanes is made up of 11 municipalities and 314 barangays, with a population of 190,000 as of 1990.

Agriculture is the province's primary sector. Major crops are rice, corn, coconut, and hemp. Most agriculture is rain-fed, and yields are generally low. There is little processing of crops, and most are exported raw to Luzon. The industrial sector is limited to small grain mills and a chicken-feed mill. The island's reasonably active service sector includes general shops, restaurants, tailors, and small workshops. Incomes are low, and many families rely on contributions from relatives working in Manila or abroad.

Demand for an electricity supply is high. Private suppliers are common in unelectrified barangays. A spot survey in 1993 found that two-thirds of the households in one barangay near a market center were connected to a private supplier who charged 13 U.S. cents (P3.50) per night per 20-W light bulb. The range of monthly expenditure on kerosene was between US$1.00 and US$2.50 (P30–70).

Formally incorporated in 1971, FICELCO was a recipient of surplus U.S. Army generating equipment. It became operational in 1972, when it entered into a lease agreement with the provincial government to operate the existing franchise covering the adjacent towns of Virac and Bato. Following is a description, taken from FICELCO's newsletter, of its eventful early years:

> After coasting along for 19 months with just enough revenues to sustain the cost of operations, FICELCO was suddenly jolted by the fuel-price increase in October 1973. Fuel bills began eating up bigger and bigger chunks of

the collections, demand forecasts nose-dived precipitously, and consumer complaints mounted. Expansion plans had to be reined back, and the use of alternative energy sources started to be conceptualized.

Amid the financial and operational woes, the coop persisted in the attainment of its corporate aim to give electric service to more and more households. Construction of its power plant and headquarters facilities was completed in May 1975. The two 500-kW generating sets were put in operation in January 1976 at the new Marinawa Power Plant, replacing the two 250-kW units at Camp Camacho, which had been rendered inadequate by the connection of added loads. FICELCO was then ready to serve the power demands of its growing membership.

FICELCO incurred losses during most of the 1980s, but following a rigorous reform program, moved into a surplus in its operations by 1992. A steady reduction in system losses, increased sales, and a doubling of the domestic tariff accounted for the cooperative's improved performance.

Before an area is electrified, FICELCO makes a demand assessment and organizes meetings in cooperation with barangay leaders. It is required that at least 80% of all households in the barangay sign up for a connection if a supply is to be provided. The minimum number of households required for electrification of an area is 30. FICELCO charges a sign-up fee, depending on the distance from the take-off point, and the usual fee is in the range of US$35 to US$46 (P900–1,200). The cooperative relies heavily on the use of BAPAs for its billing and collection, and almost 100 have been formed.

In 1995, FICELCO had 22,894 consumers. Of these, almost 90% were residential, with 854 commercial and no industrial consumers. All 11 municipalities and 80% of the barangays were connected, and the proportion of potential consumers with a supply was 72%. In 1995, average residential consumption was 30 kWh per month. Although low, this level was a notable increase from the 13 kWh per month average consumption in 1988. Commercial consumption averaged 130 kWh per month. The domestic tariff was 17 U.S. cents (P4.67) per kWh, with a minimum monthly charge of 10 kWh. Collection efficiency was 96%, and system losses were 14%. FICELCO's A rating in 1995 was upgraded to A+ the following year.

Conclusion

Final judgments on the Philippines rural electrification program at this stage are not possible. The program has experienced many changes in the past and, no doubt, will face new challenges in the coming years. For example, the restructuring of the electricity sector may bring significant changes in attitudes and operating modes of many RECs.

Nonetheless, the program provides a number of important lessons, which are highlighted below. Together with questions regarding the implications of power-sector restructuring, these lessons can help not only the Philippines move forward in rural electrification, but can assist other developing countries that are grappling with extending electricity supplies to their rural residents, particularly within complex political, social, and geographic environments.

The cooperative system has effectively delivered reliable electricity supplies to low-income rural areas on a financially viable basis. Although direct involvement of members in the running of the RECs tends to be minimal, the cooperative concept appears to bring a greater degree of local responsibility because local people, not a remote bureaucracy, manage the operation. Thus, it is easier for members to accept the idea that cooperatives in more remote areas must charge considerably higher prices than those operating under more favorable conditions.

Effective cooperative management requires careful and objective choice of franchise areas, professionally competent design of distribution systems, and effective management thereafter. Because cooperatives are small organizations with limited technical and administrative capacities, their personnel need to be provided regular training and updating of skills. They also need access to the skills and resources required for dealing with emergencies and more specialized problems.

Proper supervision and accountability are critical because cooperative managements tend to be relatively fragile and isolated, making them susceptible to local corrupting influences. As long as the central support structure remains effective, such problems can be detected and dealt with, or at least kept within acceptable limits. Where leadership and discipline are lacking at the core, the system inevitably begins to break down.

An adequate tariff system and a clear framework of financial responsibility are keys to success. Many problems encountered by the Philippine RECs sprang from the fact that tariff increases failed to keep pace with rising costs of the early 1980s (World Bank 1988). Maintenance was neglected, leading to deterioration in service. Financial discipline was undermined, and the effects spread throughout the RECs and into the consumer base, leading to pilfering and nonpayment of bills. It is notable that the large, but obviously necessary, tariff increases of the early 1990s were accepted virtually without protest by consumers.

The path of rural electrification in the Philippines has been difficult at times, in many cases because of forces outside the control of the RECs. The promising start eventually got bogged down with multiple political demands for the cooperatives to do much more than just distribute and sell electricity to rural people. In the Marcos era, the rural electric cooperatives were directed to provide many different types of services to many different masters. The ability to deal with all of the problems encountered over the decades and to continue to expand and provide high-quality rural service into the new millennium bodes well for the future of rural electrification in the Philippines.

Note

1. Configuration of the distribution system is known as three-phase, four-wire, multigrounded. This name means that, in addition to the three phases in the backbone system, there is a neutral conductor, which is grounded at each pole. The primary distribution system voltage

is 13.2 kV (phase-to-phase) and 7.62 kV (phase-to-ground). The secondary distribution voltage is 240 V for single-phase consumers and 415 V for three-phase consumers. The secondary distribution lines are placed under the higher voltage lines (underbuilt) or on their own poles (open secondary lines). A single-phase transformer always provides the secondary supply; three-phase consumers are supplied through three transformers. Overhead conductors are aluminum-core steel reinforced. Service drop conductors are of the duplex type with an isolated live conductor and a bare neutral twisted together. The average service drop is about 20 m on wooden poles.

CHAPTER 4

Rural Poverty and Electricity Challenges in Bangladesh

Daniel B. Waddle

*I*N 1976, BANGLADESH WAS A NEWLY independent country grappling with
the challenges of creating its national policies and programs amid underdeveloped infrastructure, a rapidly growing population, and frequent natural disasters.
Few would have predicted that 25 years later, this poor South Asian country would
have succeeded in providing electricity to more than 50% of its total population,
including 28% of its rural households. The rural cooperative distribution system
losses of 15–17% are quite low, and the revenue collection rate of about 95% is
high by developing country standards.

There are several reasons that the story of electricity development in rural
Bangladesh is important. The first is that despite the current relatively low coverage rates compared to other countries reviewed in this study, every year 500,000
new rural customers are receiving electricity service. This yearly increase exceeds
the total rural population of many countries. Second, Bangladesh is in a unique
position on the South Asian continent with its well-run and well-managed rural
electricity program. Last but not least, many people have argued that rural electrification cannot succeed in extremely poor countries. Bangladesh is among the
poorest countries in the world, so it provides an example of how rural electrification programs can make progress even in adverse economic conditions.

Since it was formed in 1978, the country's Rural Electrification Board (REB)
has funded 63 rural electric cooperatives, known as Palli Bidyut Samities (PBSs).
Fifty-four PBSs are operative, and several more will soon begin operations. The
REB program has received more than US$900 million in funding from 16 inter-
national donors and the government of Bangladesh. However, the program has
also faced it share of problems. Desiring to serve as many rural residents as possible,
many newer PBSs have accelerated service expansion into remote areas where
industrial and commercial load growth are minimal. Resulting shortfalls in operating income now risk these cooperatives' future ability to finance infrastructure

improvements. Valuable lessons can be learned from this country's remarkable rural electrification program and how it is grappling with its problems as it approaches the crossroads of sustainability.

Background on Bangladesh

Bangladesh was established as a country in 1971, when Bengali East Pakistan seceded from its union with West Pakistan. Bordered by Myanmar (Burma) to the west, India to the east, and the Bay of Bengal to the south, this South Asian country is slightly smaller than the U.S. state of Wisconsin. Terrain consists mostly of flat alluvial plains, with the exception of the hilly southeast. Much of the country is routinely flooded during the monsoon season, and droughts and cyclones are common.

One of the challenges for rural electrification in Bangladesh is that it is one of the world's poorest and most densely populated countries, with an estimated population of almost 130 million people. Despite ongoing domestic and international efforts, Bangladesh still has 35.6% of its people living in poverty. The country's labor force of 56 million is predominantly agricultural (63%); rice cultivation is the most important economic activity. Major industries include cotton textiles, jute, garments, and tea processing. In 1996, unemployment was 35.2%. Electricity in rural areas is used widely for pumping water in regions that have access to service because most rural people in Bangladesh depend on agriculture.

As is the case with electricity, the other infrastructure in the country is modest. Of the country's 201,182 km of highways, only 19,112 km are paved. Access to potable water is limited, and water-borne diseases are prevalent. Use of commercial pesticides results in water pollution, especially in fishing areas. Water shortages are common, caused by falling water tables in the north and central regions, deforestation, and overpopulation.

The country's culture dates back thousands of years. Bengalis make up the largest ethnic group (98%), with fewer than 1 million tribal people and 250,000 Biharis. The predominant religion is Islam; Hinduism, Buddhism, and Christianity are also practiced. In 2000, infant mortality was 71.66 per 1,000 live births, and life expectancy was 60.4 years for men and 59.9 years for women. In 1996, public expenditure on education represented 2.9% of the gross national product. As of 1997, 50% of men and 27% of women were literate.

History of Rural Electrification, 1971–1976

When Bangladesh gained independence from Pakistan in 1971, the country faced the daunting task of establishing a completely new set of public institutions to finance and implement national policies and programs. The growth of government over the ensuing years has been complicated by recurring natural disasters, an often marginally stable political environment, and a severely underdeveloped infrastructure.

From its beginning, the government has been supportive of developing a program for providing electricity to rural areas. In 1972, the newly formed government of Bangladesh (GOB) committed itself to this goal by stating in its constitution (GOB 1998a, Article 16) that:

> The State shall adopt effective measures to bring about a radical transformation in the rural areas through the promotion of an agricultural revolution, the provision of rural electrification . . . so as progressively to remove the disparity in the standards of living between the urban and the rural areas.

The Bangladesh Power Development Board (BPDB) was formed to operate and expand the generation, transmission, and distribution network of the former East Pakistan Water and Power Development Authority (EPWAPDA). BPDB was given responsibility for operating electric power services throughout urban population centers, including the cities of Dhaka, Chittagong, Khulna, and Narsingdi. In addition, it was charged with operating the fragile generation and transmission system it inherited from EPWAPDA. After the new directorate for managing rural electrification responsibilities was established, it soon became apparent that the program would become too large for BPDB to manage effectively, given the board's other obligations.

In 1977, less than 10% of Bangladesh's land had electricity service. That same year, the BPDB commissioned the National Rural Electric Cooperative Association (NRECA), financed by the U.S. Agency for International Development (USAID), to conduct a study on alternatives for a national rural electrification program. In response, NRECA developed a master plan, closely modeled after the U.S. rural electrification program of the 1930s, for establishing 72 rural electric cooperatives (RECs).

According to the NRECA master plan, the cooperatives would be financed and supervised by a semiautonomous government body, which would establish the electricity distribution infrastructure to enable all rural inhabitants to avail themselves of modern electric power provided through a central transmission grid (NRECA 1978). The program would emphasize providing electricity to areas whose energy demand is driven by agricultural mechanization, irrigation, and rural industries.

The program would also follow the principle of area coverage rural electrification, which was successfully implemented in the United States. The area coverage objective is to develop infrastructure that is expected to lose money in the early years, followed by profitability as load grows to cover the cost of service. The advantage of this approach is that a distribution network is established early on that forms the basis for higher levels of access in subsequent years. Moreover, the area coverage program provides a uniform tariff to all consumers in specific classes located within a specified distance from the distribution grid. The area coverage approach also promotes equal access to the grid through application of a uniform connection fee. Finally, it is based on the cooperative ownership principle, whereby consumers govern the operation of each electricity distribution system through an elected board of directors.

The NRECA master plan was adopted in 1977, resulting in the establishment of the Rural Electrification Board the following year (REB 1987). The first phase of the REB's development was financed by USAID and aimed to develop the capacity of the REB to manage the program and to establish the first 13 RECs. In the years that followed, 68 RECs, known as Palli Bidyut Samities, have been constructed and are now in operation. To date, the REB system includes more than 200,000 km of lines, more than half of which have been added in the past five years.

The growth of rural electrification under the REB and PBS systems has been remarkable. Today after 30 years, existing PBSs already constitute an extensive distribution network that covers most of rural Bangladesh. As Figure 4-1 shows, these PBSs serve almost 7 million rural households—more customers than the Dhaka Electricity Service Authority (DESA) and BPDB combined. But as indicated, service connections continue to increase at an annual rate of almost 500,000, although this number includes come customers taken over from DESA and BPDB.

REB Organizational and Program Overview

The REB is an agency of the Ministry of Energy and Hydrocarbons. It is responsible for planning and implementing all investments in rural electrification infrastructure, as well as overseeing PBS performance and regulating prices. Since its inception about 20 years ago, the REB has constructed the infrastructure for Bangladesh's entire national rural electrification system. In this capacity, one of the main responsibilities of the REB is to manage loans and grants provided by international donor agencies to finance the infrastructure development program. It also has the role of providing technical assistance to its own staff and PBS staff managers, engineers, and line workers. Another essential function is to finance

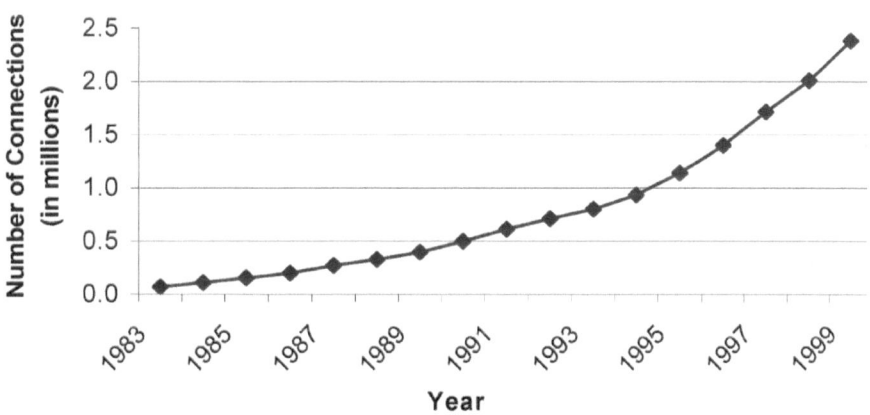

Figure 4-1. Cumulative REB Household Connections, Bangladesh, 1983-1999

Source: REB Form 550.

short-term capital needs of newly established PBSs. Finally, the REB monitors the PBSs' technical and financial performance and assists them in evolving into modern electricity distribution utilities.

To accommodate its wide range of essential program functions, the REB is divided into three management units: finance, PBS oversight and training, and engineering. Each unit is subdivided into directorates, each of which is directly responsible for a single, focused activity.

Finance

The financial management unit is responsible for monitoring, reporting, and coordinating overall program financial matters among the REB, other government agencies, and program donors. The unit is composed of six directorates: finance, accounts, procurement, the directorate for PBS loans, internal audit, and rates/contracts. Finance reports on program expenditures and results and coordinates annual budgets with the Ministry of Finance. Accounts manages payable and receivable accounts, including payments for program commodity contracts and power sales agreements. Procurement manages the bid solicitation and evaluation process, coordinating with Systems Engineering Design, which is responsible for developing bid specifications.

The directorate for PBS loans monitors payments of interest and principal for all REB loans to the PBSs. The directorate for procurement, in contrast, manages all aspects of the procurement and construction process on behalf of the PBSs. This work is done on behalf of the PBSs who do not themselves manage new construction. This procurement includes bulk purchase of equipment and materials for new projects that are being executed as the backbone and lateral lines are constructed. They also manage and finance construction contracts. Internal audit monitors all REB financial activities to ensure that they comply with program policy, as established by the Ministry of Energy and Hydrocarbons and the REB board of directors. Rates/contracts manages data and performance costs of service and revenue analyses, a function whose importance will only increase in the coming years as power generation is liberalized and bulk power rates begin to reflect market trends. In addition, after turning over the lines, the REB monitors the PBSs' technical and financial performance and assists the cooperatives in improving their performance with the goal of making them evolve into modern electricity distribution utilities.

PBS Oversight and Training

The PBS oversight and training unit provides ongoing assistance to PBS personnel and members to enhance program sustainability and to promote user participation in PBS governance. This unit is important because it provides the basis for monitoring the financial performance of the PBSs. The oversight and training unit has two directorates: development and operations, and training. The PBS development and operations directorate assists PBS membership in governance functions, report preparation to the REB, member service functions, and promotion of productive uses of electricity; and it helps ensure that board members are properly trained and oriented toward the best interests of the members they represent. The training directorate designs and provides training for REB personnel, as well as PBS managers, engineers, line workers, and administrative staff. Extensive training programs encompass all facets of utility management, procurement, maintenance, system design, safety, and overall REB program management.

Engineering

The engineering unit is responsible for monitoring technical operations at all PBSs, establishing and enforcing standards for all new construction, receiving equipment procured for new projects, and overseeing construction for the expansion program. One unit in the engineering unit manages expansion programs and, accordingly, has a large complement of staff and resources. The operations unit ensures that already constructed rural electric distribution systems are properly operated and maintained. This work is especially important for systems as they grow older and provide service to greater numbers of rural customers.

System Expansion, PBS Development, and Local Participation

In Bangladesh, the organizations responsible for program implementation are the PBSs or RECs. As described above, the REB provides extensive support to the PBSs. This support is one of the hallmarks of the rural cooperative electricity distribution system. New cooperatives are often weak and require significant technical assistance in their early stages to survive. Many of the established cooperatives run into difficulties from time to time and require assistance from the REB to straighten out their problems. The REB has also given financial support for system expansion, so it is in their own interest to monitor their investments. The following sections review how this overall system works.

PBS Development and Expansion Planning

This begins when the engineering directorate unit conducts a PBS feasibility study to determine if a new service territory will be financially viable. The first step is to complete a comprehensive demographic survey. The proposed PBS service area is divided into a series of polygons for which a census is taken on projected household, irrigation, and commercial electricity loads. The census estimates the number of households, rice mills, tubewells[1], and stores, as well as average household income (Bangladesh Bureau of Statistics 1998b, 1999). This comprehensive field survey then becomes the primary source of demand data that the REB uses to analyze the proposed PBS's ability to meet certain revenue criteria.

The revenue criteria require that each kilometer of line constructed generate the equivalent of US$400 per month to cover operating costs. This benchmark is the primary factor used to determine financial feasibility for all new projects. It is applied to the proposed PBS territory as a whole, as well as to each set of communities that will be connected to the PBS feeders. Communities that fail to meet revenue criteria may be added later if the total estimated demand for electricity in the cooperative grows at higher than expected levels.

If results of the feasibility study are positive, the REB engineering unit develops a PBS master plan that designates the location of the backbone 11-kV feeders and determines when each part of the system should be built to optimize financial viability. The plan proposes the sequence of providing electricity to all viable communities, beginning with those with the highest estimated electricity demand compared to the costs of providing service.

During the organizational phase, the engineering unit continues to develop designs for the PBS backbone system and the first set of lateral distribution lines to be constructed. The REB uses standard designs across the whole system to reduce equipment costs. This cost reduction is achieved through annual block purchases that cover many cooperatives. Thus, special procurement actions for new PBSs are seldom necessary.

PBS Construction and New Electrification

As the equipment procurement process moves ahead, contractors are hired to begin construction of the distribution system. REB executive engineers oversee the construction process by monitoring reports provided by local consulting engineers. These consulting engineers directly review and accept lines and substations built of REB standard construction units by REB-approved contractors. Service connections are made as PBS members are enrolled and pay the necessary fees.

Assisted by its PBS training directorate, the REB provides practical training to PBS employees before providing electricity. When a new PBS receives electricity, an REB employee is assigned as the interim PBS general manager for a fixed time. After the system begins operating, the PBS must monitor all operating parameters and file monthly reports with the REB. These reports are submitted in a standardized format known as Form 550, which summarizes all technical and financial data of the cooperative. This includes energy sales and purchases, system losses, number of meters, changes in plant and equipment, and other relevant data (Box 4-1).

In the years following the initial service period, the PBSs continue to expand to communities without electricity by extending lateral lines in accordance with the PBS master plan. Households and agricultural pumps are connected to intensify load in areas that already have electricity. As feeders and substations begin to reach their design capacity, the PBS upgrades conductor size, distribution transformer capacity, and substation equipment. These changes are made by PBS consulting engineers and are coordinated by REB engineering personnel.

Local Participation and Interaction with Communities

As design and construction of the infrastructure proceed, owner-members of each newly formed PBS are educated about their rights and responsibilities. Also, potential new consumers are informed about how they can obtain service. Meetings with community leaders are held to disseminate key information and to begin

---

Box 4-1. REB Form 550—The Heart of Performance Monitoring

A: Revenue and Expenses Statement. Summarizes PBS operating revenues and expenses.

B: Aging of Accounts Receivable. Provides data on receivables from PBS members for the current month, 30 days, and over 90 days.

C1: Balance Sheet. Summarizes PBS assets and liabilities, including a summary statement of long-term debt obligations.

C2: Changes in Utility Plant. Summarizes the value of all PBS assets, any assets retired for the current month, and changes for the year to date.

D: Consumer Sales and Revenue. Summarizes sales by customer category for the current month and year to date.

E: Energy and Demand Data. Summarizes energy and demand data at each substation metering point within the PBS.

F: Plant and Consumer Data Sheet. Summarizes the total number of in-service, disconnected, and idle connections. Also summarizes the kilometers of line constructed by the PBS and the length of lines taken over from DESA or BPDB.

G: Accounts Payable Statement. Summarizes payments due to BPDB for power and to REB for outstanding loans.

H: Uncollectables. Summarizes receivables likely to go uncollected.

---

the process of board elections. Meetings are also held with rural industries, farming groups, and commercial leaders to ensure that their interests are considered and that they participate by becoming PBS members.

Community involvement in PBS administration is encouraged in a variety of ways. Before a PBS's formation, members can provide direct input through focus groups and membership drives. REB personnel organize membership committees to encourage community participation in the organization and formation of the PBS. Just before receiving electricity and in all subsequent years, members participate in electing board officers during annual meetings. Awareness of key issues is promoted by distributing information bulletins along with members' regular monthly bills. Information is also circulated at consumer meetings and through village electrification committees.

The village advisory program, a relatively new arrangement, allows community leaders to meet periodically with PBS member representatives to discuss how PBS management can most effectively address customer concerns. Village advisers meet twice yearly and are most often selected from among the community's schoolteachers.

Member services also facilitate communication between PBSs and their members through meetings with local village leaders; focus groups that address specific problems, such as load shedding and voltage quality; and educating members about their role and responsibility as PBS consumer-owners. PBSs have also begun an experiment to establish a village advisory program, whereby member representatives are selected to attend regular meetings to discuss issues of common concern to consumers in their villages. Although results from taking this approach have not yet been verified, one benefit already realized is a stronger relationship between the PBSs and village leadership.

More recently, PBSs have begun to include an informational newsletter along with the monthly electric bills distributed to members. The newsletter includes special announcements, such as dates of membership meetings, board elections, and key issues that affect power availability and quality. The REB believes that this approach may improve communication and information flow, particularly in the absence of radio, television, and postal services.

PBS Bill Collection and Commercial Practices

Running a successful rural utility depends on establishing and following sound commercial practices. Many utilities run into difficulties because of poor billing and collection programs, inadequate attention to consumer needs, and the absence of control over aging accounts.

In Bangladesh, all PBS commercial practices are designed to maximize the rate of bill collection and minimize fraud, theft, and administrative losses. The methods used, which have been refined over time, have proven to be effective. Individual PBSs manage their billing and collections cycle. PBS accounting staff calculate energy bills, which are printed within each PBS, and local PBS employees deliver the bills to customers. Payments can be made either to rural banks located close to members' villages or directly to the PBS's headquarters office. A long-standing arrangement with rural banks allows the PBSs to pay the banks a minimum fee for this service. This arrangement has proven effective, and local banks manage the majority of bill collection.

Members have 30 days in which to pay their bills and are assessed a fine if they fail to pay in full within this time period. If payment has not been received within 60 days, service is disconnected. If the bill has not been liquidated within 90 days, the meter is removed and PBS membership is revoked. Reconnection may be requested, but a fine of Tk50 (about US$1.00) is assessed for residential customers.

Meter readers are hired as fixed-term contract PBS employees. To minimize risk of customer fraud, meter readers are routinely rotated among service territories. In addition, their contracts are limited to three years, which is thought to minimize administrative losses and fraud.

Systemwide losses (including technical and administrative) are less than 17%, which is relatively low in comparison to other South Asian distribution utility systems. Collection rates average well above 95% throughout the REB system, mainly because of the tight control exercised by the PBSs and oversight of the REB. The system losses and collection rates over a six-year period for all PBSs in the REB program are presented in Table 4-1. Increased losses during 1998–1999 were most likely caused by the recent takeover of DESA lines by some PBSs. Residential energy sales far outpaced any other single category. The

Table 4-1. *PBS System Losses, Collection Rates, and Sales, by Customer Category in Bangladesh, 1994–1998*

Gains and Losses	1994	1995	1996	1997	1998
			Collections and losses		
System losses (%)	16.10	15.80	15.20	16.30	16.90
Collection rates (%)	98.80	95.10	98.20	94.90	97.00
			Sales (mWh)		
Residential	237,242	310,352	402,234	455,677	569,037
Commercial	43,704	57,236	68,983	74,324	87,902
Irrigation	157,945	272,112	242,069	212,063	191,473
Institutions	9,441	11,528	13,874	14,874	17,427
Small industry	220,459	266,414	310,245	348,347	399,764
Large industry	99,204	127,322	131,847	130,701	164,498
Street lights	1,929	2,131	2,431	2,620	2,538
Total sales	769,924	1,047,095	1,171,683	1,238,606	1,432,639

Notes: The irrigation sales bounced back in 1999, 2000, and 2001. The level of sales was 312,204 GWh in 1999; 269,757 GWh in 2000; and 379,690 GWh in 2001. The reason for the falloff was probably the high levels of load shedding beginning in 1994 and continually getting worse through 1999, when the first independent power producer came on line.

Source: Data are from REB Form 550: Government of Bangladesh 1998b; REB 1998.

second fastest growing sales category was small industrial sales, followed by irrigation, large industrial, and commercial sales.

Monitoring and Evaluation, Training, Operational Guidelines, and Technical Assistance

Recognizing the need for continuous program oversight, the REB has developed a variety of tools to assist the PBSs in organizing information and in monitoring the effectiveness of their commercial viability and internal management. The principal tool for monitoring PBS performance is REB Form 550. Using this form, all the key indicators of PBS and program viability are collected on a monthly basis. These indicators include collection rates, system losses, construction performance, debt service coverage, and PBS expenses.

Monitoring and Evaluating the REBs

REB Form 550 (Box 4-1) provides monthly summaries on all operational and financial characteristics of the PBS, allowing the REB to note problems as they begin to occur. This form allows the REB to monitor PBS performance on a consistent and timely basis. Moreover, the monthly routine of filing these reports helps the PBS to establish management discipline. From the REB's standpoint, the reporting routine allows collection of systemwide information that its management can use to evaluate each PBS by itself or in comparison to other cooperatives.

The REB has developed a concise reporting format that summarizes the Form 550 data, known as the Management Information System (MIS) report. This document is processed monthly by the financial management unit's rates/contracts directorate. Using the MIS report, REB management can quickly review the statistics of all PBSs in the system for key performance indicators. These would include such indicators as system losses and collection rates. MIS reports are circulated throughout all three REB management units, whereas individual Form 550s are used only for more detailed review by REB officers in the rates/contracts directorate.

Annual performance targets are set for 21 indicators through a negotiated process between REB management and PBS managers. The indicators set benchmarks for reducing system losses, increasing sales, meeting customer expansion levels for various categories, and maintaining or improving collection rates. The negotiated agreement is aptly called the Performance Target Agreement.

The degree to which targets are met forms the basis for annual employee bonuses.[2] Although it is not perfect, this system provides employees a strong incentive for achieving annual targets. Performance bonuses range from 0 to 15% of an employee's annual salary. Senior PBS staff are responsible for monitoring progress, and oversight is the responsibility of the PBS development and operations directorate.

In addition to monitoring financial and technical performance, the PBS development and operations directorate monitors the extent to which PBSs follow their bylaws, manage internal financial matters, and manage contracts with local service providers. Administrative and institutional monitoring are accomplished through a biannual management audit supervised by the PBS executive director, controller, and the directorate's chief engineer (or these officers' designees). They are responsible for developing a checklist of items considered important at each PBS's particular stage of development; consequently, the checklist may change from one audit to another. A positive atmosphere is thus promoted in which all parties are encouraged to identify problems and solve them creatively.

REB Training: Development of Core Competencies

Over the years, the REB has developed a comprehensive training program that provides both developmental and in-service training for all positions within the REB, the PBSs, and contracting organizations that support the program. Curricula have been developed for each position, and training is tied to employees' acceptance of positions and promotion to posts of higher responsibility.

The training directorate, which coordinates the program, is responsible for course offerings and scheduling and curricula review and development. The program is divided into technical, managerial, and financial training. All courses are accompanied by standardized tests that must be successfully completed by all candidates for positions of higher authority. For many positions at the PBS level, training is offered to individuals seeking PBS employment. Candidates for such positions as line worker, meter reader, bill collector, and other entry-level administrative positions are trained before being hired. They are accepted only if they pass the exit examination.

The training directorate also offers training for board members. Although not a prerequisite to assuming board positions, the training provides incoming members a better understanding of board procedures, basics of PBS purpose and functions, and limits of board responsibilities. It also promotes a clearer understanding of the board's role in relation to administration of the PBS, as well as the PBS's role in relation to the REB.

Special training activities are developed to meet specific technical needs of the REB and the PBSs. Examples of special courses where outside expertise is often used include electronic metering and data management, hot-line training for line workers and substation technicians, design and maintenance of renewable energy systems, power generation planning and load forecasting, load flow analysis, and loss reduction. As courses are developed for special purposes, they are often included in the REB training curricula so that they can be institutionalized to train the personnel who will manage them in the future.

In addition to these special training programs, the REB provides training for participating private-sector service providers, including technicians, electricians, building contractors, and consulting engineers. The training provided to these collaborating partners is critical, allowing the REB to build capacities and competencies needed for essential services. The success of this ongoing part of the training program has over time proven its immense value.

Whereas training has been an essential and extremely effective part of the program's sustainability, growth of the training directorate and its personnel has not kept pace with the rapid growth of the REB and the PBSs. Moreover, as the PBSs accelerate the integration of advanced technology into their distribution systems, the training directorate will need to increase its personnel's level of expertise.

Operational Guidelines: PBS Instructional Series

One foundation of the REB program is its comprehensive set of guidelines for all phases of REB and PBS operations. Modeled after the *REA Bulletin* that is periodically published by the Rural Electrification Administration (REA) of the United States Department of Agriculture, these guidelines are known as the Instructional Series. The series is divided into two parts: PBS Instructional Series (Series 100–300) and REB Procedural and Oversight Series (Series 400–800). The first part, written specifically for PBS managers, project and consultant engineers, and administrators, covers all PBS activities. The second part covers all internal functions of the REB, as well as the activities it undertakes on behalf of the PBS community. The Instructional Series is periodically reviewed and updated to ensure that the REB and the PBSs can make use of policies and procedures to address their specific needs.

Series 100 provides guidance on operating distribution equipment and engineering design and analysis procedures. It includes 56 sets of guidelines on equipment specifications, management of construction contracts and contractors, transformer repair, load management, and store management. It also includes design specifications for distribution feeders, subtransmission lines, substations, and line equipment. It provides maintenance information for all facilities managed by PBSs, including re-closers, power transformers, metering stations, voltage regulators, and other associated equipment. In addition, it provides guidance on hot-line maintenance and other line worker practices, as well as shop repair information.

For analytical guidance, the series provides project and consulting engineers instructions on designing the layout of backbone lines and laterals and conducting postconstruction analysis (including load flow analysis, capacitor placement studies, load management studies, load forecasting studies, and other standard engineering analyses). The series also provides guidance on materials management (including warehousing, store management, and inventory control).

Series 200 provides guidance on the financial functions that the REB oversees. Its 31 sets of guidelines cover management of accounts payable, accounts receivable, banking functions and reporting of cash accounts, controls for various PBS accounts, property records and acquisitions, budget development and monitoring, and audit procedures. The series also includes instructions on loan procedures (from the REB to PBSs and from PBSs to their members), as well as guidelines on insurance procedures.

Series 300, the longest and perhaps most comprehensive set of guidelines, has immense programmatic value because it focuses on technical, administrative, and financial activities at the PBS level. The series includes policies governing employee selection, training, evaluation, and rights; management of PBS district offices, bank accounts, and service connections; and guidance to the board of directors on member selection, election process, functions, and training.

Series 400 focuses on the REB's administrative role in program support and oversight. This role includes its authority to intervene directly in PBS operations, its responsibility to establish nondiscriminatory loan policies toward the PBSs, financial oversight of PBS operations, and internal guidance for auditing its own financial programs.

Series 500 is composed of 20 sets of guidelines that outline the REB's role in providing engineering and materials for PBS construction and maintenance. Because the Bangladesh rural electrification program is still growing rapidly, REB continues to play a central coordinating role in all engineering, construction, and procurement activities. This means that the REB requires PBS consulting engineers to submit annual construction plans to the regional supervising engineer who, in turn, submits the proposed construction projects to REB headquarters in Dhaka. There the projects of neighboring PBSs are consolidated, and an REB-wide project list is created. (These projects may be expansions of older PBSs or construction of PBSs that are not yet energized.) The REB then develops a master materials list, which becomes the construction budget for the fiscal year.

This series also provides a comprehensive set of guidelines that governs the construction and procurement processes managed directly by the REB. The series establishes the boundaries of professional management of contractor–vendor–REB relationships to avoid conflicts of interest. It also delineates acceptance procedures for completed projects and provides guidelines for materials management procedures before distribution of equipment to the PBSs. Finally, it establishes guidelines for testing and repair of major equipment and REB supervision of these maintenance programs on behalf of the PBSs.

Series 600 provides guidance on the REB's relationship to the PBSs with respect to financing and accounting. It includes procedures and regulations for the REB's annual budgeting process and a detailed description of the uniform system of accounts that all PBSs must use. It also describes fiscal management procedures to be followed by all PBSs and the manner in which the REB will oversee them. It describes contract management procedures, how contractors will be paid, and the surety bonds that are required as a function of contracting procedures. Finally, it establishes PBS auditing procedures, clarifies the REB's role in supervising financial audits, and provides guidance on the REB's management of foreign currency accounts with donor agencies.

Series 700 provides 12 sets of guidelines on the internal functioning of the training directorate, including development of training materials, course offerings, and distribution of training information. These guidelines supplement Series 300 with respect to PBS development in that they describe how the PBS development directorate will perform monitoring functions for various PBS activities.

Series 800, which contains two sets of brief instructions, establishes guidelines for the selection of and governance by the REB board of directors.

Technical Assistance to the REB

At the outset, the REB elected to model Bangladesh's rural electrification program after the U.S. REA system. The design standards, materials specifications, and procedures that were later converted into the REB Instructional Series were all derived from U.S. REA program guidelines. Design standards were adapted to local conditions, and maintenance procedures were modified to fit the labor-intensive nature of the PBSs.

NRECA International[3] has been REB's principal technical adviser since the program began. As background, NRECA gained most of its early experience serving the U.S. cooperatives participating in the REA program. To ensure continuity, the REB decided that NRECA should continue as principal technical assistance adviser throughout the rapid-growth phase of PBS development. REB management reasoned that involving many institutions in technical assistance would result in a patchwork of as many approaches. Because the U.S. REA program and the REB were compatible, the board decided against reviewing other programs that might offer some small advantages but overall would require many systemic adjustments and procedural changes. As indicated previously, NRECA directed the initial feasibility study, designed the original master plan, and, subsequently, has provided all technical assistance.[4]

Although there is much to be said for having access to a variety of approaches to developing a rural electrification program, changes in direction during a program's early stages can be disruptive. The REB's decision to use a single source of technical assistance has resulted in clearly established and documented policies on procurement, equipment standards, and management procedures that are transparent to all concerned—the donor community, vendors, and program participants.

Financial Policies and Programs

The PBS development process is based on a demand-driven master plan. As clearly established in the REB's policies and procedures, PBS master plans assign highest priority to those areas with the highest load density and therefore the highest potential for financial viability. Only communities that meet strict revenue criteria are considered for inclusion in the master plan process. Thus, even in areas of high population density, absence of tube-wells, rice mills, and other agricultural and industrial loads may preclude some communities from being connected and served by local PBSs.

Load Promotion and Low-Cost Solutions for Rural Electrification

As the PBSs have gradually expanded service into areas that produce reasonably high loads, political influence has been kept within reasonable bounds; however, financial performance has fallen short in many cases. Load growth in many parts of the REB system has been sluggish, depending on domestic consumption in areas with little agricultural, industrial, or commercial demand. The desire to extend service to as many villages without electricity as possible is evidenced by the newer PBSs, which, by virtue of the master planning process, are marginal in comparison with their older, more established counterparts.

One factor that has helped increase load density in several PBSs has been the turnover of former BPDB and DESA service territories to PBSs in rural areas where the REB has blanket authority to provide electricity service. This decision resulted from years of discussion among the country's electricity service authorities. In many cases, the former BPDB and DESA service territories have been operated as pockets surrounded by PBS clients; these pockets were always intended to be turned over to the neighboring PBSs on energization of the rural electric systems. In most cases, these pockets have been characterized by relatively high levels of electricity consumption. Adding these service territories to the neighboring PBS has, in many cases, helped the PBS meet the revenue requirements projected in the original feasibility study.

Although the turnover process has been important to increasing load density and energy sales, it is not without its share of problems. Historically, these territories have experienced a high rate of losses and low collection rates. In some areas, combined technical and administrative losses exceed 50% of the energy sold, and collection rates are often less than 50% of energy billed. Reducing losses from such high percentages to the system average of about 15% poses a serious challenge. In most cases, losses can be reduced to around 20% within three months, but this often requires a complete overhaul of the distribution system and an aggressive PBS information campaign. As explained above, such turnovers can also create short-term management problems. Nonetheless, once the physical plant has been improved and billing and collection procedures stabilized, such areas can become important sources of PBS income. The turnover process, which has been gradual, has played a key role in increasing load density and in moving PBSs toward financial viability.

Engineering standards are another important element for establishing construction costs commensurate with the lower loads found in rural areas. Instructional Series 100 and 500 include a comprehensive set of construction guidelines and equipment specifications that govern the REB expansion program. The guidelines for consulting engineers provide standards for conductor size, voltage drop, criteria for specifying pole class and line hardware selection, and other critical equipment specifications. These specifications have been modified over the years to meet changing local needs.

The engineering standards promote the use of single-phase construction using wooden poles, aluminum steel reinforced conductor, and low-capacity transformers wherever possible. Also, they have emphasized minimizing distances covered by the secondary distribution systems to reduce distribution losses. Single-phase lines can be upgraded to two- and three-phase lines as loads grow. The system layout and type of construction to be used are determined by a region's load-growth estimates derived from demographic analyses and census-taking in each new PBS service territory. The tendency in some other developing countries has been to overbuild with three-phase electricity service. A recent country comparison of line-construction costs showed that Bangladesh's costs are among the lowest in the developing world (World Bank 2000).

Widespread use of engineering and construction standards has permitted economies of scale in the purchase of materials and equipment. The REB consolidates the materials and equipment requirements of the PBSs, conducts large-scale international procurements, receives and warehouses equipment, and issues it to PBSs on a project-by-project basis. All procurement is done on an open solicitation basis following strict REB equipment standards. The economies of scale resulting from this process are clear. Granting a 15% cost incentive to Bangladeshi companies encourages procurement from local suppliers; even so, more than 80% of program commodities are purchased from foreign suppliers.

Clearly, construction standards will need to be revisited in future years. The rate of PBS expansion has begun to level off because financial resources to maintain high growth are insufficient and territories that can afford service at present costs are saturated. One potential solution would be to lower the cost of service by modifying engineering standards to result in lower investment cost for expansion into areas of low-load growth. Discussions with REB project management suggest that this option should be seriously considered.

Promoting Productive Uses of Electricity

The initial impetus behind Bangladesh's REB program was the constitutional guarantee to provide households with electricity service of a quality equal to that of urban areas. Over time, it was realized that the promotion of productive uses of electricity is equally as important for rural areas. In fact, it has become a principal program theme for the REB. The promotion of productive uses is indirectly reflected in the requirement that all rural communities meet certain revenue requirements, which almost by default means that they must proactively promote productive uses of electricity. Marketing productive uses of electricity is an important business opportunity that many rural distribution utilities routinely overlook. Payment for electricity service requires cash, which is often in short supply in rural communities. Productive uses of electricity can help generate the cash needed to support lighting and other lifestyle improvements afforded by electricity. In the case of Bangladesh's REB program, most of the emphasis has been on promoting irrigation, rice milling, and household power looms. The significance of promoting these productive loads goes beyond a constitutional guarantee; the extremely low average of domestic energy consumption makes it essential to promote power sales to enhance PBS viability.

The types of special services and loans extended to PBS members perhaps best illustrate the productive-use theme. As defined in the REB Instructional Series, productive-use promotion functions are the responsibility of the PBS member service departments. The PBS programs receive technical support and training through the PBS development directorate. Member services include orientation of new PBS members to the basics of the economic advantages of electricity for agricultural processing, irrigation, and small businesses. Assistance is given in the proper sizing of irrigation pumps and other uses of electric motors. In most cases, financing is not an issue; equipment vendors provide short-term financing. Moreover, such institutions as the Bangladesh Rural Advancement Committee (BRAC) and the Grameen Bank provide businesses with short-term start-up loans for pumps, grain mills, and other applications. In some cases, member service departments may coordinate with vendors and organizations such as BRAC and the Grameen Bank to channel loan funds to PBS members for productive uses of electricity.

PBS member service departments also provide various types of technical assistance to potential small industrial and commercial clients. For example, they assist PBS members in purchasing electric motors and machinery, stressing a preference for single-phase motors as an energy conservation measure. They also help estimate service-connection costs for members required to pay such fees, which includes those who live more than 50 m from a PBS distribution line. In addition, they train electricians in house wiring, inspect all installations, and register those who successfully complete the PBS training program.

Rural Electrification Program Financing

One of the most remarkable aspects of Bangladesh's REB program is the tremendous coordination and collaboration that has been achieved among its 16 international donors and the GOB. Donor funds target specific projects negotiated to support the phased development of the overall rural electrification program. The REB functions as the conduit for all donor funds invested in the development of new PBSs, expansion of existing ones, and improvements in the REB–PBS system. Loans are made through the GOB to the REB, which, in turn, channels resources to the PBSs in the form of equipment, materials, and construction contracts, all of which are managed by the REB on behalf of the PBSs. As of early 2001, more than US$1 billion had been invested in PBS development and expansion (Table 4-2).

Before obtaining approval for development, all new PBS service areas must meet revenue requirements. Feasibility studies are conducted to determine whether these requirements are met, not only for the PBS overall but also for each service area within it. The revenue requirement standards allow for financial losses

Table 4-2. *Summary of Rural Electrification Donor Support in Bangladesh, 1975–2001*

Funding Source	No. of PBSs or Activities Funded	Funding Level (US$ millions)
International Development Association	18	413.47
United States Agency for International Development	17	210.96
Japan Bank for International Cooperation	3	92.40
Asian Development Bank	Expansion	173.23
Kuwait Fund for Arab Economic Development	11	79.98
Government of Bangladesh	10	71.36
Canadian International Development Agency	Intensification and expansion	58.00
Norwegian Agency for Development Cooperation	1	49.70
Netherlands	Intensification	74.21
Saudi Fund for Development	Intensification and cyclone grant	25.50
Organization of the Petroleum Exporting Countries (Fund for International Development)	2	23.60
Finland	1	23.02
Islamic Development Bank	3	29.21
China barter	Intensification and expansion	20.00
Germany (GT2)		2.14
Japan Debt Relief Fund	1	19.00
Government of Saudi Arabia	Rehabilitation	5.73
Government of France	Solar grant	1.08
Korea		20.00
Netherlands		74.71
Total	67	1,467.39

Source: Data are from REB 2001.

(referred to as "negative margins") during the first several years of PBS operation. Load growth typically is gradual as the system infrastructure is being developed. The electrification program is designed to support this development process by providing cash flow, as well as low-interest loans with long repayment periods. These support funds allow PBS operations to mature through the first five years after they begin offering electricity service. After this initial period, the PBSs are expected to reach financial stability. This funding is available to all PBSs, but it also is closely monitored.

Subsidies for Rural Electrification

Subsidies are common in virtually all rural electrification programs. However, in some programs the subsidies lead to problems. In Bangladesh, the subsidies have been directed toward making the rural electric cooperatives financially viable. The REB and the GOB support PBS development and operations through two main subsidy programs. The first and perhaps most important is a coinvestment by the central government with the PBSs, which involves extending low-interest, long-term financing to the PBSs through the REB; 30-year loans are provided to the PBSs at 3% interest. The second subsidy program is a form of subsidized interest that is deferred for a fixed term. That is, during the first five years of commercial operation, the PBSs do not make principal payments, and interest is capitalized into the investment cost of their original loans. Moreover, the interest charged during this time is at a reduced rate of 0.75%.

The REB receives donor funding through government loans at 2% interest. One percentage point is added to loans made by the REB to the PBSs to cover program management costs. Furthermore, the government forgives 33% of the loan charges to the REB. This means that, of the 3% interest charged, the REB retains 1% to cover its operating expenses and, in addition, collects 33% of program income, which it does not have to repay to the central government.

These interest rates are dramatically lower than market rates. However, they are similar to the rates being offered to cooperatives in the United States and to rural cooperatives in several other developing countries. Without such long-term financing, it is doubtful that rural electrification could take place in Bangladesh. One must also consider that the subsidies have been successful in making service affordable to rural consumers with low incomes. In addition, the rural electrification program in Bangladesh is aimed toward ensuring the long-term financial sustainability of the rural cooperatives.

One final subsidy is a somewhat low bulk power tariff. The REB enjoys a preferential bulk power rate from the BPDB. The board's price to industrial consumers on the system averages Tk2.45, and the REB has negotiated a system price of Tk1.77. With the recent reforms in the power sector, this is yet another risk factor that could significantly contribute to program stability; as new power providers begin to sell power to BPDB and thereafter, to the Power Grid Company of Bangladesh, the REB will begin to pay increasingly more "realistic" prices for bulk power. However, some of the new bulk tariffs from private electricity generators are similar to the preferential price received by the REB.

Five-Year Cash Flow Support to Achieve Self-Sufficiency

During the design phase of the REB program, it was recognized that the PBSs would require substantial financial support to achieve sustainability during the initial period of infrastructure development and load growth. The measures designed to provide the needed support include a five-year grace period for principal and interest payments and, perhaps equally important, a cash-flow support program designed to last a maximum of five years.

Because system loads often grow slowly and are subject to factors beyond the rural utility's control, the REB established a policy to support the PBSs' cash-flow needs during the initial years of operation while moving them toward financial self-sufficiency. Under this policy, a portion of support is provided to each PBS as a direct subsidy. In addition, the PBSs receive cash-flow subsidies, assessed annually, for periods of up to six years. By reviewing each PBS's needs on an annual basis, the REB can identify measures to achieve early financial maturity, thereby minimizing the amount of funding spent on subsidies.

The REB and the PBSs evaluate the cash-flow support subsidies during the annual performance review process. The REB uses a standard evaluation method that considers the number of years a PBS has operated, the customers it serves, its system losses, its collection rates, and other performance indicators. If these indicators show that the PBS is underperforming because of poor management, corrective actions may be taken, along with providing more cash-flow support. Once budget allocations for the subsidies are prepared, the Ministry of Energy and Hydrocarbons reviews them and decides whether to grant final approval.

Many of the PBSs have been able to recover all or most operating costs after the initial subsidy period has expired. This point is illustrated in Table 4-3, which reports the percentage of the cost of service covered for all PBSs through revenues derived from tariffs. Shortfalls reflect deficits that are internally financed by using a depreciation allowance to cover that portion of the cost of service that is not recovered directly through energy bills and REB-provided subsidies.

Table 4-3. *Summary of Cost of Service Covered in Bangladesh*

PBS Phase	No of PBSs	Source of Financing (%)		
		Revenue	Subsidies	Shortfalls
New (0–5 years)	6	74	26	0
Maturing (6–10 years)	8	86	11	3
Mature without margins	22	83.5	0	16.5
Mature with margins	17	100	0	0

Note: Data are aggregated across all PBSs systemwide.

Source: Data are from REB Form 550.

The data given in Table 4-3 reflect reported performance for 53 PBSs ener-gized through September 1996. Only 6 PBSs are included in the new group, for which service was initiated during 1995–1996. The maturing group includes 8 PBSs. The 2 mature groups included PBSs energized between program inception and 1990. Of the 39 PBSs in these two groups, 17 showed positive margins during FY1998 and 22 did not.

Interestingly, mature PBSs without margins are similar to maturing PBSs, except that the subsidies that are applied to the maturing group are unavailable to the older PBSs. Therefore, any unrecoverable costs are now being financed from internal reserves or depreciation funds recovered through the tariff. This group of mature cooperatives without margins is problematic in that unless tariffs are raised or loads are increased, they will continue to lose money.

Tariff Policies: A Case for Considering Reform

The tariffs that the REB sets for the PBSs are influenced by the Board's mandate to promote rural economic development by encouraging agricultural and indus-trial production and to provide electricity access to as many rural households as possible. In setting tariffs for each PBS, the REB attempts to balance customers' ability to pay for service with the program's need for economic sustainability.

The REB's general approach has been creating a cross-subsidy between indus-trial and commercial clients and residential and agricultural PBS members. This approach assumes that each PBS can achieve financial stability through successful marketing of service to potential industrial and commercial clients. In practice, however, this strategy has proven problematic. Load growth in new PBSs has been slow, and most PBSs have not yet achieved financial viability.

The situation is serious but can be solved in several different ways. One of the simplest would be to increase the residential tariff. The existing tariffs are presented in Table 4-4. Most of the PBSs that are accumulating operating deficits could yield positive operating margins if the residential tariff were increased from Tk2.45 to Tk3.2. If residential tariffs were increased by up to 30%, all but six PBSs would generate sufficient revenue to cover all operating expenses (REB Form 550). The remaining 6 PBSs would need to increase tariffs by up to 50% of their current levels to cover operating expenses.

Can rural consumers pay for such increases in their electricity bills? Two recent studies conducted in villages with and without electricity show that many rural households would be willing to pay higher residential tariffs. Field surveys of a socioeconomic impact evaluation found that 49% of households with elec-tricity would be willing to pay up to 25% more for the service they receive, and 7.9% would pay 10% more (Prokaushali Sangsad Ltd. 2000; Unnayan Shamannay 1996). This means that there seems to be a willingness to pay for higher electric-ity costs.

Perhaps a more revealing statistic is the penetration rate that many mature PBSs have achieved in areas that have had electricity for more than a decade. Whereas rates for residential customers vary greatly, in most cases, they level off at about 40% of available households in the more mature PBSs. This figure, however, may

Table 4-4. *REB Tariffs for Most PBSs with Comparable Energy Consumption in Bangladesh*

	Consumer Class								
	Residential								
Tariff	<300 kWh	300–600 kWh	>600 kWh	Commercial	Institutional	Irrigation	General Power	Large Power	Streetlights
Tk/kWh	2.5	3.6	5.4	4.60	2.85	2.55	3.55	3.45	3.3
US$/kWh	0.05	0.07	0.11	0.092	0.057	0.051	0.071	0.069	0.066
% Energy	39.6			6.2	13.4	1.2	27.9	11.5	0.2

Note: Tariffs vary slightly for some PBSs.

be misleading because it implies that only 40% of the population that has access to service chooses to electrify. Due to system configurations, it may be that many fewer households have access to service and that the penetration rate of those with access is higher than the entire population of households in villages that are counted as having electricity. No existing studies illustrate the specific factors that contribute to such low penetration rates, but they are likely a function of lower household income, effects of load shedding, and nonresponsiveness of the PBSs to requests for connections. This last factor could be an important determinant, given the high degree of emphasis placed on system expansion.

Considering the results of these studies, along with the fact that many PBSs cannot yet cover the cost of service to their members, it becomes clear that tariff reform is a measure that should be seriously considered in the near future. Because the rate of demand growth for nonresidential customers will not be achieved in most PBSs, tariff increases could safeguard REB's investment in those PBSs that continue to struggle financially.

This pricing issue illustrates why it is important to have an agency like the REB involved in overseeing the financial records of the PBSs. Thus, the original cross-subsidy plan has not worked to the REB's advantage because industrial load has grown at low rates in all but a few PBSs. However, the positive aspect of this is that if residential tariffs are adjusted, then most PBSs can cover their operating costs. Although such increases are politically difficult to implement, they will be required in the near future to ensure program sustainability. Without an oversight agency like the REB, the task would be more difficult.

A Tale of Success and Struggle in Two Cooperatives

When the REB program was first designed, it was projected that the PBSs would require six years to achieve financial stability. During this period of infrastructure development and demand growth, the REB helps the PBSs achieve financial stability through various means. For example, the PBSs are not required to make payments on the principal of outstanding REB loans, and they can incur negative margins, using the depreciation allowance to finance revenue shortfalls to cover operating costs.

Early on, it was hoped that loads overall would grow much faster than they have and that industrial consumption would compensate for low levels of residential and agricultural demand. However, such expectations have been realized only for those PBSs located close to the capital city of Dhaka, where the cooperatives benefit from other development activities, such as construction of all-weather roads.

To understand better the factors that have influenced the cooperatives' financial stability, the following section compares two dramatically different PBSs. The first is a large, fast-growing PBS in an area with a proportionately heavy industrial and irrigation load. Narsingdi PBS-1 initiated service in 1986 and is one of the REB system's best performing PBSs, with a healthy annual profit margin. By contrast, Barisal PBS-1 is a 10-year-old cooperative that no longer receives a cash-flow subsidy from the REB and yet continues to lose money. The sections below provide insight into the causes of Narsingdi's success and Barisal's ongoing struggle to reach financial viability.

The Cooperatives of Narsingdi PBS-1 and Barisal PBS-1

Narsingdi PBS-1 started service on October 29, 1986, and is responsible for an area of 736 km^2, including six subdistricts and 1,328 villages. Of these villages, the cooperative has provided service to 776 to date. This cooperative is located close to the town of Narsingdi, a bustling center of commerce only 1.5 hours from Dhaka. It is one of the oldest PBSs in the REB system and has grown dramatically in recent years. As a consequence, with the assistance of the REB it has made the transition from a medium-size cooperative with an unstable financial position to one of the largest, most financially profitable PBSs in the country.

Narsingdi has recently taken over significant service territory from DESA in population centers around which Narsingdi has historically provided services. This hallmark event has resulted in both challenges and benefits. First, there were extremely high losses of more than 50% in the service areas it inherited from DESA. As expected, this caused overall PBS system losses to rise to almost 20% after the takeover was completed. The benefit of the takeover was that the former DESA service areas have a much higher load density and contained a number of relatively large industrial clients. As a consequence, Narsingdi's customer base increased dramatically, and since 1998 it has added more than 60,000 new customers and 50 MW to its system load.

Two more problems were that the service areas formerly managed by the DESA had poor collection rates and their physical plant was in almost total disrepair. Together, the REB and Narsingdi responded to this challenge by rebuilding the entire distribution system. This rebuilding means normalizing service connections, taking a census of the customers to be served, and introducing a program to reduce technical and nontechnical losses. Although a tremendous amount of work, taking such measures will improve its financial viability once the takeover areas are fully integrated into the system. In recent months, system losses were reduced to 18% and are expected to fall even further.

Barisal PBS-1 serves an entirely different type of area than Narsingdi. It is in a relatively isolated area along Bangladesh's southern coast, and it often experiences harsh weather, including occasional cyclones and heavy flooding. It initiated service on September 1, 1990, to an area of 1,749 km², made up of 847 villages. As of June 1998, electricity service had been provided to about 309 of these villages. The area has few small industries and little commercial activity. Its poorly developed infrastructure prevents transporting products to distant markets. Consequently, Barisal's sluggish economy has not profited nearly as much as have communities with better developed infrastructure located closer to major population centers. Barisal also began taking over service territories from BPDB shortly after it was energized, but this process slowed after the fourth year of operation. Thus, it is unlikely that Barisal will increase its customer connections in the dramatic way that Narsingdi did.

Cooperative Performance Differences

Narsingdi's huge advantage over Barisal and many other PBSs in the REB system can be attributed to two major factors. The first is the high growth rate in nonresidential energy sales. This difference is vitally important to PBS profitability because it provides the means to cover cross-subsidies from industrial and commercial consumers to residential customers. The second factor is the high growth rate in residential sales per kilometer of energized distribution line, perhaps the most important factor affecting the viability of rural electric distribution systems.

During its first six years of operation, Barisal's rate of growth in sales to residential customers was similar to that of Narsingdi. However, growth in nonresidential sales lagged far behind. Between the fifth and sixth years of operation, nonresidential energy sales increased more than 150% in Narsingdi, but they fell more than 2% in Barisal (Table 4-5). Narsingdi's high growth in nonresidential load contributed significantly to the cooperative's early graduation from the REB subsidy program.

As mentioned above, one of the more indicative measures of PBS sustainability is the specific load per kilometer of energized line, also known as *energy consumption density*. In this respect, Narsingdi's growth far outstripped that of Barisal, as Figure 4-2 illustrates. In Narsingdi, nonresidential customers contribute a disproportionately high source of income to the PBS with respect to real cost of service. Because the cross-subsidy favors residential clients, increased sales to nonresidential clients is required for program sustainability. The case of Barisal shows that low growth in nonresidential sales can result in serious financial stress as a PBS matures.

By far, the most important factor that affects PBS sustainability is the tariff applied to all customer categories. The REB's approach has been to charge commercial and industrial customers relatively high rates so that residential and irrigation rates can be held at affordable levels. As the case of Narsingdi shows, this

Table 4-5. *Performance Comparison of Bangladesh's Narsingdi and Barisal PBSs over the First Six Years of Operation*

Service Year	Fiscal Year	Service Connections	Residential Energy Sales	Nonresidential Energy Sales	Residential kWh per km	Nonresidential kWh per km
			Narsingdi PBS-1			
1	1987	1,733	94,399	111,724	609	721
2	1988	3,965	689,634	470,918	2,611	1,783
3	1989	5,578	1,025,009	1,711,593	2,956	4,936
4	1990	7,790	1,614,054	3,479,809	2,287	5,578
5	1991	10,974	2,226,759	9,200,467	2,887	11,928
6	1992	15,570	4,097,404	23,003,992	3,995	22,429
			Barisal PBS-1			
1	1991	3,826	669,435	233,417	2,270	792
2	1992	7,012	1,300,425	542,161	2,714	1,132
3	1993	10,202	2,141,726	868,936	2,794	1,134
4	1994	16,107	3,632,281	1,873,591	3,204	1,653
5	1995	21,326	5,725,049	2,557,195	4,521	2,020
6	1996	25,156	6,438,546	2,496,933	4,668	1,811

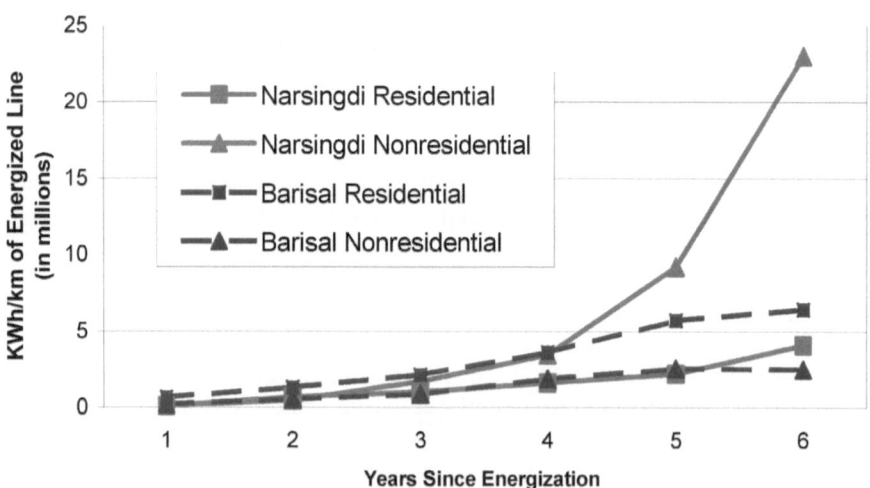

Figure 4-2. *Comparison of Growth in Energy Consumption Density in Bangladesh*

cross-subsidy strategy works well where there is sufficiently high commercial and industrial load growth to compensate for the subsidized cost of service to residential customers. Narsingdi's commercial and industrial load growth makes up about 70% of total energy sales, whereas residential and irrigation growth accounts for 30%. Conversely, Barisal's commercial and industrial load growth is slightly less than 30% of total consumption.

For the majority of PBSs, commercial and industrial loads are insufficient to cover operational deficits. To achieve sustainability without significantly increasing energy sales, Barisal's residential tariff would need to be raised by almost 100%. Such a measure, although impracticable in the short term, indicates the extent to which residential tariff adjustments may be needed to ensure long-term sustainability. Clearly, if Barisal and the REB system's many other PBSs with similar consumption characteristics are to reach financial sustainability, they must change their pricing structure and coordinate actively with other rural development programs to generate more commercial and industrial load growth.

Lessons from the Bangladesh Experience

Twenty-five years ago, Bangladesh was a newly established nation without a rural electrification strategy. Today, however, the country's REB program extends electricity service to an estimated 2.5 million rural households. Bangladesh's cooperatives operate with system losses among the lowest in the developing world and revenue collection rates among the highest. This section highlights the major factors that have contributed to such dramatic early results, as well as recommendations to ensure the program's future sustainability.

Centralized Supervision, Decentralized Operations

The REB program is characterized by centralized planning, design, and construction and decentralized operational responsibility. Centralized supervision authorizes the REB to monitor and evaluate the cooperatives' performance using standardized, objective tools and to make and enforce changes as needed. Decentralized operational responsibility through the PBSs ensures that the personnel most knowledgeable about specific problems are empowered to make day-to-day operational decisions.

Standardized Procedures and Practices

The REB has established carefully considered and clearly stated planning, engineering, administrative, and business procedures. They have consistently been put into practice throughout the entire program, covering all aspects of the development and operations of an electricity distribution system. Standardization has allowed growth of the construction program to accelerate while giving operations engineers the opportunity to share plant and technical resources.

Performance-Based Measurements

To evaluate performance and ensure quality control, a management system was established that links pay and promotion to measured compliance with clearly stated expectations. Measures of success or failure include indices that allow monitoring of loss reduction, collection performance, and other business goals.

Effective Commercial Practices

An important factor contributing to Bangladesh's rural electrification success has been the effective implementation of day-to-day commercial practices. In particular, effective billing and collection procedures have resulted in collection rates exceeding 95%. In many developing countries, the most critical reason for program failure is the poor rate of bill collection. Other measures implemented by Bangladesh's REB program, such as rotating meter reader routes and centralizing collections in rural banks, have effectively limited fraud and theft, thereby contributing to high PBS performance.

Prioritized System Investment

At the outset, the REB program established a clearly defined master planning process that prioritized system investment according to revenue generation. This model has been used almost universally. Political pressure has influenced the selection of some projects, resulting in poor performance, but it has not been a major factor in overall program implementation.

The program in Bangladesh also profited greatly from the consistent technical advice of an international firm that has been involved in the program since its inception. The long-term commitment of the government of Bangladesh also has a significant role in solving problems as they arise and making program adjustments as the program matures and faces new challenges.

Conclusion

The challenge of providing electricity to some of the poorest populations in the world should not be underestimated. Overall, the program is up to the challenges. By slowly building capacity to expand service and paying attention to the planning details, the program has grown from providing fewer than 20,000 new customers a year with electricity to about 500,000 new customers a year. This chapter provides the details of how this was accomplished in the hope that other countries will accept the challenge of providing electricity to their rural populations.

The challenges have been many. They involved the development of a subsidy strategy that aims at financial independence rather than dependence. New institutions were developed specifically to promote rural electrification. Diverse donors and the government were convinced that the program is a worthwhile investment for the future of the country. Line workers were trained in a professional manner, bills were collected, and losses were minimized. However, Bangladesh still has the future challenges of continuing to expand service to poor populations and encouraging local economic development to make electricity service affordable to greater numbers of rural people.

Acknowledgments

I would like to express appreciation to officers of the Rural Electrification Board and colleagues at NRECA International in Dhaka, Bangladesh, who gave extensively of their time and expertise to complete this study. Special thanks go to James Ford, NRECA/Dhaka team leader, for sharing his invaluable insights from REB program experience in Bangladesh. Allen Inversin and James VanCoevering, of NRECA headquarters in Arlington, VA, reviewed early drafts and provided critical comments that helped refine the analysis and focus the message. Najmul Hossain of Data International assisted in reviewing the economic effects of the REB program. Throughout the process, Douglas Barnes provided valuable guidance and practical advice and support.

Notes

1. In Asia there are two types of wells: dugwells and tubewells. Dugwells are as the name implies; a large hole is dug in the ground, water seeps in, the hole fills up, and then water is pumped from it. Tubewells are more like an artisian well (but not as deep). A tube is sunk into the ground and is placed at the bottom of the well to pump water up. This generally is not possible without electricity, as a submersible electric pump is used.

2. Although PBSs are autonomous in theory, REB retains much financial and managerial control over them, including the power to set performance goals and allocate annual bonuses. The means by which bonuses are set is determined directly by the ability of PBSs to meet their targets. A formula has been established to weight performance indicators, with bonuses granted to those PBSs that score high on an absolute scale in the PTA review.

3. NRECA is the acronym for the National Rural Electric Cooperative Association, but several years ago NRECA International adopted the acronym as its formal name as well.

4. NRECA subcontracted Gilbert Commonwealth, an engineering consulting firm, to provide engineering design services during the first phase of USAID-supported technical assistance.

CHAPTER 5

Public Distribution and Electricity Problem Solving in Rural Thailand

Voravate Tuntivate and Douglas F. Barnes

*W*HEN THAILAND'S RURAL ELECTRIFICATION PROGRAM began in 1972, the country as a whole was at a fairly low level of development, with rural residents among the least advantaged. Notably, only 10% of people living outside the Bangkok metropolitan area had access to electric power. Thirty years later, however, more than 99% of Thailand's villages have electricity service. The country's dramatic growth in gross national product over these decades stemmed, in large part, from the government's commitment to developing and improving living standards in rural areas.

The organization of the electricity industry in Thailand had significant implications for the way the rural electrification program was implemented. The country's power generation and distribution are divided among three different public companies. These include a company responsible for power generation, a company responsible for electricity distribution in the Bangkok metropolitan area, and the Provincial Electricity Authority (PEA), which was responsible for providing electricity to all other regions of the country. The PEA is an autonomous government agency for electricity distribution, and it was given responsibility for developing and implementing the program. Its ability to focus solely on public distribution in areas outside the Bangkok metropolitan area contributed significantly to program success.

The organizational structure of distribution, however, was not the only factor that contributed to the success of Thailand's rural electrification program. As the implementing agency, the PEA remained financially sound throughout the entire program, despite large capital outlays it had to make for system expansion. This was a significant accomplishment, given the comparatively higher costs and lower revenues inherent in providing electricity to rural areas. The PEA accomplished this through careful planning for system expansion, diligent attention to keeping costs low, a unique billing program to ensure revenue collection, and a system of cross-subsidies that helped the company serve high-cost rural consumers at reasonable prices.

The program gained momentum and credibility from the PEA's flexible, highly innovative, and pragmatic approach to the problems it encountered in its almost 20 years of implementation. Finally, the company paid careful attention to customer service, service promotion and marketing, and local community involvement—all factors that enhanced local support for the rural program.

Background on Thailand

Thailand is a Southeast Asian country comprising an area about the size of Texas. Most of Thailand's 62 million people speak a dialect of Thai and share a common culture. Chinese and Malay-speaking Muslims make up the largest minorities, followed by Khmer, Mon, and Vietnamese. Despite urban growth, concentrated principally in the capital region of Bangkok, Thailand remains predominantly rural. More than 60% of Thailand's labor force is employed in agriculture, and rice is its most important crop. Thailand is a constitutional monarchy with a multiparty system, and it has three major branches of government.

Most of the topography in Thailand is suitable for rural electrification. Most rural people live in the rice-growing areas of the central, north, and northeastern regions. The Chao Phraya River and its tributaries run through the country's central plain. The Khorat Plateau rises in the northeast, bordered on the east by the Mekong River. The more difficult region for rural electrification involves the area of the sparsely populated mountains. Numerous hill tribes, including the Karen, Hmong, and Mein, represent about 2% of the population.

At the end of the rural electrification program, Thailand's economy had made some impressive gains. Assisted by the government's strategy to develop infrastructure for its exports and to invest in rural development, by 1999 Thailand's gross domestic product was about US$124 billion, and gross national product per capita was US$1,960. Major exports include textiles and footwear, fishery products, rice, tapioca, rubber, jewelry, computers, and integrated circuits. The United States is Thailand's largest export market, followed by Japan and other Asian countries. In 1997, life expectancy was 66 years for men and 72 years for women. In 1999, infant mortality was 30 per 1,000 live births. An estimated 93% of Thailand's people are literate.

History of Rural Electrification in Thailand—1970 to the Present

The planning for rural electrification in Thailand began several years before the first project was implemented. Plans for Thailand's rural electrification began in 1970 with the help of a consulting reconnaissance team from the U.S. Agency for International Development (USAID). The next year the government of Thailand approved a plan based on the results of this work for the total coverage of the country with grid electricity. USAID subsequently provided the PEA with a grant to conduct a prefeasibility study, which was completed in 1972 (Middle West

Service Company 1973). In 1973, the government of Thailand officially approved the prefeasibility study report and adopted it as the National Plan for Thailand Accelerated Rural Electrification. This plan was to begin an intensive program of rural electrification for the country that would result in virtually all parts of the country having access to electricity service by the year 2000.

Preliminary Steps in Program Implementation

Like other major infrastructure investment proposals, the national plan was developed in close cooperation with Thailand's National Economic and Social Development Board to ensure that rural electrification was an integral part of the country's overall development strategy. The board is responsible for drafting Thailand's National Economic and Social Development Plan. Each plan, which covers a five-year period, is considered the country's blueprint for development activities. The inclusion of rural electrification in the development plan would have important implications for the coordination of this program with other rural development activities taking place in the country.

In this regard, the national plan set forth the full scope of the rural electrification program. This comprehensive document outlined the justification for and objectives of the expansion program, as well as specific guidelines for implementing each step for successful completion. The plan was sufficiently detailed to serve as a blueprint for the government, the PEA, and officials directly responsible for the program. It included guidelines and criteria for designating priority regions for electrification, criteria for selecting villages for electrification, organizational requirements, and many other details. Many governments have made grand pronouncements that providing electricity to rural populations is a responsibility for the country, but Thailand backed this up with program details not usually found in such communications. In other words, the government was serious about carrying out this program in coordination with other development programs.

The next major step was preparation of a detailed feasibility study, which reviewed components addressed in the national plan and provided explicit analyses of the economic, financial, and technical issues of the proposed individual electrification investment projects. In addition, the study addressed the scope of these projects and detailed the methodology for allocation, selection, and setting of priorities for electrifying individual villages. The general idea was to first target those villages that would have the lowest costs and the highest demand for electricity. Thus, the consensus was to develop a plan for communities with the highest financial and economic rate of return.

The government adopted a 25-year timetable for program implementation, based on recommendations of Thailand's National Economic and Social Development Board. This timetable consisted of five consecutive five-year stages. As the national plan had outlined, regional priority was given to the more economically underdeveloped and politically unstable areas. The highest regional priority therefore went to the Northeast. Each subsequent five-year period would concentrate on a specific region, and finally, the country as a whole. The implementation of

Table 5-1. *Original Timetable for Thailand's National Rural Electrification Plan*

| Stage | Fiscal Years | Project Villages | | Region |
		In Program	Accumulated	
I	1977–1981	5,200	5,200	Northeast
II	1980–1984	8,000	13,200	South
III	1983–1987	13,500	26,700	North
IV	1986–1990	14,500	41,200	Central
V	1990 onward	5,800	47,000	Countrywide

Notes: The total number of villages in 1977 was 47,000, 12,000 of which were already electrified. The remaining 35,000 unelectrified villages were projected to increase to 47,000 over the approximately 15-year time frame, assuming an annual growth rate of 2%.

Source: PEA 1978.

the program was to be carried out by the Office of Rural Electrification (ORE), a new agency created within the PEA.

However, given the government's expectations and rising rural anticipation of improved living standards, the original timetable proved too conservative. In 1975, a new government took office and decided to accelerate progress. The program would still rely on the national plan, but the five-year stages, formerly consecutive, would now overlap. The timetable was thus compressed from 25 years into 15 (Table 5-1).

The revised national plan, as approved by the Council of Ministers in 1975, required that there be an estimation of the magnitude of physical and financial requirements and of the time frame in which to complete the program. The criteria for both regions and villages would have to be worked out in detail. Also, there would be an estimation of the management requirements of the program. Among other things, attention would be paid to the technical standards, pricing, and other aspects of the rural electrification program.

The actual implementation of the program was not that far off from the original rural electrification plan. Between 1977 and 2003, within the framework of the larger rural electrification plan, the PEA implemented 23 separate electrification projects (Table 5-2). However, 20 of these projects were implemented between 1977 and 1997.

Institutional Choices for Rural Electrification Implementation

The government of Thailand's decision to create a separate agency responsible for providing electricity to rural areas and provincial cities was important for the program. Conventional, vertically integrated electricity companies typical in many developing countries tend to concentrate on the more urgent and profitable business areas, such as power generation and transmission, and on distribution to urban, industrial, and large commercial customers. As a consequence, such

Table 5-2. *Summary of Rural Electrification Projects in Thailand, 1977–2001*

Project Plan and Name	Years	New Villages with Electricity	Project Budget (millions US$)
Fourth National Economic and Social Development Plan (NESDP) (1977–1981)			
Accelerated Rural Electrification I	1977–1985	4,667	74.9
Village Electrification in Four Principal Provinces	1977–1984	746	12.6
Village Electrification in Three Southern Provinces	1977–1981	362	6.2
Village Electrification in Ten Northeastern Provinces	1977–1981	478	5.7
Tambon Electrification	1979–1984	4,692	74.0
Subtotal		10,945	173.4
Fifth NESDP (1982–1986)			
Accelerated Rural Electrification II	1981–1989	7,948	232.4
Village Electrification, Phases 1 and 2	1981–1989	4,749	117.1
Normal Rural Electrification I	1983–1988	4,307	98.1
Subtotal		17,004	447.6
Sixth NESDP Plan (1987–1991)			
Normal Rural Electrification, Phase 2	1988–1992	1,645	69.5
Village Electrification, Phase 3	1988–1992	2,005	67.5
Normal Rural Electrification, Addition 1	1988–1989	2,014	36.3
Normal Rural Electrification, Addition 2	1989–1990	2,020	52.3
Normal Rural Electrification, Addition 3	1990–1991	2,004	60.8
Remote Village Electrification Project, Phases 1 and 2	1991–1993	214	17.1
Village Electrification Doi Tung	1988–1990	34	1.0
Village Electrification Tung Kula Rong Hai	1987–1991	300	6.2
Village Electrification Koh Chang	1991–1992	7	6.5
Subtotal		10,243	317.2
Seventh NESDP Plan (1992–1996)			
Village Electrification, Phase 4	1992–1996	1,000	100.0
Village Electrification Project for Settlement of Population in Deteriorated Forest	1992	35	2.2
Remote Village Electrification, Phase 3	1992–1993	126	8.1
Subtotal		1,161	110.3
Eighth NESDP (1997–2001)			
Rural Household Electrification Project First Stage[1]	1995–1998	400,000 HHs	384.8

(continued)

Table 5-2. *Summary of Rural Electrification Projects in Thailand, 1977–2001 (continued)*

Project Plan and Name	Years	New Villages with Electricity	Project Budget (millions US$)
Island Electrification Project[1]	1998–2001	6,196 HHs	25.0
Rural Household Electrification Project Second Stage[2]	2001–2003	150,000 HHs	93.8
Subtotal		556,196 HHs	503.6
Total		39,353 Villages	1,552.1

[a]The Eighth NESDP involved intensification, and few new villages received electricity because virtually the whole country had access to electricity.

[1]The Projects were planned for the Seventh National Plan, but the implementation was delayed until during the Eighth National Plan.

[2]Households under this project will be electrified using the solar PV home system. Under this project PEA is only an implementing agency for the government; the government bears full cost for providing the solar PV home system. Households will qualify for off-grid electrification using the government-sponsored solar PV home system if the average cost per household connection using grid electricity is more than 30,000 baht (US$750; based on exchange rate of US$1=40 baht).

Notes: Budgets are based on an exchange rate of 25 baht to US$1.00, and from 1998 to 2003 budgets are based on an average exchange rate of 40 baht to US$1.00.

Source: PEA Annual Reports 1975–2001.

companies often find it difficult to accord high priority to rural electrification, which is often unprofitable for them. One way to deal with this problem is to hand over the responsibility of implementing rural electrification to an organization solely dedicated to this task and guaranteeing the organization a high decree of autonomy.

At the initiation of the rural electrification program, Thailand's electricity sector involved three public institutions (Figure 5-1). The company producing electricity for virtually the whole country was the Electric Generating Authority of Thailand (EGAT). Two other companies were also responsible for electricity distribution in Thailand. The first was the Metropolitan Electricity Authority (MEA), a distribution company servicing the Bangkok metropolitan area. The second was the PEA, which is a public distribution company responsible for serving all customers outside greater Bangkok, including rural areas. EGAT sells virtually all of its electricity to MEA and PEA.

In addition to having a clear operational mandate, the PEA had full autonomy and complete control over the budget for program implementation. As part of the national power supply system, EGAT charged the PEA a lower rate for bulk power purchases. The lower rate was designed to compensate for the higher operating and investment costs of rural electrification and to help ensure the company's financial independence. In addition, PEA customers in large and medium provincial cities provided a good revenue base to help finance expansion

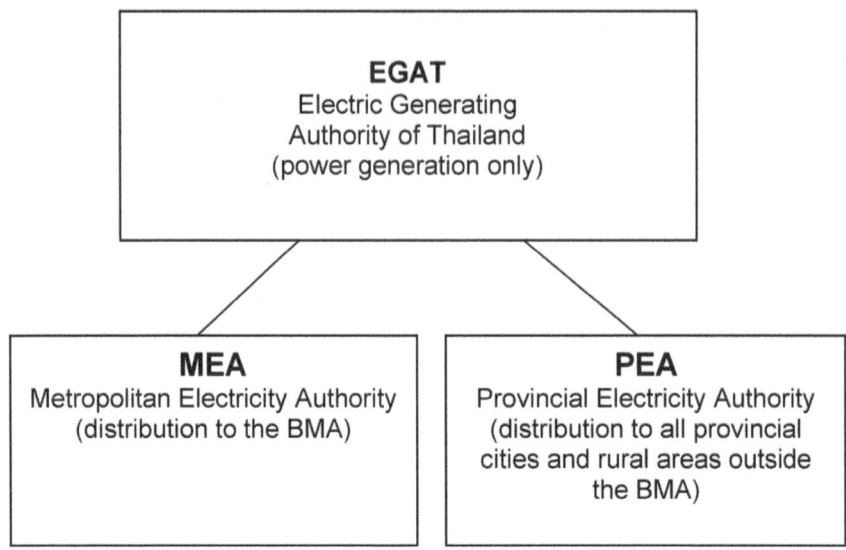

Figure 5-1. *Division of Responsibilities for Electricity in Thailand*

of electrification in rural areas, as well as to compensate for losses sustained in some rural areas and villages.

To further concentrate attention on the task of providing electricity to rural areas, the PEA's board of directors created the ORE specifically to carry out the expansion of rural electrification. Operating within the PEA, ORE was responsible for constructing grid extensions and providing electricity services to villages under the national plan. It was also responsible for followup and conducting load promotion and demand management in the villages that had received electricity. By creating the ORE, the PEA focused a distinct set of resources on the task of serving rural customers, allowing the company to concentrate on providing electricity to provincial urban areas. Once the task of extending the grid to most villages had been completed, the PEA dissolved the ORE and resumed responsibility for serving villages and other nonurban provincial customers.

The result of the institutional choices made by the government and the PEA was a highly successful program that steadily built coverage in provincial areas. Over some 20 years, the PEA methodically expanded the number of rural villages with access to electricity (Figure 5-2), and by 1993, more than 97% of rural villages had been reached (Table 5-2).

Village Selection Method Mitigates Political Interference

Rural electrification programs are particularly subject to lobbying by politicians eager to demonstrate their ability to provide low-cost electricity to their constituents. Once a program succumbs to such pressures, however, it is virtually impossible

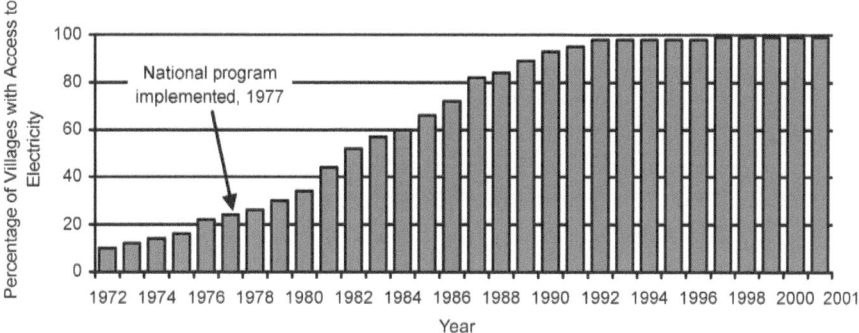

Figure 5-2. *Percentage of Thai Villages with Access to Electricity, 1972-2001*

Note: Figures between 1972 and 1980 are estimates.

Sources: Electric Utility Data Book for Asia and Pacific Region, Asian Development Bank, Manila, 1993; PEA Data Book (internal document), December 2001.

to maintain financial viability because of the expense of providing electricity to remote locations. To avoid such interference, PEA staff developed an objective, well-thought-out methodology for village selection. The full scope of the rural electrification program, including the village selection plan, was then integrated into Thailand's five-year National Economic and Social Development Plan, which further inhibited politicians from interfering with the program's implementation.

From the outset, the PEA realized that it would not have the resources to connect all rural villages throughout the country. Its management knew that providing electricity to less developed villages would mean higher program costs, low potential productive use of electricity, and risk of financial losses. It also believed that electrification was one among many interconnected components of successful rural development. Unless other favorable factors were present, the distribution company reasoned that the incremental effect of electric power would be mainly social. In this section, we discuss in some detail the methods used at the early stages of electricity planning as an example for other countries just beginning the process of providing electricity service to their rural populations.

Quantitative and Qualitative Assessment for Distribution Planning

The general framework for setting priorities for villages to be included in electrification projects was based on three major principles. These principles were maximizing potential benefits while minimizing project cost, integrating rural electrification into a broader national development strategy, and giving consideration to the social and political requirements of less stable areas. The third criterion was entirely political, and it was accepted that including remote, less stable areas would mean higher costs. However, even within the provinces included for largely political reasons, the PEA applied the standard village selection criteria.

The PEA used three quantitative steps, along with a final qualitative assessment, to determine which villages would receive electricity under its projects. The first two steps involved allocating the number of villages to receive electricity in each province, followed by selecting the number of villages in each district. After these steps were completed, villages were ranked according to seven major factors relevant to both the costs and benefits of providing electric power. The result of this procedure determined which villages would receive electricity under a specific project.

Village Allocation for Each Province. The process of village selection in a country without significant levels of rural electrification is an important issue. Availability of funding under the PEA's early rural electrification projects was limited.

The first step in allocating the number of villages within provinces involved an analysis of each province's socioeconomic conditions. Secondary data were collected on 30 variables and then analyzed (Box 5-1). Each of the 27 provinces covered was assigned a score for every variable. The score was standardized according to the mean of the variables; 0 was assigned for a province at the mean, and ±1 was assigned for provinces one standard deviation (approximately 34%) above or below the mean. The sum of these scores was then used to determine the number of villages within a province that would receive electricity under the project based on a formula.[1] Even though this method favored the more advanced provinces, the overall program goal was to reach the largest number of beneficiaries within the project's budget.

Village Allocation for Districts within Each Province. The next stage in the planning process involved the selection of the number of villages to receive electricity in each district within each province. For this purpose, a quite different methodology was used. The goal of this stage was to predict the level of electricity use by villages in individual districts. The nature of the information necessary to make such predictions is some knowledge of household characteristics. However, no secondary data were available at the household level in Thailand at that time. As a consequence, the PEA conducted a detailed household survey of close to 10,000 households, including those with and without electricity. In addition, at the village level, surveys were conducted in some 1,500 villages to measure their level of infrastructure development.

For the survey, information was collected on many different household characteristics, such as income, expenditures and household assets (Box 5-2). With this information at hand, for each province they used standard regression techniques to estimate household electricity use for households with electricity. They then used these estimates within each province to estimate a predicted level of electricity use for households that did not have electricity at that time. The PEA needed some measure of electricity demand in communities without electricity, so they estimated electricity use in villages with electricity and used these results. After this, the actual and the predicted electricity use for a district were then combined to determine the number of villages to receive electricity in a district. In general, districts with the highest predicted levels of actual and

---

Box 5-1. Variables Used to Determine Viability of Electrification for Village Selection

Access of Rural Households (%)
Public well as source of water supply
Private well as source of water supply
Electric lighting
Radio
Television
Sewing machine
Refrigerator
Electric fan
Water pump for agricultural use

Households Using Cooking Fuels (%)
Charcoal
Wood
Gas
Other modern fuels

Dwellings Constructed in Past Five Years (%)

General Health Indexes
Rural population density
Population birth rate
Population growth rate
Ratio of population to local physicians

Source: PEA 1978.

Ratio of Students
Upper elementary to lower elementary school
Lower secondary to upper elementary school
Upper secondary to lower secondary school

Agricultural Assessment (%)
Gross area in agricultural use
Arable land under rice cultivation
Arable land under field crop cultivation
Total land under fruit trees and tree crops

Baseline Electricity Data
Consumption in villages already electrified
Ratio of electrified households to total households
Households with electricity (%)

Overall Village Characteristics
Average household size
Average village population

---

potential demand for electricity were allocated a larger share of villages to be covered under the program.

Ranking Villages for Final Selection. The final step in the selection process was to rank the villages based on certain characteristics. At this point, the process of village selection was simplified so that anyone interested in the allocation plan could easily understand it. Before the ranking was completed, obvious candidates for exclusion were eliminated, including villages on islands or in remote areas. These adjustments were based on such factors as technical feasibility, spatial distribution, administrative and security requirements, and other ongoing or planned development efforts of the national government's various agencies and ministries.

To obtain a preliminary ranking, the villages were rated according to seven equally weighted factors. The factors are familiar to anyone involved in rural electrification planning. Again, the emphasis for the PEA was on holding down costs and generating the highest amount of revenue possible for those

---

Box 5-2. Village Socioeconomic Factors for Forecasting District Electricity Demand

Household Characteristics
Household size within village
Net income per expenditure

Agriculture
No. of households that own lowlands
for rice cultivation
No. of households that own upperland
for other crops
No. of households with cattle and
water buffalo
No. of households owning small livestock

Household Income
Average village income from agricultural
sources
Average village income from livestock
sales
Average village income from wages and
salaries
Average village income from other
sources
Total annual income

Household Expenditures
Fixed expenditure
Variable expenditure
Subsistence expenditure
Social expenditure
Total annual expenditure

Source: PEA 1978. Vol. I.

---

communities to receive new electricity service. The factors are explained in detail below.

Proximity to the Grid and Roads. Village proximity to a distribution line or to other electrified villages was given important consideration because it would be less costly to connect villages close to the grid. Villages were rated according to their accessibility to roads or highways, which generally meant lower costs for transporting equipment, and thus less time for project implementation.

Village Size and Number of Expected Customers in First Five Years. The electricity distribution construction costs per customer are lower in larger villages because these villages have more potential customers than smaller villages. Also, larger villages usually function as local commercial, administrative, and social service centers. For this reason, larger villages merited a higher priority for electrification. Villages were evaluated based on the expected number of customers. Unlike cost-per-customer considerations, this factor is important for the prediction of the amount of revenue that the company would receive from its new customers, thereby increasing the project's economic viability.

Productive Uses: Agriculture, Small Industry, and Commercial Establishments. The total kilowatt equivalent of power load based on the presence of diesel irrigation pumps and energy use in local industries was evaluated for each village. Rural industries not only would use significantly larger quantities of electricity than average households, but they also would increase productivity and generate more income for the rural population. Given the lower costs of electricity, it was predicted that electrification would prompt a rapid conversion from diesel engines to electric motors.

The number of commercial establishments in a village was used as an indicator of the village's relative importance in the hierarchy of village settlement in the vicinity. Such establishments as general-purpose commercial shops, tailoring shops, and barbershops would consume significant amounts of electricity. Almost without exception, these establishments were among the initial consumers in the early years of electrification.

Number of Public Infrastructure Facilities. The number of public infrastructure facilities located in a village was also used as an indicator of the village's importance in the overall hierarchy of village settlements in the vicinity. Presence of a health clinic, for example, suggested that the village enjoyed a position of centrality in the spatial distribution of settlements in the area. Presence of grade schools, potable water supply systems, and other public facilities similarly indicated centers of rural development. Supplying electricity to such villages would enhance the effect of existing infrastructure facilities and would improve the overall utility of the villages to the surrounding areas.

Depending exclusively on quantitative information to determine which villages should receive electricity can sometimes lead to an anomalous situation. Therefore, final selection of villages was based on qualitative information as well. For example, some villages with highly dispersed settlement patterns were excluded because of the high costs associated with reaching potential customers, even though they might not be located far from the electricity distribution system. Villages included in other rural electrification projects, such as Tambon and Normal Rural Electrification, which required local contributions for electrification, were excluded. Some smaller villages, excluded initially, were later added because they were located along the direct path of the distribution line and could be served with relatively small incremental investments.

The net result of the rigorous selection process was a list of villages for which the delivery of electricity service would have the highest initial impact in both social and economic terms. In addition, by eliminating guessing or haphazard approaches, which will be discussed in the next section, much of the political pressure to select particular villages or regions that would develop slowly and produce low revenue for the distribution company was deflected by this process.

Options for Accelerating Village Selection

The PEA also developed an incentive program by which individual villages could accelerate the pace at which they received electricity, depending on their willingness to pay some or all of the associated construction costs. Though developed to mitigate the capital costs of constructing the distribution infrastructure, the program functioned as an additional set of criteria for village selection.

In effect, the PEA provided opportunities for villages to "jump the queue" based on local ability to contribute to the costs of electrification. It did so by classifying the program into separate projects depending on the funding sources. If the villages were covered by one of the accelerated rural electrification projects, villagers were not required to contribute to the extension costs; the PEA would bear full construction costs for grid connection. These villages were scheduled for electrification in accordance with the standard village selection criteria.

However, if a village not in the normal accelerated program was willing to pay 30% of the construction costs in cash or labor contribution, it would be moved higher on the list of villages to receive electricity under the accelerated projects. Finally, the PEA would respond immediately to any villages willing to pay 100% of the construction costs. In practice, the villagers themselves did not pay much of these costs involving either option. Rather, politicians were encouraged to secure local social development funds or individual contributions to finance the costs of accelerated village selection. Those instances in which villages were willing to pay part or all of the capital costs for their electrification had another important simultaneous advantage of reducing the company's own costs.

The flexible financing program described above was not the only mechanism for minimizing political interference. The PEA established a reputation as a well-managed government enterprise with a sound financial standing—one that was financially and politically independent. This reputation further helped to insulate the rural electrification program from political pressures. In addition, the king of Thailand issued a decree that publicly supported rural electrification throughout the country. Although the king is a constitutional monarch, his public endorsement gave the PEA management and its staff confidence in their capacity to act independently. Unique to Thailand, the king's public degree worked to reduce political interference in PEA management's policy and procedural decisionmaking.

Lowering Costs and Increasing Local Participation

A critical aspect of the rural electrification program's success was the distribution company's ability to reduce costs in many areas of program operations. As an autonomous agency, the PEA had a clear incentive to manage costs because it was fully accountable under its mandate for its own financial performance. Moreover, it had the authority to do so because it had been granted full control over its own budget. In the face of intensive capital investment that the initial years of the program would entail, PEA management made a concerted effort to ensure that the rural electrification effort would not jeopardize its own financial standing.

The result was that the PEA developed and implemented highly effective cost-cutting strategies for rural distribution that simultaneously improved efficiency. It implemented a productive use-promotion strategy to increase revenues from rural areas. To defray some of its own expenses, it required economically well-off villages to pay for some construction costs. Finally, it concentrated on building consumer confidence by providing responsive customer service.

Strategies to Minimize Construction Costs

To help minimize system construction costs, the PEA instituted a far-reaching policy of system standardization, including technical, equipment, and other components. It also relied on locally manufactured materials to the greatest extent possible and spread the large capital costs of system expansion by providing incentives to allocated villages to contribute to those costs wherever possible.

Several approaches were taken to reduce costs through standardization. During the early planning stage, PEA engineers worked closely with EGAT's engineers to develop standards for the program's electrical system. As the agency responsible for Thailand's electricity generation and transmission, EGAT would ultimately work closely with the PEA in this area. It was thus critical to establish a general systems plan and distribution system that complemented existing or planned transmission networks. After consulting with EGAT, the PEA's engineers selected 33 kilowatt volts (kV) and 22 kV as the standards for its distribution system throughout the country. The company chose the 33-kV standard primarily for the South, which frequently experiences severe thunderstorms, and decided on 22-kV lines for all other regions of the country.

To further control construction costs and increase efficiency, the PEA opted to standardize all equipment and components used for constructing the distribution systems of all electrification projects. This decision enabled project engineers to minimize the variety of equipment and components used. As a result, the PEA was able to reduce procurement, materials handling, and purchasing expenses through bulk purchases, which gave them control over an important component of the overall project costs. It also significantly reduced the risk of equipment and component shortages, providing project managers an efficient way to handle materials required for field construction. Because they were familiar with the equipment and technical standards, field construction crews were better able to complete assigned tasks on time and within the allocated budget. Standardization also reduced the difficulties or uncertainties involved in the tasks of field engineers and construction crews. Field engineers solved most technical difficulties simply by referring to the detailed handbooks derived from the national plan. Most field construction was therefore completed without delays that could have increased project costs significantly.

Another strategy to minimize costs involved the reliance on locally manufactured materials. PEA management realized that imported materials were costly and might require a longer procurement time, with consequent delays in the construction schedule. Whenever possible, therefore, the company relied on locally made or assembled equipment and components, not only to reduce costs but also to ensure a steady supply and timely delivery of construction materials. As an example, the PEA purchased concrete poles and prestressed concrete-spun cross-arms from local concrete pole and pipe manufacturers located in the major provinces of all regions.[2] Another example was the development of the local capacity to produce wires and cables. It purchased aluminum ingots from abroad but also hired local contractors to process the imported ingots into aluminum wire for project use. In this case, material costs were dramatically reduced without compromising quality.

As a result of the cost savings, the PEA was able to provide electricity to an additional 837 villages under the Accelerated Rural Electrification Project, 22% more than had been estimated when the project began (World Bank 1978, 1980).

The involvement of the local population in paying for part of the construction costs also helped PEA to reduce its financial costs of providing rural electricity service. The PEA recognized that most villages would be unable to afford the full cost of electrification. Likewise, the company's own ability to finance the entire program was limited. Thus, to maximize the number of villages that could receive service, PEA management set up a voluntary system that provided villages who could afford to and were willing to do so an incentive to contribute to the capital costs of extension. The system was considered voluntary because, in theory, the PEA would bear the construction costs for all villages in the country. In practice, the incentive-based policy consisted of three approaches. Over the course of the program, almost 75% of the designated villages waited their turn to receive electricity based on the standard village selection plan. Some 25% agreed to contribute 30% of construction costs, and only 1% was willing to pay the full cost of construction.

These incentive-based contributions helped relieve the PEA's financial burden and enabled it to extend power to some of the poorer, more remote villages. In addition, the system let politicians focus their energies more on raising funds to finance construction than on pressuring the PEA into connecting their constituents' villages ahead of schedule.

Strategies to Increase Local Participation

The PEA actively sought and received strong local support and cooperation during all phases of system development, beginning with the initial surveys and continuing through construction, electrification, operation, and maintenance (United Nations 1990). Local support and participation enabled the company to develop and implement projects at a satisfactory pace and carry out the entire program efficiently through innovative cost-cutting measures and by reducing technical and nontechnical difficulties. With full support from local communities, the PEA was able to obtain tangible benefits—both direct and indirect—some of which are described in this section.

The PEA was conscious that good relations with local communities were important for gaining the confidence of its new consumers. As a result, after completing the village selection process, the PEA informed and initially sought cooperation from the provincial governor, who, in turn, informed and requested cooperation from the *amphor* (district), *tambon* (subdistrict), and village heads. Developing a strong rapport with local district officials was important because village heads report directly to them. Thus, the connection guaranteed the PEA easy access to local communities.

As part of its public relations strategy, the PEA assigned its representatives to attend local government meetings. Because the PEA presented itself and was generally regarded as a government agency, its representatives were permitted to attend regular administrative meetings. Held at district headquarters, these monthly meetings also included all tambon and village heads. Through these meetings, the community leaders learned in advance whether their villages had been selected for electrification.

An important component of the rural electrification extension program was to obtain village assistance in carrying out some of the tasks necessary during the extension of the new lines to the community. The PEA actively sought assistance from village heads and residents at all program stages. When its team was ready to conduct an area survey, the PEA asked the village head to lead the survey team, help identify land demarcation, and, eventually, assist in certifying the village plan.

The town meeting was used as a vehicle to introduce the new service to community residents. Once construction was imminent, the PEA asked the village head to organize a town meeting, at which a company representative conducted a public hearing on the upcoming electrification effort. During such a meeting, the PEA representative typically sought general public support on such issues as the need for land rights of way or the cutting of certain trees along the planned route for electric lines and poles. The representative estimated the number of possible connection requests, explained hookup policies and connection fees, distributed application forms and discussed how to fill them out, and notified attendees of the approximate date the construction crew would arrive in the village. In this way, they also educated villagers about the uses of electricity and promoted the adoption of electricity by people in the village.

The PEA also was able to secure important in-kind contributions from villagers at the time of construction. This would not have been possible without extensive community support for the program. For example, the program had no budget for acquiring land rights of way or financial compensation for cutting trees with economic value. For grid connections or distribution lines, villagers typically granted rights of way and the right to cut trees without any financial compensation. Without such contributions, program costs would have skyrocketed.

The PEA was also conscious of creating a positive image during construction. One method of creating a positive image was to provide some wage-paying jobs at the time of construction. Typically, the PEA hired or contracted unskilled laborers from the village to be electrified or from those nearby. They were paid at market rates to install low-voltage poles, and local firms were contracted, also at market rates, to install high-voltage poles. In the case of house wiring and meter installation, the village head was enlisted to collect home-connection fees and to remit these in one installment before the village was electrified. Through this process, the PEA knew in advance the exact number of households and businesses wishing to be connected.

On many occasions, villagers provided free labor and animals to transport materials and construction equipment. In areas with difficult terrain, they used various means—elephants, horses, manpower, boats, and barges—to assist in transporting poles and construction materials (United Nations 1990). During the construction of high- and low-voltage distribution lines, they often donated their time to work as unskilled laborers. Following training, they worked as skilled laborers to dig holes, set poles, and lay conductors. The PEA's construction crews either rented rooms in the village or stayed in the local temple during a project's construction phase, which enabled them to become part of the village and avoid imposing burdens on villagers.

Commitment to Customer Service after Construction

Once villages were connected to the grid, PEA management adopted specific performance measures and practices to ensure high-quality, responsive service. Targets were set to increase the number of connections and reduce the number of interruptions and system losses. PEA technicians from the company's local offices regularly monitored quality of service. Each month, for example, they measured the voltage at specified locations during peak hours and submitted the results to the main office.

Requests for service connections were acted on promptly. For example, it was a company policy to respond to requests of less than 30 amperes within six working days. In 1993, 94% of such requests were completed within this time limit. The company also established a customer-service training program for staff who worked in direct contact with customers. This plan ensured that each office had trained staff on hand to provide services and offer advice on safe and efficient use of electricity.

Electricity companies in developing countries often have problems collecting bills from rural consumers, and the PEA was keenly aware of the implications of problem bill collection for the finances of the company. Consequently, it developed some innovative practices to ensure that customers paid their bills on time. One PEA innovation was to hire respected local individuals to collect on bills (Asian Institute of Technology 1992). Typically, the PEA hired a schoolteacher, village head, or village elder, who was required to post a security deposit or bond. The bond was based on land holdings or on future salary (for government employees) and guaranteed that the bills of the village would be collected and paid. In the mid-1990s, 6,035 people were collecting monthly electricity bills in villages throughout the country. Compensation for this activity was 5% per bill.

The use of local bill collectors has proved highly successful, although only about 10% of the villages in the country pay their monthly bills this way at present. The system has reduced the company's costs for bill collection and accounts receivable. It also has reduced nontechnical losses from delinquent accounts, nonpayments, and illegal connections. Currently, the company's cycle for accounts receivable across all customers is only 34 days, and nontechnical losses are only 6.1%.

The PEA's commitment to customer service created a positive environment that benefited the rural electrification program substantially. Among community members, it helped develop a sense of program ownership, which was reflected in low incidences of theft and billing problems. As a result, the PEA could provide better service at a lower cost than might otherwise have been possible.

Active Promotion of Rural Electricity Use

From the beginning, the PEA regarded load promotion and demand management as integral parts of rural electrification development. As part of its strategy, the company regularly monitored electricity demand load and conducted billing analyses of its customers. The PEA also relied on extensive market research. These analyses permitted assessments as to whether rural villages were using electricity productively or at its highest potential.

Load promotion activities concentrated on three major areas. One area involved promoting the use of electricity for revenue growth. Another was promoting the efficient use of electricity. Finally, the PEA wanted to expand connections so that the largest number of people could enjoy the benefits of electricity. Not incidentally, with a high number of households adopting electricity, the cost of service to these people would be reduced. The responsibility for load promotion was taken seriously, and a Load Promotion Unit was created as a part of the Office of Rural Electrification. The company's load promotion program began in 1979 as part of Accelerated Rural Electrification Project I.

Typical Load Promotion Strategies

The company's load promotion strategies were aggressive and focused on villages in which electricity use was below planned expectations. After a village received electricity, the PEA closely monitored the load growth, connection rates, and revenue. On a monthly basis, each regional billing department submitted a summary monthly billing report to the ORE's Load Promotion Unit. The load promotion staff conducted a billing analysis and identified those villages and provinces where it was felt that electricity use ought to be higher. A load promotion team would be sent to the region and was instructed to work to increase productive uses of electricity in these areas. The teams also were trained to help villagers understand how they could make greater use of electric power.

After the initial visit, a local load promotion team was formed, consisting of members of the Load Promotion Unit's own staff and staff from the local customer service office. This team functioned to coordinate the company's services and efforts and capitalized on the strengths of both the PEA and the local administrative personnel. The teams were familiarized with local conditions, services strategies, and customers. This knowledge, combined with the technical know-how and promotion strategies of the unit, allowed the teams to make direct contact with local villagers.

The Load Promotion Unit would engage in several types of activities. Before making direct contact with potential customers, the load promotion team visited local electric motor dealers or representatives. They assessed the current supply, retail prices, and availability of electric motors in the area and discussed with dealers the types and sizes of motors the company was promoting. This visit put them in a better position to provide direct technical assistance to rice mill owners, as well as to provide information regarding the advantages and disadvantages of switching to electric power from both a financial and engineering point of view.

The PEA's indirect approaches typically relied on the mass media, such as television, local radio, and newspaper advertisements. The Load Promotion Unit also worked with the PEA's public relations department to produce brochures and publications that provided technical information to both rice mill owners and local contractors. Unit staff and their local counterparts often set up exhibitions at local agricultural fairs and at cultural and religious celebrations to demonstrate electric motors for rice milling and other applications. All of these activities were part of the technical information dissemination and education campaign. The unit also worked and cooperated with other government agencies and their efforts, such as the Department of Industrial Development and rural drinking water programs, to promote the use of electric motors by these agencies.

Promotion of Productive Uses for Revenue Growth

Under the initial electrification projects, about 90% of rural electricity use was for household lighting. Thus, the amount of energy consumed by customers during the first year of connection was minimal, ranging from an average of 11 kWh per household per month under the first project and increasing to only 22 kWh per household per month under the second project. The level of household energy consumption after five years increased to about 50 kWh per month per household (PEA 1986, 1990). These low consumption levels, in tandem with the concentration of lighting load during the evening hours, were a financial burden for PEA. Because of the rapid and large investment involved in the rural electrification program, the company's management decided early to explore ways to improve the program's economic rate of return. It thus began to put more emphasis on promoting productive uses.

The most successful load promotion program involved direct incentives to those willing to convert from diesel engines to electric motors. Realizing that any conversion required high initial investment costs, the company worked with Thailand's Bank of Agriculture and Agricultural Cooperatives to make loans available to rice mills and other commercial concerns. The company also assisted rice mill owners to obtain these loans. Load promotion techniques encompassed no direct subsidies or low-interest loans to customers. The only financial assistance program was the PEA's credit scheme for initial house wiring and other connection costs for domestic customers. A main incentive was allowing payment of the connection fees by installments, without interest, in 12 equal monthly payments.

The PEA initiated its productive uses promotion activities during the preconstruction phase and continued until either a satisfactory connection rate was achieved or it was no longer sustaining any financial losses for the village or region. The PEA's load promotion plan to increase productive uses in the villages was systematic. It relied substantially on the market research the company conducted on potential power loads in the rural villages. The research confirmed that rice mills, water pumps for irrigation, and motorized applications for cottage industries had the greatest potential and should be targeted for promotion activities.

The PEA determined that electric motors in rice mills were the best candidates for load promotion and demand management and would provide the company with the best return on its load promotion investment. This was because almost all of the project villages had one or two small- or medium-size rice mills, amounting to approximately 50,000 rice mills of these sizes for the country. Typically, most small- and medium-size rice mills use 5- to 15-hp motors. Converting from a diesel engine to an electric motor could reduce a mill's operating costs by a minimum of 20% (PEA 1989). Moreover, the conversion to electricity of just one rice mill would increase crucial daytime load in the village, significantly improving the load factor, which had been skewed by nighttime use for lighting, thus improving the financial rate of return for investments in electricity distribution for the village as well.

In addition to rice mills, the PEA included other powered devices used in cottage industries in the lists for load promotion. These included simple power tools, small- and medium-size water pumps for irrigation, motor applications for rubber tree plantations, and applications for cassava shredding and palm sugar processing.

Lowering Costs per Customer by Promoting Household Adoption

The number of people who adopt electricity service have a significant effect on the costs of the distribution systems. In general, a higher number of electricity connections means that the cost per customer for the distribution system will be lower. Thus, the PEA's representatives capitalized on the direct, personal contact with villagers during the construction phases to achieve the highest possible connection rates.

To provide an incentive for villagers to opt for connections, the PEA established low connection fees and provided a credit line for homeowners to cover both connection fees and house wiring costs. In the first instance, the PEA launched an incentive program providing credit of approximately 2,000 baht (US$98.00) per household for the initial house wiring and other connection costs. A few years after implementing this program, the company decided that there was little need to continue it because local contractors replaced the need for the PEA to provide such incentives. As discussed in the following sections, the PEA used local contractors for the lower level technical tasks to improve and build up local private sector capability and to cut its own costs. The system's high degree of standardization made it easy for local contractors to learn and apply the newer wiring and connection technology, and the company's field engineers were available to provide technical assistance and guidance. This assistance resulted in an increased number of local private electrical contractors providing services in the villages. With an incentive to secure an expanding market, the contractors created their own individual marketing strategies. Some contractors even provided various types of credit or installment schemes to customers for house wiring and connections, thus obviating the need for the PEA to provide its own financing incentive program.

Despite personal contact and intensive campaigns during the preconstruction and construction phases, however, the average initial connection rates for the Accelerated Rural Electrification Projects I and II were only 55% and 53%, respectively. Even though these rates are higher than for most other developing countries, to PEA this indicated a need for load promotion activities as a continuing, long-term process. Household connection rates increased over time, and today most households in rural Thailand have electricity.

Cross-Subsidies and Concessional Loans Permit Uniform National Tariff

Twenty years ago, at the peak of the rural expansion program's implementation, Thailand's financial markets were limited, and most financial analyses without including subsidies yielded no prospects for an acceptable financial return on these electrification investments. However, a combination of cross-subsidies and concessional loans enabled the PEA to expand service to rural areas without undue financial stress. As a government enterprise, the company was able to secure low-cost capital and long-term loans that greatly reduced program costs and allowed it time to develop a strong financial base. In addition, the government's electricity pricing policy provided a price cross-subsidy and a lower price for bulk power purchases by the company, both of which helped finance the rural electrification program.

The government's pricing policy for electric power enabled the PEA to realize a reasonable return on investments and sufficient funds to finance system expansion to meet growing demand for electricity. The government adopted a uniform pricing policy for the country, which included a price cross-subsidy from urban to rural customers and from large to small customers. EGAT, the public company for supplying electricity in Thailand, was required to charge the PEA 30% less for bulk power sales than it charged the MEA, which serves the Bangkok metropolitan area. Retail rate structures were also set up to charge larger users at much higher rates than smaller ones. Finally, both the PEA and MEA were required to use the same retail rate structures.

These tariff and subsidy policies, along with the PEA's mandate to remain financially viable, spurred the company to focus on ways to provide efficient, reliable rural service without jeopardizing its own financial health.

Uniform Pricing Policies and Urban-to-Rural Subsidies

Typically, a uniform pricing policy leads to poor access by rural people: The more a utility company expands service, the more it loses money because costs per customer increase. In Thailand, this policy did not cause losses because of the system of price cross-subsidies between the large pool of Bangkok metropolitan area customers and PEA customers. Although the revenues resulting from the uniform pricing policy were insufficient to cover the costs of rural electrification, the PEA was compensated with a lower electricity bulk rate for power that it purchased from EGAT. In practice, the PEA justified the price cross-subsidy to the government by referring to the rural electrification program's contribution to Thailand's overall development and political stability.

As indicated, both the MEA and the PEA use the same retail electricity tariff structure approved by the government, which consists of seven customer classes and 20 tariff schedules. The tariff schedules can generally be characterized as an increasing block structure (Figure 5-3).

In practice, the average electricity retail price for all customer classes is close to the long-run marginal cost of the country's power sector. With respect to

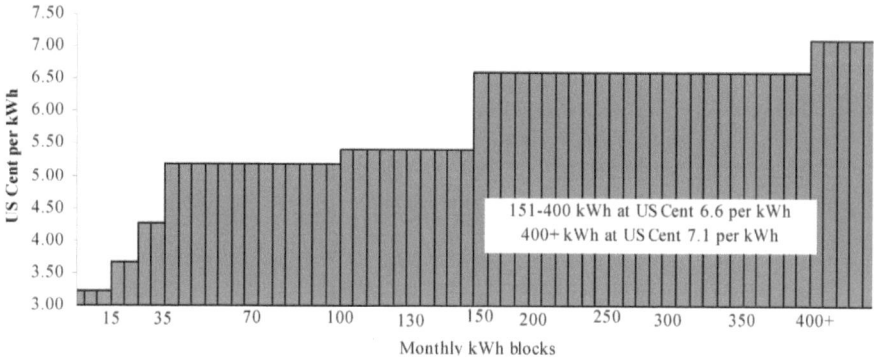

Figure 5-3. *National Uniform Residential Tariff Schedule, Thailand, 2002*

Note: Tariff schedule includes a fixed charge and the increasing block rate: < 150 kWh a fixed there is a service charge of US¢ 20 for the first 5 kWh or consumption less than 150 kWh per month less; and a service charge of 0.97US$ for consumption more than > 150 kWh a fixed charge of US$ 3.56 for the first 35 kWh or less per month. Exchange rate 42 baht = 1 US$.

Sources: Electric Power in Thailand 2002, Department of Alternative Energy Development and Promotion Efficiency, Ministry of Energy 1992.

residential classes for both urban and rural customers, the retail tariff structure consists of fixed charges that increase according to blocks representing increasing ranges of consumption, measured in kilowatt-hours per month. The tariff schedules for the first 35 kWh can be characterized as a lifeline rate for rural customers. The lifeline rate is designed to make electricity affordable to poor people, who use electricity mainly for lighting. Because this class of customers in aggregate uses so little electricity, the effect of this low rate does not adversely affect the PEA's financial performance.

Decisions of the national government's cabinet on tariff adjustments for the PEA, MEA, and EGAT are based on the tariff rate committee's recommendations for setting the country's wholesale and retail electricity tariffs. Tariff levels are set to allow each utility a reasonable return on its investment. During the past 25 years, Thailand's electricity pricing policy has provided sufficient returns and adequate financial support for the investment programs of all three utilities.

The PEA has always been keenly aware of the national government's important role in establishing a favorable regulatory environment for expanding the rural distribution system. At various times, its management has worked actively to ensure that this role was maintained, including support for price cross-subsidy policies (Box 5-3).

During the past two decades, the PEA's revenue from its industrial and commercial customers has been increasing significantly, providing an additional financial base to support system expansion in rural areas. In 1993, the company's revenue from electricity sold to industrial and commercial customers accounted for about 70.9% of its total income. Its profit margin from these customers was larger than the MEA's margin from the same class of customers because of the lower

---

Box 5-3. Price Cross-Subsidies: Avoiding a Policy Shift

In the late 1970s and early 1980s, Thailand's electricity sector policy frameworks were in line with its national goals—a financially sound power sector, a steady pace of power development, and accelerated rural electrification. At times, however, the political climate and other factors dictated a policy shift, which necessitated compromises. For example, the severe droughts of 1977 and 1978 adversely affected the thermal–hydropower generation mix, and in 1979, fuel oil prices increased some 115%. To prevent rising energy costs from driving inflation higher, the government chose to delay passing on the increased costs of electricity to consumers.

Faced with mounting criticism of its handling of the economy, the government in mid-1979 introduced direct budget subsidies for electricity tariffs, at an estimated cost of US$9 million per month. In 1980, however, it agreed to a plan of action to remove the subsidies and restore the power sector's financial independence. Wholesale and retail tariffs were raised, but the price cross-subsidy between MEA and PEA customers and between large and small users remained untouched.

PEA management realized that any shift in these price cross-subsidy policies would jeopardize its own financial viability and that of the entire power sector, eventually affecting the rural electrification program and the other utilities' expansion programs. Therefore, PEA, MEA, and EGAT worked actively with the national government to maintain these policies.

---

bulk rate at which the PEA purchased power. The PEA's effective average revenue per commercial and industrial customer was almost the same as the MEA's. Viewed another way, only 14% of the total electricity sold by the PEA in 1993 was sold to unprofitable residential customers—those with monthly consumption of less than 150 kWh.

In addition to industrial and commercial customers, the PEA's urban customers also generate a good financial base for its operations (Figure 5-4). These customers include residents in all urban centers outside Bangkok, many of which have grown steadily. Large provincial urban centers, such as Amphor Maung and large districts in Chiang Mai, Nakorn Ratchasima, Songkla, and Pathumthani, have provided healthy profits, thus helping to bolster the PEA's financial stability. The company's ability to transfer profits from these urban, commercial, and industrial customers to subsidize service to its rural consumers was particularly important during the program's expansion phase.

Effects of Bulk Rate Concession

The considerable disparity between MEA and PEA revenue from residential customers (Figure 5-5) is the primary reason for the bulk rate subsidy that PEA has received for more than 25 years. The 30% lower bulk rate resulted in significant cost savings for the company because almost 100% of the electricity it purchased came from EGAT (EGAT 1992-2001). This savings helped alleviate problems associated with the PEA's higher operating costs, combined with the fact that the

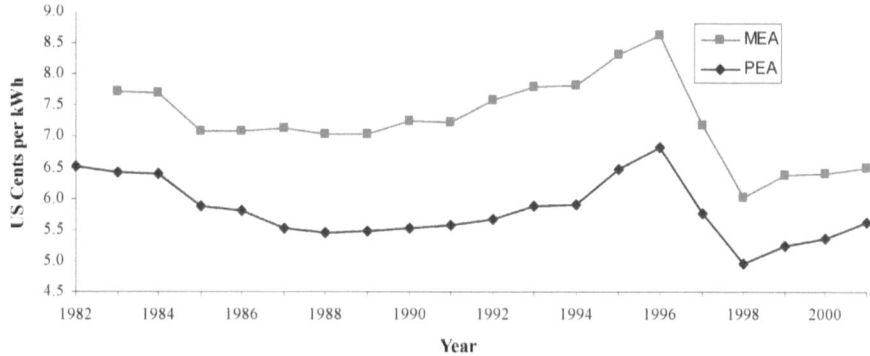

Figure 5-4. *Average Revenue from Residential Customers, Thailand, 1982-2001*

Note: Exchange rate from 1982–1996 is based on 25 baht=US $1; from 1997–2001, exchange rates are based on 31.37, 41.37, 38.0, 40.0, and 43.0 baht = US $1 respectively.

Sources: PEA Annual Report 1982–2002; MEA Annual Report 1982–2002.

company could not charge higher electricity rates to its more widely dispersed consumers. In addition to helping finance the rural electrification program, this lower bulk electricity rate afforded rural customers a lower tariff rate. In 1993, for example, the PEA spent about 29.9 billion baht (US$1.12 billion) to purchase bulk electricity. The concessionary rate it paid for bulk power was equivalent to a cross-subsidy, or cost savings for the PEA, of about 12.8 billion baht (US$0.51 billion) for that year.

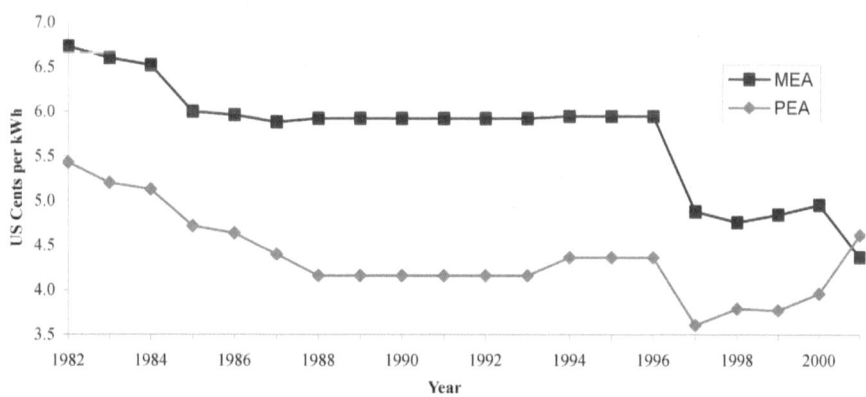

Figure 5-5. *Average Bulk Rate Charged by EGAT, Thailand, 1982-2001*

Note: Beginning in 2001, the bulk rate charged by EGAT is the same for both PEA and MEA. Exchange rates from 1982–1996 are based on 25 baht=US $1; from 1997–2001, exchange rates are based on 31.37, 41.37, 38.0, 40.0, and 43.0 baht = US $1 respectively.

Sources: PEA Annual Report 1982–2002; MEA Annual Report 1982–2002.

Recently, as part of the continuing evolution of rural electrification policy, the MEA-to-PEA cross-subsidy was deemed obsolete because the company's customer base and revenues have grown significantly over the past 20 years. In recent years, the PEA's energy and electricity peak demand have surpassed those of the MEA (MEA 1982-2002). As a consequence, in 2002 the bulk rate concession to PEA was eliminated, and both companies are charged the same rate for electricity. However, the government recognizes that the cost of services for PEA is higher than for MEA. To alleviate some of the burden, the government instructed MEA to pay 8.5 billion baht a year to PEA. Actually, the 8.5 billion baht per year is less than the price break PEA had received before the reforms.

Blending of Concessional and Commercial Loans

From the 1970s through the mid-1980s, Thailand's government had limited ability to fund rural electrification feasibility studies, not to mention the projects themselves. Therefore, the PEA sought grants and concessional loans from bilateral and multilateral development agencies. Concessional loans generally are similar to normal loans, but they have lower than market interest rates and sometimes delayed repayment schedules. As a consequence there is a subsidy component in such loans. To fund feasibility studies, it requested help from bilateral aid organizations in Finland, Norway, the United Kingdom, and the United States. Results from the studies that were subsequently funded were used by PEA to make a case to the government and to international lending agencies for concessional loans and grants. The concrete plans laid out in the feasibility studies helped the PEA to secure funding from such international lending agencies as the World Bank and from such bilateral agencies as those of Canada, Germany, Japan, Kuwait, the Organization of the Petroleum Exporting Countries (OPEC), and Saudi Arabia (Table 5-3).

Most concessional loans were long term and carried below-market interest rates; in some cases, no interest was charged. The loans significantly reduced financing costs, allowing the PEA time to build its revenue base before the loan repayment period began. The favorable grace period provided a major relief from the financial burden experienced during the period of intensive system expansion. Not only did the PEA build its financial base during this time; it was also able to mobilize revenue and other financial resources to fund individual projects and spread out payments over a long period.

In addition, the company's management blended concessional loans from various sources with commercial loans because commercial loans with market interest rates were available for a few large projects. For example, the total international funding for the Accelerated Rural Electrification Project I and the Village Electrification Project in Four Principal Provinces amounted to US$43.2 million, of which 42% was derived from concessional loans. For the Accelerated Rural Electrification Project II, total international funding was US$113.6 million, of which approximately 34% came from concessional loans. Total foreign funding for the Normal Rural Electrification Project I accounted for US$59.2 million, about 48% of which was from concessional loans.

Lessons from Thailand's Experience

The remarkable accomplishments of the rural electrification expansion program carried out by the PEA and its ability to remain on sound financial footing throughout two decades can be attributed to a number of factors. Although some of these factors are unique to Thailand, others could be quite easily duplicated in other countries. One of the main reasons for the program's success was the decision to develop an institutional structure that allowed the PEA to concentrate solely on electricity distribution in rural areas. Given the focus on program expansion, the company made conscious efforts to involve local communities, which resulted in local support and, in some cases, capital and labor contributions to system expansion.

The PEA devised a well-conceived distribution plan that, with few exceptions, focused on the most dynamic areas before proceeding to more remote ones. This strategy contributed to building a sound financial base for later expansion. In most cases, politics were not involved in village selection. Finally, the program was able to maintain good financial standing through conscientious cost-reduction measures, a system of price cross-subsidies between urban and industrial consumers to rural ones, and solicitation of concessional loans and grants from bilateral and multilateral donors.

Dedicated Distribution Company

The PEA, a company dedicated solely to electricity distribution, carried out the rural electrification program. Unlike electricity companies in many other developing countries, the PEA did not have to concern itself with power generation and transmission or providing services to megacities with high-growth, high-demand urban centers. Furthermore, within the PEA, the ORE was created to assume sole responsibility for rural electrification. This logical division of functions created an effective system wherein a committed group of professionals with control over their own budget were eager to solve the problems of developing a national distribution system to cover the country's rural areas.

Commitment to Financial Soundness

As an autonomous agency, the PEA had complete control over its budget and activities and was mandated to achieve and maintain solvency. Many innovative and imaginative approaches were devised to cut costs. The overall strategy was to first provide electricity to rural areas that were already economically advanced, with high potential for productive uses and good returns on initial capital investments. The company conducted careful market research and identified agriculture-based and other local industries that it believed should be targeted for its aggressive load promotion efforts; the rice milling industry was a particular target of these efforts. In this connection, the company also devised strategies to ensure that bill collection rates were extremely high.

Table 5-3. *Loan Sources for Thailand's Rural Electrification Projects, 1976–2000*

Country or Organization	Funding Source/Project(s) Funded	Date of Agreement	Currency & Principal (millions)	Interest Rate (%)	Payment Period
Germany	KFW		DM		
	Village Electrification	4/12/83	14	2	1993–2013
Japan	JBIC (formerly OECF)		Yen		
	Tambon Electrification	6/14/79	7,050.000	3.25	1989–2009
	Village Electrification	4/28/81	4,173.078	3	1991–2011
	Normal Rural Electrification I, Phase 2	9/22/83	3,350.508	3	1993–2013
	Village Electrification II, Phase 2	10/4/85	3,588.000	3.5	1995–2015
	Normal Rural Electrification, Phase 2	2/20/90	6,826.789	2.7	2000–2020
	Village Electrification, Phase 3	9/18/91	6,635.000	3	1998–2016
	Village Electrification Project, SWAP loan to US$	3/5/95	41.184	5.4	1996–2005
	Normal Rural Electrification Project, Third Stage, Phase 2, SWAP loan to US$	3/5/95	33.067	5.03	1996–2005
Canada	CIDA		C$		
	Accelerated Rural Electrification I and Village Electrification in Four Principal Provinces	3/2/78	12.987	0	1988–2027
	Accelerated Rural Electrification II	3/9/82	13.983	0	1992–2031
OPEC	Fund for International Development		US$		
	Accelerated Rural Electrification I and Village Electrification in Four Principal Provinces	10/7/77	7	0	1982–1997
	Accelerated Rural Electrification II	10/24/80	8	4	1985–2000
Saudi Arabia	SFD		SAR		
	Accelerated Rural Electrification II	6/1/81	68	3	1986–2001
Kuwait	KFAED		KW$		
	Village Electrification in Three Southern Provinces	4/10/76	1	3.5	1981–2000
World Bank	IBRD		US$		
	Accelerated Rural Electrification I and Village Electrification in Four Principal Provinces	3/9/78	25.0	7.45	1983–1998
	Accelerated Rural Electrification II	11/14/80	75.0	8.25	1986–2000

(continued)

Table 5-3. *Loan Sources for Thailand's Rural Electrification Projects, 1976–2000 (continued)*

Country or Organization	Funding Source/Project(s) Funded	Date of Agreement	Currency & Principal (millions)	Interest Rate (%)	Payment Period
	Normal Rural Electrification I, I	9/28/83	30.6	7.58	1990–2003
	Transmission System, Second Stage, Phase 3, and Village Electrification, Phase 4	9/27/96	100.0	Managed float	2001–2007
Asian Development Bank	ADB		US$		
	Rural Household Electrification Project	4/17/96	93.3	Managed float	1997–2007

Note: ADB = Asian Development Bank; CIDA = Canadian International Development Agency; GOT = Government of Thailand; IBRD = International Bank for Reconstruction and Development; JBIC = Japan Bank for International Cooperation; KFAED = Kuwait Fund for Arab and Economic Development; KFW = development bank; OECF = Japanese Overseas Economic Cooperation Fund; SFD = Saudi Fund for Development.

*The World Bank's first rural electrification loan of US$3.6 million is not listed here because the GOT assumed this loan on behalf of the PEA.

Source: PEA Annual Report 2000.

Avoiding Political Interference

The PEA also realized in advance that political interference in village selection could cause serious escalation in system expansion costs. The company largely avoided this hazard through the use of objective criteria for village selection and a contribution incentive program that permitted villages that could help defray costs to "jump the queue" for being connected to the grid.

Gaining Strong Local Support

By working closely with local villagers and their leaders from the outset, the PEA was able to gain considerable local support for the expansion program; increase enthusiasm for obtaining connections; and obtain a range of valuable, in-kind contributions from villagers that helped substantially in defraying program costs.

Financing Expansion through Cross-Subsidies and Concessional Loans

Extension of distribution networks in rural areas is capital-intensive. To help finance rural expansion, the PEA relied heavily on price cross-subsidies, as well as grants and concessional loans. Throughout the course of the rural electrification program, the PEA sought and received both grants and concessional loans to extend the capital costs of system expansion.

Bulk Tariff Subsidies as Compensation for Universal Electricity Pricing Structure

With a national policy of uniform retail pricing firmly in place, the rural electrification expansion effort was assisted by a series of cross-subsidies that were approved by the government to assist in system expansion. The PEA was permitted to purchase bulk electricity at 30% below the rate paid by the company serving Bangkok and approximately 15% below normal bulk-price costs. In addition, there was a cross-subsidy between urban and rural consumers within the company's service territory.

Conclusion

Thailand's rural electrification program faced all of the major problems typically posed by extending the grid to rural areas, and they found solutions. This problem solving approach was made possible by having a dedicated unit within a well-managed distribution company that dealt with the many problems that evolved during the more-than-20-year course of providing new electricity service to all of rural Thailand. In hindsight, the rural electrification success would not have been possible without the dedication, independence, and innovation of the team in the Office of Rural Electrification.

The solutions were many. The problems of lower load in rural areas were dealt with through targeting highest load villages first, developing low-cost connections, and promoting load development. The problems of high cost were handled by standardizing systems design; seeking low-cost solutions; implementing a series of cross-subsidies from urban, commercial, and industrial customers to rural ones; and concessional financing. Bill-collection problems that plague many rural electrification programs were prevented by hiring respected local officials to collect bills from the community, thus minimizing the problem of collecting from individuals. Institutional problems were solved by placing responsibility for rural electrification in a publicly owned distribution company. Political interference, which can cause serious financial problems, was dealt with by using sound methods for village allocation, selection, and priority setting. The program also demonstrated flexibility by providing alternatives for villagers and other individuals, including politicians, to receive electricity on a priority basis, contingent on their contributing to the capital costs of the electrification expense.

With its task completed, the Office of Rural Electrification has been eliminated; providing continuing service to rural customers now lies with the larger distribution company. The bulk power subsidies, which were justified because of the rural electricity expansion program, also have been eliminated, but there is still a cross-subsidy between the two distribution companies. The innovative way that the company developed solutions to the problems of rural electrification are not forgotten by the people in rural Thailand and now can provide lessons for other developing countries as well.

Acknowledgments

We thank Norma Adams for editing different versions of this chapter. We also thank Willem Floor, Elizabeth Cecelski, Karl Jechoutek, and Joy Dunkerley for comments on earlier versions of the chapter.

We would like to thank Dr. Chulapongs Chullakesa (First Director of the Office of Rural Electrification and former Governor of PEA), Mr. Vibul Khoohiran (First Chief Engineer of the ORE and former Governor of PEA), Mr. Reungvith Vechasart (Director of Policy and Strategic, PEA and former Staff Engineer of the ORE), and former staffs of the Office of Rural Electrification for sharing their personal experiences in implementing Thailand's rural electrification program and allowing us to review the project records and documents.

Notes

1. $D_i = (S_i - S_m) \times [(0.975 - D_m)/(S_x - S_m)] + D_m$, where: D_i = degree of electrification under project in province $_i$; S_i = composite score for province; S_x = maximum composite score; S_m = minimum composite score; and D_m = minimum degree of electrification achievable within project scope.

2. PEA management decided to purchase these materials locally only after it discovered that it was cheaper to buy from local manufacturers rather than produce the materials itself.

CHAPTER 6

From Central Planning to Decentralized Electricity Distribution in Mexico

Luis E. Gutierrez-Poucel

P UBLIC INTEREST IN RURAL DEVELOPMENT began shortly after the Mexican Revolution in 1920 and was fueled by ideological and economic ideals result-ing from land injustice, rural oppression, and discrimination against a burgeoning indigenous capitalist class. From 1911 to the late 1920s, as pioneering revolutionary leaders arrived from the countryside and later from the nationalistic middle class, Mexico was the scene of extended political and civil unrest. To pacify the country, revolutionary governments responded to the people's desire for a better economic and social life—a basic premise of the revolution. When this premise was neglected, no government survived long; thus, important tenets of the revolutionary govern-ments were land reform and rural development (Sterrett and Davis 1928).

The drive toward rural development carried through to later years, when the government perceived rural electrification to be an important component of the rural development strategy. Among the key reasons for the success of Mexico's rural electrification program (REP) has been the social compact between federal gov-ernment and society that considers rural electrification indispensable for social and economic betterment. Because of this commonly held view, significant resources have been steadily allocated to the REP, despite the frequent economic downturns that have punctuated the country's development. The result of this process has been a successful program. In 1938, only 38% of Mexico's 18.5 million people were electrified. In 2006, almost 97% of the country's 104.4 million people—some 101.1 million residents—have electricity, leaving an estimated 3.3 million people without electricity in predominantly rural areas, where the average electrification coverage is 89.7 percent.[1] However, coverage is not homogenous. In some states, more than 10% of their inhabitants do not have electricity. Also, while 99% of urban house-holds are electrified, only 85% of rural households are.

In addition to its sustained political commitment to rural electrification, Mexico has benefited from a technically competent and motivated national utility. Created in 1937, the Federal Power Commission (Comisión Federal de Electricidad, CFE) has evolved from its initial role as the sole rural electrification decisionmaker to one of coordinating the funding decisions that are increasingly entrusted to local governments and communities. This gradual shift toward decentralized decisionmaking attests to the CFE's ability to adapt to Mexico's ever-changing rural electrification needs.

This chapter examines the important features of this interesting process of electrification. A brief background on Mexico and its energy sector is presented because understanding of the important institutional and financial drivers within their correct historical perspective helps to explain both the public drive and the enthusiasm of the electrification effort. This background is followed by an explanation of the evolution of the program as it emerged from a central planning approach to more of a focus on decentralized distribution. Finally, we present a case study in the state of Guerrero and review the remaining rural electrification challenges and the role of renewable energy options.

Background on Mexico

After 300 years of Spanish rule, independence was proclaimed in Mexico in 1810. Socioeconomic tensions gave rise to the Mexican Revolution (1911–1920), and the country's constitution was written in 1917. Mexico is the most populous Spanish-speaking country in the world and the second-most populous country in Latin America. About 76.8% of its 104.4 million people live in urban areas. Some 60% are *Mestizo* (Amerindian–Spanish); 89% are Roman Catholic. About 21 million people live in Mexico City, the country's capital. Population growth is rising sharply in the border cities of Tijuana and Ciudad Juarez and in the interior cities of Guadalajara, Monterrey, and Puebla.

Mexico is a federal republic with three branches of government. Its president is elected for one six-year term. The country has 31 states and 1 federal district. The Institutional Revolutionary Party has historically dominated national politics; it was defeated in 2000 for the first time in 73 years by the second largest party, the National Action Party. Other relevant parties are the Party of the Democratic Revolution, the Green Ecological Party, and Convergencia.

Parallel to the electrification program, the social conditions in Mexico have been improving. Life expectancy is 75 years for women and 71 years for men; infant mortality is 30 per 1,000 live births. The country's literacy rate is 91%. Education is mandatory for ages 6–18; over the past 20 years, both secondary school and college enrollments have increased dramatically.

The energy sector is extremely important to Mexico because it is one of the world's largest oil producers. Petroleos Mexicanos (Pemex) is the national oil company in charge of all hydrocarbon activities in the country. It is the largest company in Mexico and one of the 10 largest in the world. Mexico holds around 4% of the world's petroleum resources, equivalent to 32% of Latin American reserves, with 45 years average duration of reserves. About half of Mexico's oil output—almost 3.1 million barrels a day—is exported, totaling around 600 million barrels a year.

Mexico also has significant reserves of natural gas; it is twenty-first in the world in terms of proven reserves. As a consequence of this abundance of petroleum, Mexico consumes around 11.2 barrels of oil per capita, the highest per capita energy consumption rate in Latin America. Nevertheless, this figure is significantly lower than the consumption rates of developed countries—about five times less than the rest of North America, and half the rates for Europe.

In rural areas, the main use of energy is for cooking, accounting for about 95% of all end-use demands, which explains the heavy firewood consumption in rural Mexico. The use of electricity in rural areas is mainly for household use and pumping for irrigation. For most households, lighting is provided mainly by electricity, and this makes up about 4% of rural energy use.

The Historical Context of Rural Electrification

The electricity industry in Mexico had modest beginnings. From 1880 to 1910, some 200 small generation plants were installed in Mexico to serve mainly mining, metal refining, and textile providers. Over these 30 years, the important inducement for these companies was the economic growth brought about by the fiercely held peace imposed by the Porfiriato (named after General Porfirio Diaz, who was president during most of the time between 1877 and 1911). These generating businesses found potentially attractive side markets, extending their distribution grid to serve commercial, public lighting, and household users.[2]

The market's rapid growth attracted several experienced foreign companies, which bought out most of the small companies operating at the time. The first of these foreign utilities, the Mexican Light and Power Company of Canada, settled in Mexico City and then extended its services to the country's heartland. Next, the Electrical Company of Chapala, formed by a group of foreign entrepreneurs, established itself in Guadalajara, the country's second major city at the time, and soon extended its services to western states. Finally, the American and Foreign Power Company arrived to serve northern Mexico.

The Mexican Revolution, which started in 1911, created a climate of uncertainty and civil unrest. The electric utilities reacted cautiously, keeping their market shares, but foregoing any new expansion investment. With the end of the revolutionary struggle in 1920 and the takeover by the nationalist governments that emanated from it, foreign companies continued their policy of caution, while population growth accelerated and the economy started to expand.

By 1937, Mexico's population had grown to 18.3 million, an increase of almost 35% over the previous 40 years. The foreign utilities provided unreliable service, characterized by frequent power outages and high tariffs. It is estimated that between 69 to 88% of the population lacked electricity.[3] Electrical facilities, suffering from lack of maintenance, were at the end of their useful life. Moreover, the foreign utilities, more interested in the high-yielding revenues of urban areas, largely disregarded the rural countryside, where at that time 67% of the population resided. Thus, supply shortages, combined with the poor quality of available service, frustrated an expanding population and constrained economic growth.

Growing tensions between the foreign utilities and the revolutionary governments, coupled with people's rising expectations, led the Mexican state to create the CFE in 1937. Dependent on the Secretary of the Economy, the CFE was charged with expanding electricity generating capacity and coverage. Three years later, it was decentralized and established as a state-owned enterprise. The CFE represented the culmination of decades of revolutionary ideology. In light of nationalistic priorities, the state's relationship with foreign companies, including the electric utilities, was strategically reexamined.[4]

The conflict between the state and foreign utilities was further exacerbated by two factors. First, the regulatory system, introduced in the 1930s, was based on the experience of developed countries. Second, the rapid inflation of the 1940s undermined the regulatory system's effectiveness.[5] The conflicting needs of the utilities for financial revenue, consumers for cheaper tariffs, and the state for more and better supply caused existing tensions to escalate.

From the beginning, the CFE had focused explicitly on electrification projects that the three foreign utilities would not do on their own because of the negative financial rates of return. The CFE continued to demonstrate technical prowess by developing a pioneering spirit and professional excellence among its staff, building large hydropower schemes in difficult terrain, and extending public electricity service to remote rural communities. By 1950, however, CFE realized that regional authorities would need to become more involved in rural electrification, given the challenges of spreading limited federal funds across multiple national priorities, combined with the rising expectations of a growing rural population.

In 1952, the federal government made the REP official and explicit. The CFE was charged with planning, programming, and electrifying rural communities. The practice of regularly preparing rural electrification plans began. The federal government also requested that state and municipal governments, as well as local communities, contribute to rural electrification. Through their joint efforts, 915 communities—about 1.4 million people—were provided with electricity by the end of the decade.

Sentiment against the foreign electric utilities culminated with the nationalization of the industry in 1960. The political success of nationalization signaled a renewed effort to improve systems development and expand rural electrification coverage. In 1960, the systems were interconnected, a single distribution voltage and frequency were established, the electricity supply industry was consolidated, and a more aggressive electrification campaign was launched. Once the Mexican state consolidated its presence in the electricity sector, political will maintained the momentum of rural electrification.[6]

Nationalization of the electrification industry in 1960 brought a new dynamism to the REP. The annual rate of connections rose from 3.6% (1952–1960) to 9.5% (1960–1970). Coverage climbed from 25.0% (1960) to 58.9% (1970) of the total population; meanwhile, population growth continued to accelerate, rising to 3.3%, an increase of 0.2% over the previous period (1952–1960). Access to electricity more than doubled from 1960 to 1970 (see Figure 6-1).

The structure of the nationalized electricity sector was characterized by two vertically integrated state monopolies: the Comisión Federal de Electricidad, which

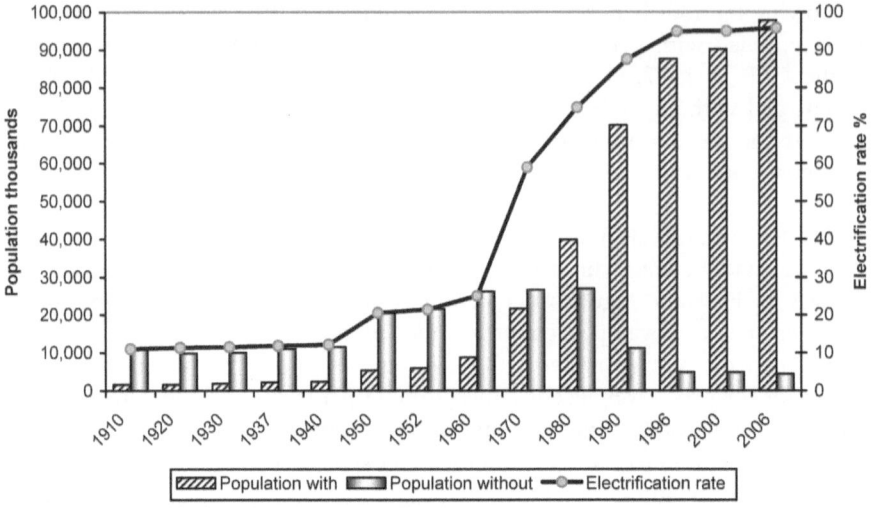

Figure 6-1. *Growth of Rural Electrification in Mexico, 1910-2006*

Sources: Based on Lara-Beautell (1953); CFE (various years), Obra realizada Grado de Electrificación, and Informa da Labores.

serves most of the country, and Luz y Fuerza del Centro (LyFC), which covers the central states (Mexico, Morelos, Hidalgo, and Puebla) and the Federal District. These two firms carry out generation, transmission, distribution and marketing activities in a monopolistic fashion (see Figure 6-2). This structural arrangement strengthened the state power company and gave new vigor to the electrification effort by removing the entry obstacles to the service areas of the private utilities and by gaining overall program control.

The benefits of rural electrification were apparent to all. With the integration of rural communities into the grid, the process of nation building could be accomplished. Rural electrification raised agricultural productivity through pumping and irrigation, and higher productivity raised rural incomes, extending and strengthening the domestic market. Investment and tariff subsidies were granted to electrification projects within a context in which agricultural products were readily absorbed because of World War II, the reconstruction of Europe, and the rapid global growth that followed. Rural electrification also enhanced community life, permitting better schooling, health services, public lighting, and entertainment (Rodríguez-Rodríguez 1994). The revolutionary governments considered education among the most crucial policy objectives, believing that access to electricity would improve rural education, which, in turn, would strengthen the nation-building process.

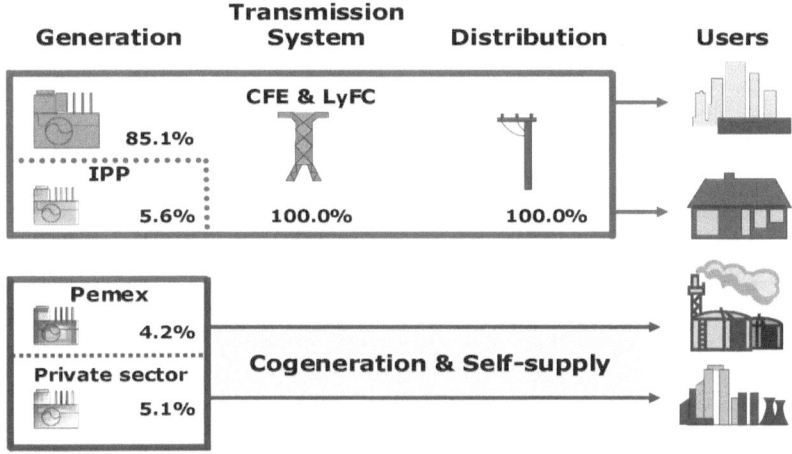

Figure 6-2. *Structure of the Electricity Industry in Mexico*

Source: Secretaría de Energía, "La Industria Eléctrica Mexicana" (www.energia.gob.mx).

The period from 1970 to 1996, marked by several economic downturns, saw coverage rates rise from 58.9% to 94.8%, a considerable achievement considering that population was growing at a rate of 2.6% per year. The average annual increase of customers grew by more than 0.5 million, an increase of 0.3 million over the previous period (1960–1970).

Over time, the responsibility for deciding the direction of and funding for electrification projects shifted toward local governments and communities. Gradually, they began to define their own priorities and decide the best uses of their resources. Today, regional infrastructure programs that are federally funded are subject to state and municipal decisionmaking as a matter of public policy, and budgetary control is becoming increasingly decentralized.

Since 1990, the program has been integrating various renewable energy technologies into its methodology. Interestingly, the CFE acknowledged the appropriateness of renewable energy technologies as complementary to, rather than in competition with, the standard grid supply. Overall, the program has succeeded in stimulating the involvement of private individuals and rural communities in electrifying areas located far from the main transmission lines.

After 1995, the state entrusted the municipalities with social development funds and decisionmaking over their allocation. From 1996 to 2000, electrification kept pace with the 1.7% population growth. Service was provided to an average of 602,000 new residential customers per year; however, the annual rate of connections fell from 5.7% (1970–1995) to 3.2% (1995-2000), and the rate of coverage remained flat. Several factors explain this slowdown. The remaining areas without service were becoming more difficult to reach and more expensive to electrify, and public funding was becoming scarce. However, the key reason was the transfer of the financial and programmatic responsibilities from the central government to the states and municipalities, a transfer that was not accompanied by a parallel build-up of local capacity to identify electrification needs/uses and plan cost-effective solutions accordingly.[7]

In 2001, the Fox Administration set out the goal of achieving 97 percent electrification coverage by 2006 (Energy Sector Program, programa sectorial de energía, PROSENER), outlying the need to implement a rural electrification program based on renewable energy options. However, as of 2006, there was no national electrification program in place and the government remained shy of attaining the 97 percent coverage target. The causes were due to a series of institutional, programmatic, and fiscal constraints, among which the lack of institutional mechanisms and know-how at the local level were the chief culprits. However, despite these shortcomings, the 2001-06 administration fared better than its predecessor (1995-00). While the rate of new connections declined from 5.1% during the Salinas government (1989-94) to 3.3% during President Zedillo (1995-00), the rate increased slightly to 3.8% during the Fox administration (2001-06).

As of December 2005, CFE estimated 3.7 million people without access to electricity. Reaching them is challenging, since most of them are in small, remote, isolated communities. CFE estimates 71,822 unelectrified communities, 68,686 of which have less than 100 people, and 2,989 with populations between 101 and 2,499 (see Table 6-1).

From the outset, Mexico's REP enjoyed numerous advantages that enhanced program success over time. The social compact between the government and society made national coverage indispensable for social and economic development. At the same time, sustained political commitment at both national and regional levels provided secure and significant funding; this commitment allowed for multiyear planning and execution, despite frequent economic downturns. Mexico also enjoyed a technically competent and motivated national utility, which lent authority to providing program guidance. In addition, construction and procurement guidelines had been established, and national construction and materials industries had proven experience. The country also had an integrated and reliable infrastructure for electricity generation, transmission, and distribution services. However, after the decentralization policies introduced in 1996, the electrification effort slowed down, especially in the remote rural communities, largely due to a manifest lack of capacity to identify infrastructure needs and execute projects in a transparent and cost effective manner.

Toward Decentralized Responsibility

One unique feature of the rural electrification program in Mexico is that in its initial stages everything was centrally planned. As the program grew over time, the REP's institutional setup and funding mechanisms gradually evolved into a more decentralized model, with local communities taking on increasingly greater responsibility. The overall context of the program has been described in

Table 6-1. *Degree of Electrification in Mexico, 2005*

Communities by size	Existing	Electrified	%	Pending	%
1–99	146,345	77,659	53.1%	68,686	46.9%
100–499	34,005	31,016	91.2%	2,989	8.8%
500–2,499	14,999	14,852	99.0%	147	1.0%
2,500–9,999	2,487	2,487	100.0%	0	0.0%
10,000 or more	833	833	100.0%	0	0.0%
Total	198,669	126,847	63.8%	71,822	36.2%
Communities by location					
Rural	195,349	123,527	63.2%	71,822	36.8%
Urban	3,320	3,320	100.0%	0	0.0%
Total	198,669	126,847	63.8%	71,822	36.2%
Residents by location					
Rural	26,113,289	23,131,494	88.6%	2,981,795	11.4%
Urban	80,722,335	79,967,234	99.1%	755,101	0.9%
Total populace	106,835,624	103,098,728	96.5%	3,736,896	3.5%

Source: CFE (2005), Grado de Electrificación.

the previous section, so here the focus is on the institutional decentralization of certain aspects of the program. As Figure 6-3 illustrates, this shift occurred in four stages. They include a period in which rural electrification was entirely the responsibility of CFE and the more recent period when the municipalities and local communities have a significant say in their infrastructure programs. A case study later in this chapter describes the more recent, decentralized approach to funding rural infrastructure.

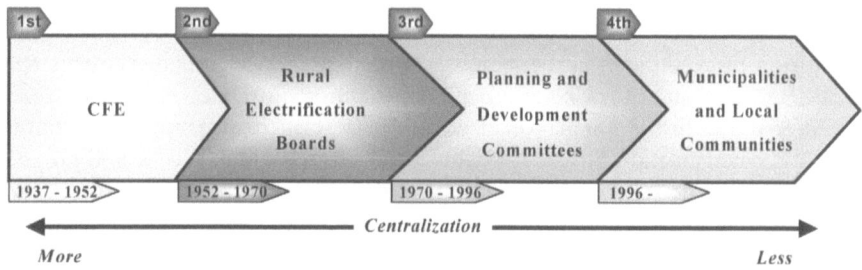

Figure 6-3. *Decentralization of Rural Electrification in Mexico, 1937–Present*

First Stage: Central Planning by CFE

The first stage began in 1937, when the government of Mexico created the CFE. At that time, one of the government's important aims, rooted firmly in the populist ideals of the Mexican Revolution, was to extend electricity service throughout the country, without explicit consideration of financial returns. National efforts focused initially on expanding generation capacity and electrifying nearby populations.

Backed by the federal government's political will and supported by state and municipal governments, CFE was made the sole entity responsible for designing, implementing, and achieving the goals of national electrification. Through its strategic leadership and political visibility, CFE played a central role in electrifying Mexico's rural areas, although no specific "rural electrification" distinction had yet been made. Mexico's federal budget provided most of the funding for electrification, supplemented by CFE's own resources.

Second Stage: Evolution toward Shared Responsibility

The second stage started in 1952, when the Mexican government formally institutionalized the REP and explicitly entrusted the CFE to extend electricity service to towns and communities in rural areas. Given the shortage of federal funds for the magnitude of the task at hand, the federal government summoned state and local governments, as well as local communities that desired electricity service, to assist in the effort. To this end, rural electrification boards were established in each state and charged with the task of preparing electrification plans and programs.

The federal government provided electrification funds to the CFE, which in turn, distributed them to the country's 31 states, based on plans prepared by the rural electrification boards. Both state and municipal governments contributed to these funds to raise the targets for the number of communities and connections made. Local communities that desired service also contributed, according to their ability to pay.[8]

Before nationalization in 1960, electrification proceeded slowly. CFE still had full responsibility for rural electrification, although it lacked total control. Coordination problems and conflicts frequently arose between CFE and the private utilities. The state had difficulty implementing the REP on a national scale, especially attending to rural customers in the private utilities service areas. At the same time, the private utilities were using different technical standards, which made electrification more difficult and expensive. Nonetheless, modest achievements were made, and a culture of cooperation and community effort was established.

After the industry was nationalized in 1960, REP management was centralized and electrification procedures were standardized, which accelerated program development. The rural electrification boards developed an efficient, effective system that responded equitably to communities' requests. Towns within the concession areas of the three foreign utilities were electrified, extending service to 9,091 new towns, representing some 7.5 million people during 1961-1970. Thus, during this stage, the REP became a sustained, well-managed effort.

Third Stage: Cooperation and Conflict Resolution

The third stage began in 1970, a benchmark year toward decentralizing the REP. New budget allocation and appraisal procedures had gradually improved program efficiency. High-density communities located close to the grid that had higher productive potential had already been connected. The challenge now was to reach the rest of the country, especially those communities whose electrification would be justified more on social merit, rather than on the benefits of productivity gains and higher incomes.

At this time, Mexico's federal government introduced a new plan for distributing social development funds. The plan consisted of aggregating all resources for social infrastructure development, including those for roads, schools, health clinics, drinking water and sewerage systems, and electrification. The amounts allocated to each state were inversely proportionate to their income levels and degree of infrastructure development. The government invited all of the country's 31 states to form planning and development committees, which were charged with analyzing requests for distribution of federal and state funds and determining the contribution level of each participating community.

The states' rural electrification boards continued to operate alongside the planning and development committees. The electrification boards dedicated most of their efforts to micromanaging electrification projects, organizing communities, and managing the implementation of projects; the planning and development committees focused more on planning and appraising the socioeconomic aspects of the communities and reconciling the development priorities of competing infrastructure projects within the state.

Over time, however, some unavoidable duplication of effort and responsibilities developed, leading to political conflict and infighting. Given the planning and development committees' wider social perspective and greater operational experience, the rural electrification boards were eliminated. The committees continued as the planning and conciliation organs charged with distribution of federal, state, and municipal resources.

Because CFE's executive representation took place at the level of the rural electrification boards, under the new institutional setup, it established rural electrification departments in each state to represent itself before the planning and development committees. These departments were created within the organizational structure of CFE's 13 distribution divisions and charged with the responsibility of programming and executing the electrification projects.

Fourth Stage: Local and Regional Autonomy

The fourth stage began in 1996, when the national Congress and federal government agreed to grant greater financial and executive autonomy to Mexico's 2,400 municipalities, entrusting them with the social development funds. As part of the decentralization policy, the government created a budgetary brand—Ramo 33—for annual distribution of federal funds among states and municipalities based on allocation rules that considered key social indexes such as marginalization, poverty, human development, and infrastructure development. The municipalities were thus responsible for selecting the social infrastructure projects to be developed each year, authorizing the allocation of available funds, and determining the amount each participating community contributed.

REP targets are annually formulated, examined, and agreed on by CFE, state, and municipal governments within each state's planning and development committee. The REP's funding structure is divided, according to origin of funding, into three subprograms (Figure 6-4). Although there is some overlapping of these categories, a description of each type is provided in this section.

Direct funding comes from federal, state, and municipal governments, as well as private sources, including local beneficiaries. In 2004, this subprogram accounted for 16% of the total REP funding, enabling the electrification of 219 rural villages and 96 suburban towns, benefiting 59,764 residents in 15,003 households, with a total investment of US$11.1 million.

Shared resources are derived equally from CFE and state and municipal governments, with contributions from beneficiary communities. This subprogram funded the electrification of 693 rural villages and 304 suburban towns serving 188,938 residents in 47,429 households. The total investment amounted to US$35.0 million (50% of the investment funds).

Finally, for the *federal funding,*[9] the federal government provides funding to states and municipalities, who then evaluate and select the best electrification projects. CFE provides the technical expertise for project design and investment budget, and it executes the project when requested by local authorities or monitors implementation if another party is selected. In 2004, this subprogram enabled the electrification of 477 rural villages and 209 suburban towns serving 130,253 residents in 32,697 households. The total cost was US$24.1 million (34% of the total funds).

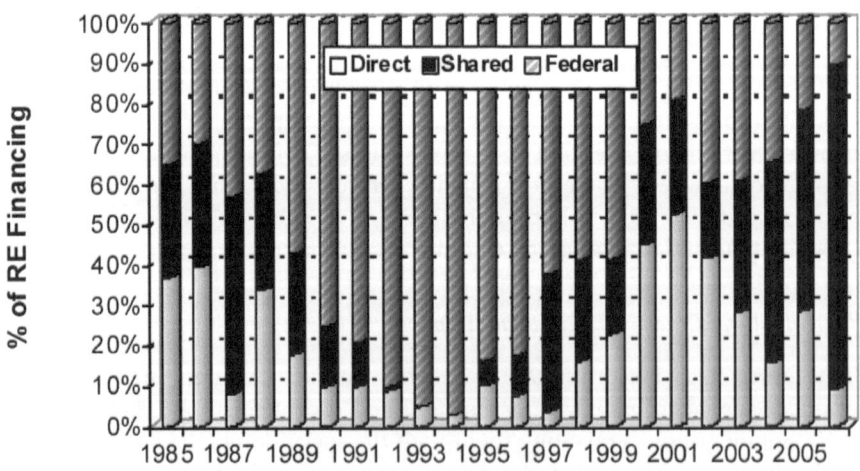

Figure 6-4. *Funding Sources for Rural Electrification in Mexico, 1985-2006*

Source: CFE (1985 to 2006), Informe de Labores.

As Figure 6-4 shows, federal funding steadily increased from 1988 until 1994, after which a new government took over, an economic downturn took place, and a policy change occurred in 1996 which transferred from the central government the responsibility for the allocation of electrification funds to the states and municipalities. Since that year, federal funding has declined in percentage terms, while shared (state and municipal governments) and direct funding have taken up the slack.

Thus, today decisions for infrastructure investment are made, for the most part, at the regional level. Federal funds for social and economic infrastructure development are given to each region in accordance with projected needs and relative income. State and municipal governments are responsible for developing the infrastructure project portfolio; each state's planning and development committee and staff from the state's finance secretary is responsible for final approval. Electrification projects in the portfolio are prepared jointly with the area's electrification department. The federal government does not usually exercise direct, official influence in these decisions.

Challenge and Design of Rural Electrification

Many national demographic standards have established a direct linkage between access to basic public services (e.g., potable water, primary education, primary health care, and access to electricity) and poverty alleviation. Without access to such services, rural families are considered to live in poverty. From the start, Mexico's public officials acknowledged the links between electrification, productivity, and economic development in rural areas. However, rather than quantify the real economic gains achieved through expansion of service, REP officials focused exclusively on the number of connections made, thus obviating the need to measure the specific economic effects of electrification. Thus, design of the Mexican REP rests on the belief that electricity provision per se is necessary for rural development and poverty alleviation.

It is a well-established fact that private utilities are not interested in rural electrification projects because of the low levels of income and electricity consumption in rural areas, which produce poor financial returns. To guarantee private utilities a reasonable rate of return, Mexico's government, through the CFE, decided to reduce initial barriers by subsidizing capital costs and providing subsidized tariffs for rural use and low residential consumption. In this section, tariff policies, subsidies, and costs issues are examined.

Standards to Achieve Lower Costs

When the REP began, it had to overcome various technical problems, including discrepancies in technologies, voltage frequencies, and operational and construction procedures. Nationalization of the industry, which began in 1960, helped to standardize service. There was now one voltage frequency, three voltage transmission levels (13.2, 22, and 33.5 kV), and two voltage distribution levels (127 and 240V).

Standardization stimulated the domestic electricity industry to manufacture most of the materials and equipment the REP required. Not only did the national industry benefit, but also industrial, commercial, and residential users saved significantly, thanks to more economical equipment and appliances. The electricity supply industry also benefited from the economies of interconnection in the operation and expansion of the system.

Tariff Policies that Provide Financial Returns

After nationalization a new tariff schedule was published in 1962, which established 11 tariff groups. Of these, corn mills, public lighting, pumping for water and sewage, and agriculture and mining (high-tension voltage) were considered socially deserving and were charged tariffs below the cost of supply.

These tariffs, with government assistance, remained in effect until 1973. They enabled the timely expansion of the industry and nationwide electrification. However, the constraints embedded in the tariff schedule gradually reduced the sector's ability to finance system expansion and reduce debt levels, resulting in the need for more subsidies.

Currently, Mexico's end-user tariff schedule is extremely complex, with 112 different billing possibilities that draw from 7 basic tariffs, 2 seasons, and 8 billing options by consumption level. Tariff rates vary according to geographic, climatic, and use criteria. Tariffs are classified by type of user (residential, commercial, temporary, services, water uses, agricultural, and industrial), supply-voltage tension (low, medium, and high), and regional temperature. Cross-subsidies significantly distort residential and agricultural tariffs. Distribution of subsidies among households is highly regressive, benefiting the "better off" households, those who consume more.

There are seven pricing structures for residential users (including rural), plus one for high consumers. These vary according to the average temperature in the users' geographic area. During the summer, residential tariffs are differentiated according to an area's mean temperature; throughout the rest of the year, a single rate is applied countrywide (Table 6-2).

Each residential pricing structure has a set of increasing block tariffs. The underlying policy aims to support low-income households, which in practice distorts the tariffs and benefits the better-off consumers. The residential tariff structure embodies two types of subsidies: support for low-income groups and those in areas

Table 6-2. *Tariff Schedule in Mexico, 2006*

Sector	Tariff	Supply-Voltage Tension and Type
Residential	1, 1A, 1B, 1C, 1D, 1E, 1F, DAC	Low
Commercial	2, 3	Low
Services	5 and 5A (public lighting), 6 (water pumping)	Medium
Temporary	7	Low
Agricultural	9, 9M, 9-CU, 9-N	Low, Medium
Water Uses	EA	
Industrial	Medium and high tension rates, time-of-day rates, backup and interruptible for low and high demand rates	

with extremely high temperatures that require air conditioning. The basic residential tariff (Tariff 1) sets a low rate for the first block of usage (up to 75 kWh), from there on the rates increase as consumption goes beyond, 75, 125, 140, and higher. The air conditioning subsidy increases the size of the lower blocks and lowers the price per block in warmer areas.

For industrial users, there are hourly (time-of-day) tariffs with low- and high-load factors, backup tariffs, and interruptible supply tariffs. Optional electricity tariffs based on marginal costs and time-of-day metering are now in effect for all high-voltage and some medium-voltage customers, which together account for over half of total electricity sales.

To maintain the real value of the tariffs, the tariff policy applies specific monthly inflationary adjustments. However, the policy has been erratic. For residential tariffs, the monthly adjustment factor led to an increase of 1,610.7% from 1995 to 2000, while the consumer price level increased by only 226.0%. During 2001 to 2006, the adjustments lead to a tariff increase of 41% against an expected increase in the CPI of 25.9%. Agricultural tariffs increased 134.3% during the 1995-2000 period and 198.9% from 2001 to 2006. The problem is that these adjustments have not really decreased the level of subsidies.

Subsidies and Distortions: Residential and Agricultural Tariffs

Traditionally, the Mexican government has used electricity tariffs to achieve various policy goals. Electricity price subsidies have contributed to the significant growth in electricity demand. Although power-sector reforms of the early 1990s eliminated many tariff subsidies, residential and agricultural tariff subsidies have persisted.

Given their low consumption levels, rural households fall within the consumption block of 1–140 kWh per month, which has an average price of US¢6.15 per kWh, close to 30% of the supply cost. However, for residential consumption levels exceeding 315 kWh per month, the average price a Mexican household pays is higher than a consumer in California, which pays among the highest tariffs in the United States.

In many tariff structures around the world, customers who use more electricity pay a lower rate per kWh consumed (Figure 6-5). The underlying tariff policy reflects cost of supply because the unit cost declines as consumption rises and the load factor improves. In Mexico, conversely, the residential tariff schedule responds more to distributional, fiscal, and political criteria than to economic efficiency. In Argentina, Canada, Peru, and the United States, the residential tariff follows a declining cost curve, whereas, in Mexico, it rises with consumption. For example, the average residential price for 120 kWh (without taxes) is US¢10.1 per kWh in Argentina, 18.3 in Canada, 6.1 in Mexico, 9.7 in Peru, and 17.8 in the United States, making Mexico considerably cheaper than the other countries for low consumption levels. However, the unit prices for a consumption level of 450 kWh are US¢8.7 in Argentina, 10.8 in Canada, 16.2 in Mexico, 9.3 in Peru, and 14.7 in the United States, making Mexico the most expensive of the five countries for higher consumption levels.

Obviously, a significant distortion exists in the residential tariff, but the problem is also one of fairness because distribution of the tariff subsidies is regressive.

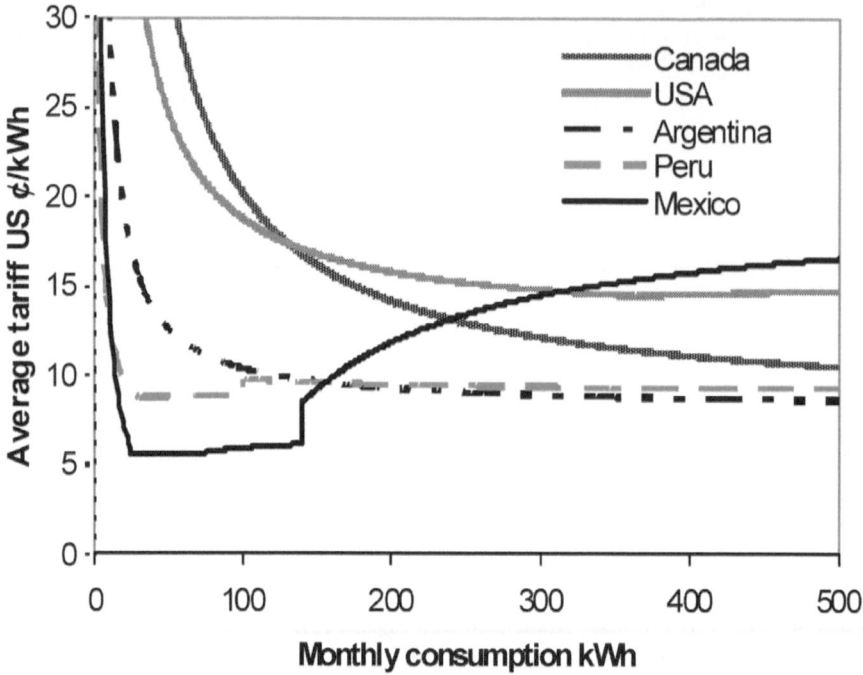

Figure 6-5. *International Comparisons of Residential Average Prices, 2006*

Source: Author's elaboration based on the published 2006 residential tariff schedules: Argentina, Edesur; Canada, Hydro One Ontario; USA, California, Sierra Pacific Power; Peru, Coelvisac; Mexico, CFE.

High-income households, which consume more electricity, receive the lion's share of the subsidies (see Figure 6-6). Thus, 40% of low-consumption users receive only 24% of the total subsidy, whereas 40% of high-consumption users receive close to 60% of the subsidy (deciles 6 to 9).

The rural communities that are being reached by the electricity grid are required to pay an average US¢5.4 per kWh (or 0.595 pesos/kWh). This amount already includes the subsidy. The following subsidies are provided to grid-connected customers:

• Connection: high cost of line extension but flat connection fee
• Consumption: tariff below cost (cross subsidy)
• Consumption: no meters in poor households/ low billing collection (with no disconnection policy)

In 2005, the average agricultural tariff for irrigation was US¢4.0 per kWh, which was insufficient to recover the cost of supply (US¢14.1 per kWh). Despite the large subsidy, collection of this tariff is difficult. Farmers, for example, complain that prices for their products are so low that they can barely afford public electricity, let alone pay on time. Given rural producers' low incomes, their consumption and arrears continue

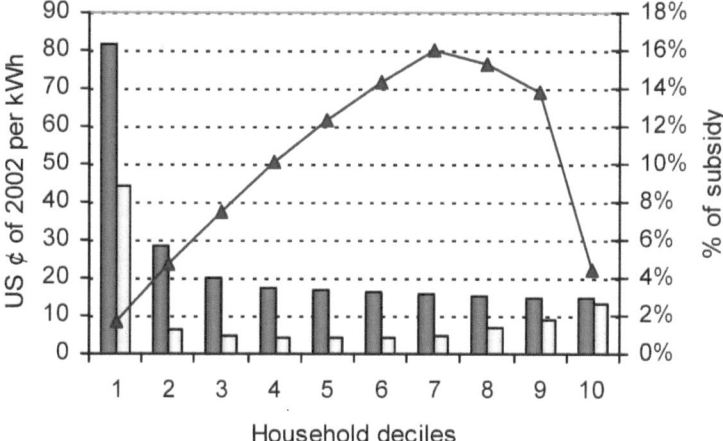

Figure 6-6. *Average Residential Price, Cost and Subsidy, Mexico, 2002*

Source: World Bank (2003), Mexico Public Expenditure Review.

to increase. They fail to pay on time within each agricultural cycle, which leads to lower revenue for the utilities and higher fiscal subsidies from the government.

The subsidy to agricultural tariffs causes several distortions. The low tariff permits an artificial profitability of some uneconomic crops, while inducing the overexploitation of underground water reservoirs. Also, electricity demand grows faster, not only because of the price elasticity of demand, but also because it is cheaper to consume more electricity than to buy a more efficient pump or properly maintain an existing one. Thus, equipment ages and becomes less efficient. The fiscal load of this subsidy has remained fairly stable over time, representing about 0.1% of gross domestic product (GDP) from 1990 to 2005.

In Mexico, energy subsidies have a long, uneven history. In the past, the Mexican government increased tariff levels and successfully eliminated several subsidies. However, during difficult times, such as the political clashes of 1994, subsidies were reinstated. At times, Mexico's tariff levels have approached those of the United States and Canada. Historically, a low point was reached during the early 1980s. For example, in 1983, Mexico's prices for agricultural electricity were only 10% of the rates paid by U.S. agricultural users. Similarly, in 1987, Mexico's residential electricity prices were only 37% of U.S. levels. By 1990, much of this gap had been bridged, only to widen again in real terms, as nominal prices remained constant while inflation soared.

From 1990 to 2006, the average electricity price has increased in nominal terms at an annual rate of 14.3%, with the largest absolute increases in agricultural (17.7%) and commercial rates (14.8%). Inflation-adjusted prices have risen by 1.9% per year. Despite these increases, the price-cost ratio has remained around 66% since 1990.

As might be expected, these low electricity charges are made up for through the allocation of tax revenues as a subsidy for the electricity industry. During 2000, the national electricity sector received a fiscal transfer of 56.7 billion pesos (US$6.0

billion), representing about 1% of GDP. This operating subsidy was composed of residential subsidies (62%) and agricultural and services subsidies (12%); the remaining 26% was used mainly to fund inefficiencies in the utilities, especially Luz y Fuerza del Centro. The fiscal support for CFE was 37,454 million pesos (US¢11 per kWh sold), whereas for LyFC it was 17,764 million pesos (US¢30 per kWh sold).

To summarize, because the current policy sets tariffs based in part on political and fiscal objectives, rather than on industry costs alone, residential and agricultural tariffs are insufficient to allow CFE and LyFC to recover their costs. Subsidies are distributed without a social criterion, which promotes an inefficient and irrational use of electricity, a growth of demand above the optimum, and greater investment requirements to expand the system. In addition, the fiscal load of the subsidies reduces public funding for other important social and economic expenditures, reducing national welfare.

Competition for Social Funds

The REP strategy is designed to promote competition between the municipalities by using federal and state funds to support the best distribution systems or to finance the cost of renewable energy systems. Mexico's state governments receive project proposals from their respective municipalities, which compete for limited social development funds. In general, those projects that provide new connections at the lowest cost per connection (net of municipal and community contributions) are funded progressively, until the funds for that year are exhausted. The system promotes the participation of the municipalities and communities in funding, to the extent possible, to reduce government funding per connection.

Annual investment in the REP has tended to follow national economic trends, falling when the economy slows and rising when economic growth picks up. This relationship held true from 1952 until the end of the Salinas administration in 1994, after which time the annual electrification investment dropped significantly (Figure 6-7). The average annual investment in electrification (at 2000 prices) was US$11,527 million (1952–1960), increasing to US$ 91,816 million (1961–1970), and then US$132,027 million (1971–96), after which time it declined to US$61,930 million (1997–2000), and US$51,156 (2001-2006).

Electrification connection costs have risen as coverage has increased, and the marginal effort to connect the next community has risen progressively. Thus, before privatization of the industry, the average cost per kilometer of line was about US$8,513 for the period 1952–1960, and it now stands at US$13,108 on average for the 2001–2006 period (Figure 6-8). From 1952 to 2006, this cost has varied dramatically, averaging US$10,385, with a low of US$5,896 in 1988 to a spike of US$26,019 in 1961.

Overall, Mexico's electrification costs are higher than those of comparable Latin American countries.[10] The most important reason for this difference is clear: The country's existing high level of coverage, combined with its mountainous topography and the nature of its remaining unelectrified communities, has determined that more effort is required to reach the next new user. Although Mexico's electrification industry is fairly efficient, it is reasonable to expect that costs per user will continue to rise as the marginal unelectrified community becomes more difficult to reach.

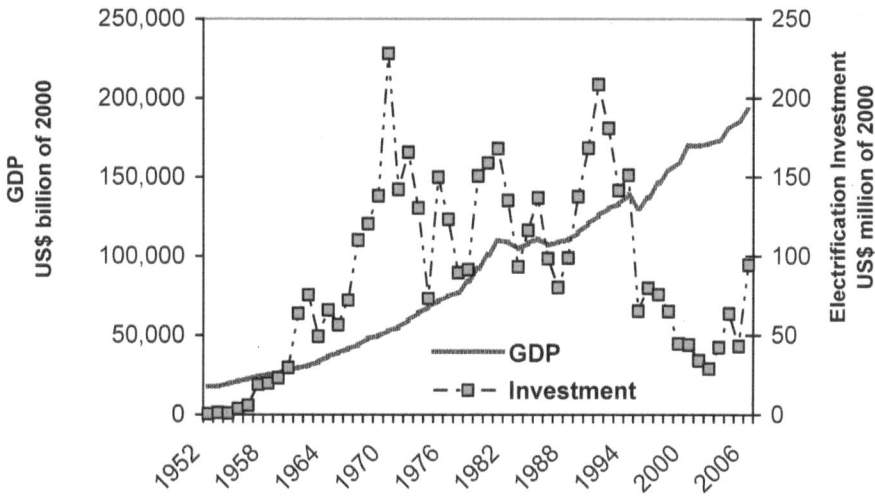

Figure 6-7. *GDP and Investment in Electrification, 1952–2006*
Source: INEGI and CFE.

As an example, the total REP investment during the 1997–2000 time frame was approximately US$231 million (in 2000 dollars), which resulted in the electrification of 330,000 households serving 1.6 million people, at an average cost per connection of approximately US$700. The investment represented a mix of federal funds (55%); state and municipal governments, supplemented by local communities (30%); and direct agencies (15%).

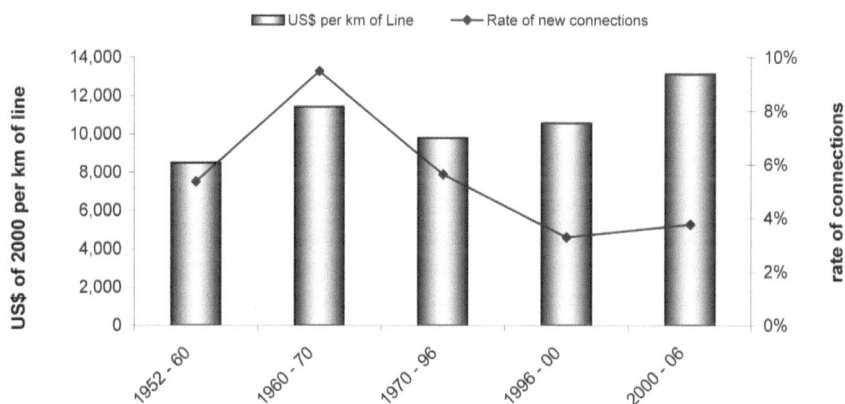

Figure 6-8. *Investment per Kilometer of Line and Connections, Mexico, 1952-2006*
Source: CFE (1995 to 2005), Obra realizada and Grado de Electrificación; and CFE (2001 to 2006), Informe de Labores.

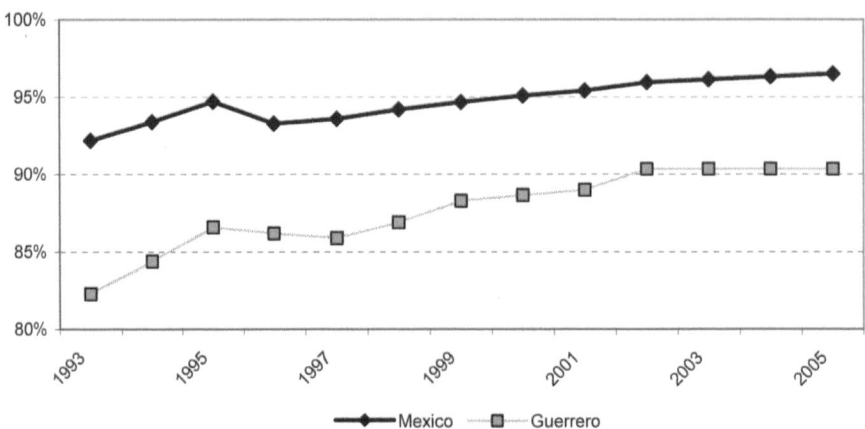

Figure 6-9. *Rural Electrification Coverage, Mexico and Guerrero, 1993–2005*

Source: CFE (1993 to 2005), Grado de Electrificación.

A View of Five Communities: The Case of Guerrero

To illustrate how the rural electrification program works in Mexico, we will examine the process through which five rural communities in Guerrero attained electricity service. In 1970, Guerrero's electrification rate was only 37.5%, significantly lower than the country's average of 61.2%.[11] By 1999, the state's rate of coverage had increased to 88.3%, still well below the country's average of 94.7% (Figure 6-9).

Despite its well-known tourist attractions, including the beach resorts of Acapulco and Ixtapa, Guerrero is one of Mexico's four poorest states. About 45% of Guerrero's almost three million citizens are rural (20% above the national average), and only 70% are literate (20% below the national average). In 1999, the state's GDP per capita was US$2,560, almost half the country's average of US$4,634. Tourism and agriculture are Guerrero's major economic activities. For many people, agriculture is their major means of livelihood. Almost 60% of the population have no steady employment; of those who are economically active, 53% earn only US$6.00 per day.

Located along Mexico's southern Pacific coast, Guerrero has over the past 30 years experienced much social unrest and witnessed the emergence of various guerrilla insurgencies. In terms of literacy, drinking water, and sewerage systems, it ranks last among the country's 31 states. In terms of electrification coverage, Guerrero lags behind all but two states. Mountainous terrain, poor access roads, and widely scattered populations increase the challenge of providing electricity to Guerrero's remote rural households.

Description of the Selected Communities

Beneficiaries were interviewed in five of Guerrero's recently electrified rural and suburban communities: El Quemado, La Venta Vieja, El Plan Liebre, El Naranjo,

Table 6-3. *Characteristics of the Case Study Communities in Mexico, 2000*

Community	Municipality	No. of Residents	Average Age of Household Head	Average Household (Income/Year in US$)	Year of Electrification
El Quemado	Acapulco de Juárez	540	37.8	2,375	1999
La Venta Vieja	Eduardo Neri	325	32.2	2,980	1996
El Plan Liebre	Eduardo Neri	295	31.8	3,059	1998
El Naranjo	Metlatónoc	480	38.2	2,191	1992
Villa Xochitl	General Heliodoro Castillo	420	35.4	2,159	1993

Sources: INEGI and field survey.

and Villa Xochitl. For the most part, these communities are similar in terms of size, age of household head, and household income (Table 6-3).

Mexico's federal and local governments created the community of El Quemado several years ago to relocate families displaced by Hurricane Paulina in 1996. It took authorities almost two years to relocate community members and three years to provide them basic infrastructure services. Along the way, the community had to pressure authorities, including blocking an important highway, to ensure that they honored their commitment. In 1998, families were given small brick and cement houses.

La Venta Vieja and El Plan Liebre, located adjacent to each together, are about 7 km from Chilpancingo, the capital of Guerrero. According to one interviewee, these communities are at least 130 years old. Their indigenous residents speak the Náhuatl language. La Venta Vieja has about 60 houses, and El Plan Liebre has 55. All houses are made of adobe and have thatched roofs. Most heads of households practice subsistence agriculture, with corn accounting for at least half the food consumed by community families. (The *milpa* or cornfield functions as the center of these rural villages.) None of those interviewed used pumping for irrigation, given the region's water shortage. Some community members work as craftspeople, stonemasons, brickmakers, and laborers at construction sites in Chilpancingo. These two communities waited more than 20 years to receive electricity, having made their first formal request in 1974.[12] La Venta Vieja was finally electrified in 1996, and El Plan Liebre was supplied two years later.

The towns of El Naranjo and Villa Xochitl are located within a 20-minute walk of each other, and both are about 110 km from Chilpancingo. Dating back to the late 1880s, French intervention soldiers intermarried with the local population in these mountainous, rural settlements. Though the two towns belong to different municipalities, they are similar in appearance. El Naranjo has 80 houses, and Villa Xochitl has 70. Most houses are made of adobe, but a few are constructed of brick and cement. Most roofs are thatched.

Most residents practice subsistence agriculture. Corn, beans, pumpkins, and chilis are the main crops. Among the poorest of the villagers interviewed, the residents of El Naranjo and Villa Xochitl earn an average monthly income of about US$180. They overuse the land the way their forebears taught them: burning to clear it, overworking the soil until productivity drops, and then moving on to another plot. The aggressive chemical fertilizers they now use serve only to compound the problem, causing soil erosion and contaminating water resources. El Naranjo was provided with electricity service in 1992, and Villa Xochitl was supplied the following year.

The Decentralized Process of Electrification

The process of providing electricity to the five Guerrero communities was similar in all cases except El Quemado, which followed the federal government's disaster relief funding procedures. In the other four cases, electrification was a joint effort of beneficiaries, local and municipal authorities, and the CFE.

The communities requesting electricity service created an electrification committee, charged with the following responsibilities. With CFE's assistance, each committee formulated the electrification project and budget and made a formal request to the municipal authorities. They also promoted the project and collected contributions from community members. The committee also arranged for any assistance needed to complete the work by the community. The monitoring of the contractual work was also a responsibility of the committee. This included monitoring work one year after the completion of the project to ensure that CFE's service met the standards agreed on in the contract.

In the case of La Venta Vieja and El Plan Liebre, the municipal government requested the final project design, cost estimates, and funding from CFE. Then, CFE studied and approved the projects. It agreed on their priority and timing and funding mechanisms with the state government and also implemented the electrification work. By contrast, in El Naranjo and Villa Xochitl, the municipal government contracted private firms to carry out the project, using CFE monitoring and technical specifications. Once the electrification work was completed, CFE assumed operation and maintenance of the facility.

One hallmark of the program was that the communities contributed to the initial costs of the program, both through their access to subsidy funds and through contributions by people in the community. Electricity was introduced to El Quemado in 1999, at a cost of about 1 million pesos, or US$990 per user. The municipal government, grants from nonprofit organizations, and the federal government's disaster relief program provided funding.

In La Venta Vieja and El Plan Liebre, investment costs amounted to 520,000 pesos, or a per-user connection cost of US$585 in La Venta Vieja and US$603 in El Plan Liebre. Funding was provided by the community (10%), CFE (35%), and the municipality (55%). Each of the 115 households contributed 500 pesos (US$53) toward project costs. In addition, community residents participated in laying out the poles and carrying materials over difficult terrain.

Investment costs in Naranjo and Villa Xochitl totaled 500,000 pesos, or a per-user cost of US$900 and US$914, respectively. The municipal government provided all funding. Beneficiaries participated in the electrification effort by digging holes, setting up poles, carrying material and equipment, and feeding the engineers and workers.

The electricity prices vary quite a bit among the five communities. In El Quemado, electricity users pay 90–200 pesos [US$9–21] bimonthly, according to the number of appliances they own, which is below the residential tariff. Households in La Venta Vieja and El Plan Liebre consume below the residential tariff, paying 150–200 pesos (US$16–21) bimonthly. In El Naranjo and Villa Xochitl, users' consumption level is below the residential tariff. They pay 100–200 pesos (US$11–21) bimonthly, according to the number of appliances they own.

Because Guerrero's communities are small and located far apart, few permanent CFE offices are located in this region. CFE sends bill collectors to Guerrero's remote villages; however, given the difficult access, low electricity consumption, and frequent payment delays, the cost of bill collection almost equals the revenue collected. Thus, CFE views supplying Guerrero's rural customers as a social service.

The quality of service also varied quite a bit, but compared to most other developing countries the problems were quite few in number. However, residents still had strong opinions about the quality of service. Most of the interviewees in El Quemado consider the quality of CFE service acceptable, though some have complained of frequent blackouts lasting several hours or more. Residents of El Naranjo and Villa Xochitl have complained of poor-quality supply and low and erratic voltage delivery, which causes their appliances to malfunction or break down. They have also complained of frequent blackouts lasting several hours at a time. These are caused by the geological fault, which precipitates landslides that affect lines and transformers. Because residents are at risk, the government is conducting relocation studies and will fund the costs of relocation, including new houses and basic infrastructure. CFE is conducting the technical studies on providing the two towns with electricity.

The Effect of Rural Electrification in the Communities

Residents of all five communities said that electrification has enriched their family and community life. The most common use of electricity is domestic lighting, which has allowed residents to extend educational, business, and leisure activities well into the evening hours. Other major uses include powering electric appliances, artisan tools, radios and televisions, and rural telephones, as well as running audiovisual equipment in local schools. The results have been healthier, better educated, and more productive village residents.

Electricity also had a significant effect on rural productivity. Although most of El Quemado's male residents work as laborers on construction sites in nearby Acapulco, some are artisans who have become more productive since they started using electric tools. Many female residents are single mothers who are unable to work in Acapulco because they must care for their children. Electricity has made it possible for these women to earn income by establishing small businesses, such as tortilla shops and grocery stores.

Before electrification, craftspeople in La Venta Vieja and El Plan Liebre used manual tools; as a result, the few rustic furniture pieces and decorative figures they produced were of inferior quality. Once they began using electric tools, however, both their productivity and the quality of their products improved, and their incomes increased. One interviewee noted, "We used to emigrate to Chicago because our incomes were very low and we had nothing to do, but now we make more money and can listen to the radio and watch TV."

In El Naranjo and Villa Xochitl, interviewees mentioned that electricity makes it possible for them to use corn mills and set up small grocery stores equipped with refrigerators for selling soft drinks, ice cream, and dairy products.

With electricity, El Quemado residents perceive that they have improved their health and quality of life. They can refrigerate their food, which has resulted in fewer cases of dysentery. By using electric fans to cool their homes, they now suffer fewer insect bites. In La Venta Vieja and El Plan Liebre, electrification has made it possible to establish a health clinic with a refrigerator that preserves medicines, lab samples, and serum to combat venom from scorpion bites, a common occurrence in the area.

Before electrification, residents in El Naranjo and Villa Xochitl lit their homes with kerosene lamps and firewood, which caused indoor air pollution. Some interviewees pointed out that, with electricity, they no longer have to cut down trees for firewood, which has allowed some natural reforestation to occur.

Education is important to the people in the five communities, and most agreed that electricity made it easier for students to learn. El Quemado recently opened a public school, consisting of one classroom. The school's teacher said that, thanks to electricity, she now can teach both morning and afternoon classes, thereby doubling the number of students, and enhancing the quality of their education by using a videocassette player and a television.

Because of electrification, the primary school that serves La Venta Vieja and El Plan Liebre has an audiorecorder, a videocassette player, and a television. One of the school's three teachers said that, "Since electrification was introduced, the quality of learning has improved and has diversified to include art instruction, and a refrigerator now makes it possible to store free breakfasts for the school's 79 students."

The towns of El Naranjo and Villa Xochitl each have primary schools consisting of one room divided into two classrooms. Three teachers work in both schools, teaching 35 pupils in El Naranjo and 37 pupils in Villa Xochitl. The teacher interviewed said that electricity makes the use of a videocassette player possible, which enhances the teachers' instruction. More importantly, because the children must help their parents work the land during the afternoon, electric lighting makes it possible for them to finish their homework during evening hours.

Electricity has also provided village residents a bridge to the larger world. As one El Quemado interviewee said, "Now I can hear the news on the radio and know what's happening in the rest of the world." Other interviewees in El Quemado mentioned that electric lighting had improved personal safety and public security.

Thanks to electrification, the shared health clinic established in La Venta Vieja and El Plan Liebre has rural telephones, which enabled staff members to communicate with nearby hospitals. Similarly, in El Naranjo and Villa Xochitl, a shared health clinic equipped with rural telephones make it possible for workers to report emergencies to the hospital in Chilpancingo. Being able to communicate is critical, given the frequent accidents caused by perilous terrain and landslides.

Thus, electricity, in combination with education, communication, and improved business, is perceived by people in these five communities to have a beneficial effect on their quality of life.

Renewable Energy Options: The New Frontier

Since CFE was established in 1937, electricity service has been viewed as a Mexican citizen's right. The country's massive, sustained effort has delivered electricity to even the most remote communities. Nonetheless, a number of communities are still not linked to the grid, largely because they are remote from power lines, their residents are widely dispersed, their terrain is mountainous, and their consumption levels are low.

Based on CFE data, it is unclear how many communities do not have access to electricity. The estimates vary from 70,000 to 140,000 rural communities (4.6 to 6 million people) without electricity coverage in 2006. Of these communities, 69,000 to 136,000 (with 1 to 2.3 million people) have fewer than 100 people per community (the average community size is only 17 residents per community) and are located far from the grid. Renewable energy is their only logical option as an electricity source.

During the early 1990s, rural electrification using nonconventional energy sources took off in Mexico. Electrification authorities recognized that traditional electrification was becoming prohibitive in remote, unelectrified settlements. In response, the Mexican government, with CFE collaboration, started an ambitious program to electrify isolated rural communities using nonconventional energy systems. To date, about 42,000 photovoltaic (PV) modules (used mainly for domestic lighting) and hybrid (solar–eolic–diesel) systems have been installed. Since 1991, the Mexican government has devoted $US11.2 million to develop PV systems, benefiting 41,950 rural households, representing some 250,000 residents.

Solar Modules

Where grid connection is not feasible, decentralized electric power generation systems have been the most attractive option since 1991, when CFE helped to install 1,202 PV cell systems. These solar modules have been disseminated through private companies' direct sales to users and through government-assisted projects. The solar modules program focuses on two types of PV systems. They include home-use kits that generate small amounts of electricity for lighting and to power small radios or black-and-white television sets and systems designed for community buildings that demand large amounts of electricity intended for a wide range of uses, including audiovisual resources in schools and basic medical equipment in rural health clinics.

Institutional Setup and Funding: Keys to Success

Mexico's solar modules program complements CFE's grid expansion program, forming part of the national solidarity program established to extend social infrastructure facilities to poorer regions. A special budget was created to fund the program, whose key features include an internal process in which only local authorities participate, avoiding any intervention by outside institutions. Also, social program guidelines are structured to provide services for the poorest communities.

Three levels of government (federal, state, and municipal) handle the solar program's political aspects and plan its overall strategy, making funds available according to the goals established at municipal and community levels. CFE acts as the regulatory and technical agency in charge of standardization and equipment monitoring.

A key to the program's success is that public funding is not dispersed paternalistically. A counterpart funding mechanism requires that beneficiaries contribute financially and through personal effort, creating a community commitment crucial to a project's success. As a result of this policy, many participating villages have set up community funds to maintain and expand their domestic solar power systems.

Each state's planning and development committee evaluates project requests from its municipalities and grants final project approval. CFE's regional electrification department usually acts as the executive agent responsible for project planning, development, and management. Private companies handle equipment supply and are responsible for installation and after-sales services, as well as training users in basic system maintenance. Communities provide basic technical support for a project during and after installation of the domestic systems, as well as the underwriting of maintenance costs and financing of capacity expansion.

Stages of the Solar Program

Mexico's solar electrification program involves a well-laid-out process. The process includes community selection, promotion of the program, a formal application and implementation process, and finally monitoring of the results.

To participate in the program, communities must comply with a series of requirements that stipulate their size (they must have fewer than 100 residents), density (the average distance between their homes must be greater than 100 m), and location (they must be more than 1–2 km from the main power lines). Also, they cannot be included in other plans to connect to the grid over the next five years.

Communities are informed about how PV systems work, their benefits, and the differences between solar power and grid systems. Demonstration systems are installed in selected communities. Once potential users are convinced that PV cell systems are worth acquiring, a formal application is forwarded to the local government. In the application, consumers agree to make proper use of and maintain the systems and to contribute financially, according to the size of their budget, to their purchase and installation. Once the formal application is approved, the state contracts specialized companies to handle more complex systems not designed for domestic use.

Once technical approval is completed, the state government puts these projects up for competitive bidding. Each project is awarded as a package deal (from design through construction, user training, and development of the infrastructure required to operate and maintain the system). After implementation, the project is approved by comparing the performance of the installed system with its specifications. Once the system is operational, users can call on the concessionaire for help in solving problems.

Hybrid Systems

Mexico has an impressive track record in developing hybrid systems that blend wind and solar power. The combined characteristics of these two energy sources generate electricity reliably at low cost. In addition to wind farms and PV cells, these systems consist of batteries and sometimes include a diesel-powered backup system to enhance reliability. Table 6-4 shows the hybrid systems that are currently operative in Mexico.

Problems and Achievements

The solar program in Mexico is an example of the application of a well-known technology to solve problems of providing electricity to remote rural populations. After all, about 90% of Mexico's population already had electricity from the grid in 1991 when the first 1,202 solar modules were installed. At the time, it was felt that it would not be equitable to leave out the most rural and remote populations in the country. However, the difficulties in serving such remote populations meant that the program faced quite a few challenges. The result was that the program ran into difficulties at several stages of its implementation.

Mexico faced exaggerated claims of system quality by the institutions involved; these claims have frustrated new users. Improper installation of equipment has weakened system performance. Users, who are mostly illiterate and impoverished, are not used to modern technologies; thus, they frequently damage the systems

Table 6-4. *Hybrid Wind–Solar Systems Operative in Mexico*

Location	Year Established	Solar Power (kW)	Wind Power (kW)	Diesel Power (kW)
Santa Maria Magdalena	1991	4.32	5	18.4
X-Calak	1992	11.2	60	125
El Junco	1992	1.6	10	—
La Gruñidora	1992	1.2	10	—
Ignacio Allende	1992	0.8	10	—
El Calabazal	1992	0.8	10	—
San Antonio Agua Bendita	1993	12.39	20	40

Source: Huacuz 1995.

through misuse. Faced with budget shortages and rising unit costs, CFE has had trouble maintaining the impetus of the program's early years. In 1996 (the last year CFE installed modules), the average cost per system was US$1,125. Despite such problems, Mexico's solar modules program can be considered a success, proving that large-scale development of solar-powered systems is feasible at both household and institutional levels.

The various systems in use throughout Mexico have had wide-ranging results. Still a developing technology, hybrid solar–wind power is plagued by problems that likely will be resolved in the future. For example, the technology's assessment methods are not yet sufficiently disseminated, and some technical glitches must be resolved to ensure reliability.

CFE, versed in large-scale electrification and grid-based systems, has encountered some difficulties in managing this technological diversity. Although unlikely to reject renewable-energy options outright, CFE is also unlikely to pursue alternative technologies without continuing federal pressure.

There are concerns that the initial costs of renewable-energy technologies are higher than those of grid-based electrification, even at low levels of service. In addition, maintenance costs are lumpier, making it more difficult for CFE to plan ahead. This has contributed to a lack of understanding regarding the performance of renewable energy in rural distributed applications, resulting in a generalized resistance to these options. Nevertheless, CFE has gained substantial experience working with the solar modules; with equipment supplies and communities' participation, solar-module maintenance has been economical.

In contrast to public service from CFE's grid-based electrification schemes, tariffs for decentralized power systems are unregulated. The high and inefficient tariff subsidies to residential customers and agricultural uses reduce the competitiveness of renewable energy. Installation of renewable technologies, particularly those based on wind power, need to be designed with detailed attention to the geography of each site. To achieve cash flow commensurate with cost of service, each project presents unique challenges for negotiating tariffs with municipalities and communities.

Currently, no method is being used to account for the environmental benefits of solar or hybrid solar–wind technologies. Although their benefits might not be considered important in rural areas, the emissions savings from grid-electricity substitutes could be relevant because incremental grid electricity is based on thermal generation. Thus, some allowance could be made at the program level (within the social evaluation) to account for the environmental benefits of renewable-energy alternatives.

Lessons from the Mexican Experience

Mexico has relied on its state-owned enterprises far longer than many developing countries to provide energy and infrastructure services. In the REP's early years, the country benefited from a nationalistic, motivated, and well-trained staff. A strong economy throughout Mexico's postrevolutionary period, coupled with more recent market-based policies, have laid a solid foundation for continued subsidies for poverty alleviation and infrastructure development, including rural electrification.

Despite its unique characteristics, Mexico's REP has many relevant lessons for other countries, particularly those in need of rational, consistent mechanisms that can prioritize projects backed by finite fiscal resources—whether federal or state subsidies, community funds, or some kind of combination. The Mexican experience is especially relevant in countries anxious to transfer a portion of their planning activities and administrative oversight to local governments and communities.

Sustained Government Commitment

Rural infrastructure programs require long time frames in which to mature. They cannot be completed quickly, and their benefits cannot be realized within a presidential cycle—six-year—in the case of Mexico. In Mexico, rural development has been one of the federal government's top priorities for decades. Since the triumph of the Mexican Revolution in 1920, rural development, of which rural electrification has been an integral part, has been a policy goal, creating institutions with long term frameworks responsible for rural development.

Natural Resource Base

Mexico enjoys an ample energy resource base, particularly petroleum, which has allowed the electricity industry to expand and electrification to advance at a rapid pace.[13]

Decentralization and the Changing Role of the Central Stakeholder

At the outset, Mexico established a state-owned enterprise devoted to electrification. This institution, the CFE, centralized the REP, making electricity easily available to rural households. As the REP matured and electrification of remote communities became more difficult and expensive, the federal government began involving state- and local-level players, whose participation was coordinated by the CFE. The purpose was to strengthen the institutional, programmatic, and financial capabilities of the municipalities in line with a bottom-up approach, empowering communities to decide how to develop. This decentralization of social development funds and transfer of responsibilities to the local level was done without proper consideration of state and municipal capacities, resulting in a slowdown and atomization of the electrification effort. Even though CFE remains in charge of the engineering and installation of grid extension projects, it now functions only as a contractor under the philosophy and development objectives of different, and sometimes conflicting, programs. The continued involvement of the CFE in helping Mexico chart a course toward decentralized decisionmaking with a parallel build-up of local capacity to identify electrification needs/uses and plan cost-effective solutions should be carefully considered.

Community Integration

One main advantage of Mexico's REP is the way it has gradually integrated local communities into assuming responsibility for electrifying their villages. In more remote communities, where renewable-energy options are the most viable alternatives, local communities are even responsible for system maintenance. Their involvement nurtures a sense of ownership and ensures that equipment is better maintained and more carefully handled.

Training users in system maintenance has also figured prominently. In many developing countries, electrification programs fail because of the expenses involved in maintaining systems in remote locations. When repairs are delayed, government funding is constantly drained to finance long-distance maintenance schedules and consumers are dissatisfied because of lengthy power outages.

Policy Reform to Ensure Financial Viability

Although there might be apparent excellent reasons to initiate and maintain subsidies, allocating them in a general way through tariffs not only puts the financial viability of the sector at risk, but prevents new renewable options from being developed. It would be preferable for any required subsidies to be made transparent, non-distortionary, and directed toward those customers with the greatest need.

The Right Fit for Renewables

The main distribution network has been overextended to cover most of Mexico's rural settlements. Rural areas still to be electrified have special features that undermine the feasibility of further grid extension. For these remote communities, a relative successful electrification program was put in place based on the installation of PV cells that supply solar power to rural dwellings. The CFE acknowledged the appropriateness of renewable-energy technologies as complementary to, rather than in competition with, the standard grid supply.

The Success of Creating Competition

The goal of creating a competitive environment, where communities and municipalities compete for federal and state funds, has been important if not essential to the REP's success because it has involved beneficiaries in all phases of the electrification process.

Conclusion

The rural electrification program in Mexico was a massive undertaking in a country with millions of people in rural areas without electricity. At first, the program relied on a central electricity company to plan and extend the program. However, over time the politics of providing the subsidies changed from a direct financing of the central agency to the development of a combination of fund-ing sources, including central, municipal, and local funds. Also, communities could take advantage of social funds to select rural electrification as a priority for them. They did this in large numbers, sometimes selecting electricity over other social projects.

Today the electricity sector is still undergoing significant changes. The subsi-dies to the companies involved in electricity are still massive, but they cover more than just financing for rural electrification. The mechanism is a general subsidy voted through the legislature for the purpose of lowering energy costs and making their products and services more competitive in a worldwide economy. One can quibble whether these are the right policies, but the results for rural electrifica-tion mean that the electricity industry is on stable financial grounds and is able to provide service to rural people.

The legacy of this interesting mix of central planning and decentralization of decisionmaking is still to be written. However, it certainly dispels the myth that the public provision of electricity service is an endeavor performed in only one way. Certainly the rural electrification program has benefited millions of rural people in a way that would be difficult to measure. The final reality is that the significant investments have reached deep into the rural areas of Mexico to help rural people lead better and more productive lives.

Acknowledgments

I wish to acknowledge the support of Fernando Solís-Cámara, Juan Eibenschutz, Gerardo Dorantes and León Alazraky, for their support and thoughts on several issues on CFE, poverty, access, electrification, rural development financial issues, and social welfare; and, Julio Cesar Rodriguez, for his assistance with the inter-views in Guerrero.

Notes

1. Data to August 2006; CFE (2006), *Sexto Informe de Labores*.

2. Rodriguez-Rodriguez (1994) refers to this practice as mixed service: providing power for textile production and illumination needs, while selling the remainder to nearby communities.

3. The author tends to favor the lower estimate in line with Lara-Beautell (1953). It seems that the high estimates considered all users, not just residential.

4. See Comisión Federal de Electricidad (1997).

5. By the 1940s, the growth rate of Mexico's population was 2.8%, an increase of 0.8% over the previous decade. This accelerated growth reflected the socioeconomic improvements that were being made: rising productivity and incomes, improved sanitary technologies, and vac-cination campaigns. Improved social welfare and education, in turn, had generated an improved public health system, resulting in decreased infant mortality and increased survival rates. The average rural family now had seven children, and Mexico's total population had reached 19.6 million; however, only 13% had access to electricity.

6. During the 1960s, the REP accelerated into a sustained effort and extended service to almost 71,000 communities, benefiting 35 million people. During this period, CFE's leadership was critical in stimulating each region to achieve the ambitious objectives proposed by the federal government.

7. The studies of Hernandez-Trillo (2004) and Castañeda-Diaz (2004) indicate that the expenditure efficiency at the municipality level is extremely low.

8. The state governments were responsible for preparing socioeconomic studies of local communities to determine how much they could contribute, based on project costs and beneficiary income levels.

9. Federal funding refers to the Branch 33 (*Ramo 33*) composed of 7 funds earmarked for several purposes, among which the FAIS (Social Infrastructure Development Fund) is the only one for social infrastructure development: health, education, transport, water and electrification works, among others. Before 1996, SEDESOL (Social Development Secretary) managed these funds.

10. In Chile, the regional average cost per kilometer of line has varied between a low of US$4,300 in 1993 and a high of US$7,500 in 1996 (Region VII–El Maule). In Bolivia, costs are marginally lower than in Chile, whereas in Argentina, costs are similar to those in Mexico.

11. For further information on the electrification of Guerrero, see Bustamante Castañeda (1978).

12. As informed by the local authority (*comisario*).

13. For further reading, please consult CFE (1988) and Comisión Nacional para el Ahorro de Energía (1999).

CHAPTER 7

Electricity and Multisector Development in Rural Tunisia

Elizabeth Cecelski, Joy Dunkerley, Ahmed Ounalli, and Moncef Aissa

T HE GOVERNMENT OF TUNISIA HAS STRONGLY supported rural
electrification since the early 1970s. Although the program has done many
things right to accomplish the country's goal of universal electricity access, one
unique feature of the program has been the extensive consultation with other
agencies and reliance on funds from rural development programs. The country's
leaders understood from the start of the program that electricity by itself will not
have the development effect of a program that is integrated into a broader strategy
of rural development.

Tunisia's rural electrification program was launched in the mid-1970s, a time
when only 6%, or 30,000 of the country's rural households, had electricity. At
that time, about half of Tunisia's population lived in rural areas. Over the ensuing
years, the country has made impressive gains in providing electricity to its rural
population. By the end of 2004, 97% of all rural households had electricity service.
Today the country has begun a program to serve even the most remote areas with
photovoltaic (PV) systems. The accomplishment is even more remarkable because
of the conservative definition of rural areas, which involves households outside of
incorporated areas. Many populations that in other countries would be defined as
rural villages and towns are defined as urban in Tunisia. Thus, Tunisia's rural population is highly dispersed and isolated, with long distances between small groups of
sometimes scattered houses.

Tunisia has been able to artfully balance the sometimes conflicting priorities of
having substantial state subsidies, integrating rural electrification with rural development goals, and maintaining the commercial viability of a public electricity
company. As Tunisia approaches universal electricity coverage, the question arises
as to whether the experience in Tunisia is applicable to rural electrification programs in other African nations. The many factors that contributed to the program's
success—strong government policy and financial commitment, gender and social

equity, institutional esprit de corps, technical innovation, and uniquely enabling political and economic conditions—are lacking in many other African countries. Nonetheless, the Tunisian experience can provide useful lessons—even in some of the most unpromising situations.

History of Rural Electrification in Tunisia

Electricity generation in Tunisia began in 1902, when a French concessionaire that was already providing gas installed the first power plant to serve the capital city of Tunis. Various French companies rapidly followed suit, constructing power plants in the cities of Sousse (1905), Sfax (1907), Ferryville (1909), and Bizerte (1911). On the eve of Tunisia's independence from France, in 1956, seven concessionaires controlled the country's electricity generation and distribution. The largest of these companies was the Compagnie Tunisienne d'Electricité et de Transports (CTET), established in 1952, which served Grand Tunis and parts of the northwest region. The concessionaires designed their own networks and produced their own electricity or subcontracted producers to maximize the profitability of their concession areas and the duration of their respective contracts. This design resulted in companies sacrificing long-term interests for short-term profitability, making few investments in infrastructure, and alleviating shortages with uncertain solutions.[1]

CTET owned Goulette, Tunisia's oldest and most powerful steam thermal power station (57 MW in 1952). Forces Hydroélectriques de Tunisie (FHET), the country's second largest concessionaire, created in 1952, was responsible for hydroelectric power plants in Ben Metir and Neber in the northwest region. Other companies, which mainly generated diesel power, distributed electricity to various cities and urban areas, including Béja, Bizerte, Gabès, Gafsa, Médenine, Sousse, Sfax, Tozeur, and Zarzis. In addition, Tunisia imported electricity from the Algerian Electricity and Gas Company, which also had interests in FHET. The network consisted of the interconnection with Algeria and the connection between hydroelectric plants in the northwest and Grand Tunis. At this time, rural electrification was extremely limited.

After gaining independence, the Tunisian government initiated a general policy to nationalize key economic activities, including electricity and gas, water, railroads, and banks. In 1958, the government temporarily took control of the concessionaires, replacing CTET and the other companies with management committees. On April 3, 1962, the government nationalized electricity generation and electricity and gas transport and distribution. These activities were entrusted to the Tunisian Electricity and Gas Company (STEG), a commercial company. At that time, only 26% of Tunisian households had access to electricity.

In the decade that followed, owing to rapid growth in domestic customers and initial extension of the grid into rural areas, electricity consumption increased at a pace of 11.5% annually. The STEG concentrated its efforts on rationalizing the system it had inherited from the concessionaires. Electricity generation and transport were developed to meet the demand of new industrial projects, such as El Fouledh steelworks and the textile industry. In 1965, Goulette 2 was installed, including four groups of 27.5 MW each. In 1972, a power plant was built in the southern city of Ghannouch, which included two groups of 30 MW each and a 15-MW gas turbine. Part of the electricity thus generated was used in Gabès's new industrial units, and the rest was transported to other regions through a newly looped system. Electricity generation in Baves was favored through exploitation of flared gas in the southern region (El Borma oil field associated gas) and construction, in 1972, of a gas pipeline connecting the oil field with the Gabès area.

With assistance from the French utility Electricité de France and the proactive education policy of the Tunisian government, the STEG developed a cadre of highly qualified technicians and engineers. By the mid-1970s, the utility had established sound business practices and financial sustainability achieved through tariffs related to marginal costs. Just 6% of rural households and only 37% of all households in the country had access to electricity at that time.

The Tunisian government, now increasingly concerned about the exodus from rural areas caused by the lack of public services, turned its attention to expanding rural electrification. In 1973, the STEG, with the assistance of Hydro-Quebec, undertook a technical audit of distribution systems. The audit took into account the government's ambitious goal of providing universal electricity coverage for the whole country, the potentially low levels of rural energy consumption, and the high financial requirements.

The main recommendation of the audit was to study a new, lower cost means of electricity distribution that combined three-phase and single-phase lines. Based on the North American model, this system was known in Tunisia as the MALT (Mise A La Terre). Although controversial at the time, the recommendation was confirmed by technical and economic studies conducted for the master plan for distribution in 1974–1975. The studies estimated 18–24% savings using the MALT system. In 1976, the technical decision was made to begin converting to the new system, using three-phase and single-phase lines and 30 kilovolts. On this basis, the Planning Ministry, together with the STEG, set rural electrification goals that were incorporated into the Vth Plan (1977–1981) and subsequent five-year plans (Table 7-1).

The Vth Plan allocated government funds for system expansion and identified villages to be electrified, based on lowest cost criteria. The STEG's main emphasis was on converting the existing distribution system to the MALT system. During this five-year period, 70,000 rural households were connected, and investment costs were fully recovered. During the VIth Plan (1982–1986), 80,000 rural households were connected. Savings from the new distribution system made it possible to connect an additional 10,000 households under the same budget, raising the rate of rural electrification to 28%.

The majority of the financing for rural electrification came directly from the government. It was not until the VIIth Plan (1987–1991) that Tunisia's government for the first time mobilized external funding for the program from the African Development Bank, the Kuwait Fund, and the French Development Agency, Agence Française de Développement. This step initiated an intensive phase of the rural electrification program. From 1987 to 2000, 429,000 households were connected to the grid, raising the rate of rural electrification from 28% in 1986 to 88% in 2000. Recognizing that the last remaining households

Table 7-1. *Evolution of Tunisia's Rural Electrification Program, 1972–2001*

| | Five-Year Planning Periods | | | | | |
Factor	IV (1972–1976)	V (1977–1981)	VI (1982–1986)	VII (1987–1991)	VIII (1992–1996)	IX (1997–2001)
Total investment (million Tunisian dinars)		29	52	105	130	134
No. of new connections	30,000	70,000	80,000	114,000	180,000	135,000
Cumulative connections		100,000	180,000	294,000	474,000	609,000
% rural households with electricity	6	16	28	48	75.7	88.1
% total households with electricity	37	56	69	81	90.0	94.9
No. of new households with PV systems[b]					3,919[c]	3,838

[a]Implemented through the year 2000.

[b]The PV program adds about 1% to rural electrification coverage.

[c]Cumulative through the end of the VIIIth Plan.

were scattered throughout remote areas, the National Agency for Renewable Energy (ANER) launched a PV program in 1990 that has now reached about 1% of rural households. This agency is now called the National Agency for the Management of Energy (ANME).

National Commitment to Rural Electrification

Tunisia's achievement in rural electrification has been rooted in a strong national commitment to integrated rural development, gender equity, and social equality. The rationale for the national government's longstanding resolve for rural development and modernization was based on the goal of seeking to raise the living standard of its rural citizens, to promote security in outlying regions, and to moderate urban growth.

The three pillars under which rural development was initiated by the country's IVth five-year plan (1972–1976) were basic education, improved health services (including family planning), and rural electrification.

The Three Pillars of Rural Development—Education, Health, and Electricity

The first pillar of Tunisia's rural development strategy formalized in 1972 was education. Since its independence from France in 1956, Tunisia has been at the vanguard of promoting human resources development—particularly gender equity—in the region. Before independence, most Tunisian women were illiterate. The most advantaged women only had an elementary level of education. Even by the 1960s, female university graduates numbered only about 100. Nonetheless, women participated actively in the struggle for national independence. This was perhaps a factor in the keen personal interest of Tunisia's first president, Habib Bourguiba, in promoting women's rights.[2]

Immediately after independence, on August 13, 1956, a Personal Services Code (PSC) was promulgated. Among its other provisions, the PSC abolished polygamy, instituted judicial divorce, gave women the right to vote, and set a minimum age of 17 for girls to marry. The suppression of polygamy, in particular, had an enormous symbolic effect in Tunisia and throughout the world, even though it represented only 4% of marriages in Tunisia.

In the decades after the introduction of the PSC, the Tunisian government invested heavily in education to ensure that women could take advantage of their new legal rights. The emancipation of women, viewed as a struggle against ignorance, emphasized the education of girls. As a result, attitudes toward girls' education changed radically. The principle of coeducation was recognized as a fundamental means of progress and was adopted in schools run by the Ministry of Education. Today, more than 90% of girls and boys are enrolled in school, and Tunisian women have one of the highest literacy rates in the Arab world. Female students outnumber males in universities; 5,000 women head private companies; and 12% of senior business executives and 35% of doctors are women.

The second pillar in the rural development strategy was health. This program mainly involved women's right to control fertility and have access to modern means of contraception. This goal was accomplished through provision of basic health and family planning services. At the time of independence, female mortality was higher than male mortality. Subsequent improvement of women's access to health services, as well as education, allowed greater participation of women in salaried work, another positive factor for women's rights. Today, Tunisian women have an average life expectancy of 74 years and an average of only 2.2 children.

Rural electrification was the third pillar of Tunisia's rural development program, and the government understood that it was important for it to be coordinated with both education and health. Electricity permitted higher penetration of the media, especially television, which introduced rural Tunisian families to various roles for women in urban areas and in other countries.

The national commitment to developing and improving rural living standards involved considerable investments on the part of the government. Such investments were clearly a critical factor in the success of all three aspects of the development program. Government support has proven remarkably steady in weathering political and economic changes.

Ways of Financing Rural Electrification—Domestic and International

The commitment to rural electrification was demonstrated through the various budgets used for carrying out the program. The primary source of financing the program has been from the Regional Development Program (PDR, Programme de développement régional), and this source was complemented by other domestic funds and international loans.[3]

Since the 1970s, the PDR has been the primary source of state funding for rural electrification. PDR allocations for each rural development sector are negotiated between the Ministry of Economic Development (Ministère du Développement Economique) and each regional government (governorate) on an annual basis. In many governorates, rural electrification has had first priority at various times. After the sectoral allocation is negotiated, specific projects are chosen at the governorate level, thereby ensuring local input in project selection. As Table 7-2 shows, of the various rural development sectors that made up the Regional Development Program investment during the 1997–2000 period, rural electrification represented more than 21% of the total, second only to drinking water.

In 1984, the PDR was supplemented by the Integrated Rural Development Program (PDRI, Programme de développement rural intégré), which the state also funded. Although small compared to the PRD, the PDRI takes a more integrated approach to rural development. It offers beneficiaries integrated assistance across many areas, such as vegetable and small industry production, irrigation, and electrification.

In addition to the PRD, in which funding was allocated according to a strict planning criteria, a program was created to assist projects that failed to meet the usual criteria. The president of the republic created two extrabudgetary funds. The first was a special Presidential Fund (Programme Présidential, PP), which finances projects that the president selects at regional ministry-level meetings and during visits throughout the country. The second was the National Solidarity Fund (FSN), which the president created in 1992 to improve Tunisian living conditions, particularly in underprivileged areas (*zones d'ombre*). Voluntary contributions to the FSN are solicited from nongovernmental organizations (NGOs), public and private businesses, and Tunisian citizens. As a result of this financing, the rate of rural electrification has been increased by an estimated 10%. These personal initiatives of the president, supported by contributions from a broad spectrum of society, exemplify the country's strong political commitment to rural development and to rural electrification in particular.

Later in the program, the government also borrowed money to finance rural electrification. Since 1977, the state has incurred more than 200 millions of Tunisian Dinars (MTD) in external debt—not always at concessional rates—with which to finance rural electrification. The most significant source has been the African Development Bank (ADB), which has provided five lines of credit for rural electrification, or about 80% of external financing. Other funding sources have included the World Bank (7.8%); Agence Française de Développement (7.5%); and the Kuwait Fund (2.6%). Together with ADB support, these loans have helped to finance the connection of 376,000 rural residents, 61.7% of the 609,000 rural households connected during the times of these investments.

Finally, national commitment to total rural electrification has been demonstrated by the development and funding of an ambitious, high-quality PV program established in the mid-1980s to reach the most isolated households that

Table 7-2. *PRD Rural Development Investments in Tunisia by Sector, 1997–2000*

Sectoral Activity	Amount (MTD)	% of Total Investment
Drinking water	373.7	45.8
Rural electrification	172.7	21.2
Roads, bridges, and streets	150.0	18.4
Education and teaching	47.0	5.8
Post offices and telephones	32.4	4.0
Flood-protection works	19.8	2.4
Health	7.8	1.0
Youth and children	7.5	0.9
Professional training	3.7	0.5
Total investment	814.6	100.0

Source: Republic of Tunisia 2000.

otherwise would not meet the selection criteria. This program is financed primarily by World Bank credits and NGO's not by beneficiary and regional-government contributions.

Effective Institutional Structure for Project Planning and Selection

Tunisia's rural electrification program benefits from an institutional structure that has proven highly effective in achieving rapid growth and spread of rural electrification. An iterative five-year planning and implementation process balances economic and social criteria and imposes financial discipline on rural development projects, including rural electrification. The system is characterized by centralized planning, with major regional and subregional inputs and initiatives. This planning is done within the framework of a comprehensive rural development program.

Tunisia's rural electrification program is influenced by various social, economic, and technical factors. Multiple institutions are involved in developing and implementing rural electrification programs. At the national policy and planning level, these institutions primarily include the MDE and the MI, with input from the STEG and the ANER. At the regional and local levels, regional governorates and their subdivisions, called delegations, provide input into rural development planning, whereas the STEG and the ANME are responsible for implementation.

At first glance, having so many agencies involved in rural electrification appears unwieldy. However, these disparate entities are unified through well-defined roles as an official coordinating body and a planning and implementation process that guarantees continuous interaction between agencies. Clear criteria govern the selection of rural electrification projects. All cooperating agencies are aware of the criteria governing the process and can concentrate on efficient implementation. Such close cooperation continues throughout implementation.

Well-Defined Roles and Mandates for Agencies

Each agency has a clearly defined role and mandate in rural electrification. The MDE, in collaboration with its specialized agencies, including the General Regional Development Commissariat (CGDR, Commissariat Général de Développement Régional), defines overall rural development policy, and this policy provides the framework for rural electrification. The MDE is charged with disbursing a share of national revenue to subsidize rural electrification projects in a cost-effective and equitable way. It mobilizes finances and divides the national budget for rural development between the regional governments and implementing agencies. Both the PRD and the PDRI are housed within the MDE.

The MI develops Tunisia's energy policy. It is responsible for supervising the various branches of the energy sector: hydrocarbon exploration and production; refining and distribution of petroleum products; and production, transport, and distribution of gas and electricity. As part of its mandate, the MI houses the National Rural Electrification Commission (CNER, Commission Nationale d'Electrification Rurale) and has supervisory authority over the STEG. It also provides input into the five-year planning process.

The governorates, in their role as regional executive agencies of the Ministry of the Interior, are charged, together with their delegations, with selecting rural development projects, including rural electrification projects, and allocating funds disbursed from the national budget, in addition to their own resources. The governorates are also responsible for overseeing the timely and efficient completion of projects. Thus, the governorates and their delegations provide, at an official level, primary regional and local input into project selection and design. In identifying rural electrification projects in their jurisdictions, the delegations also consult with *oumdas* (leading citizens who act as spokespersons for local interests).

The two implementing agencies—the STEG and the ANME—also have clearly defined roles (Figure 7-1). As the national electric utility, the STEG is responsible for electricity generation, transmission, and distribution, as well as transport and distribution of natural gas. Although the STEG falls within the MI's jurisdiction, it enjoys considerable autonomy in practice, especially in technical matters. However, decisions of a broader social nature, such as changes in tariffs, are made in consultation with the MI.

The STEG is responsible for implementing the major part of the rural electrification program—that based on grid extension. It maintains a regional organization that parallels the governorates. Thus, STEG districts largely coincide with governorates, facilitating regular consultations between the two bodies. The STEG is the direct counterpart of the governorates in rural electrification projects. As a statutory government corporation established by Decree Law No. 62–8 of 1962 on nationalization, the STEG is responsible not only for grid-based rural electrification projects, but also for the entire electricity sector, including generation, transmission, distribution, and export, as well as distribution of gas under the MI's supervision.

The ANME, which is under the administrative supervision of the Ministry of the Environment, promotes energy conservation and development of renewable

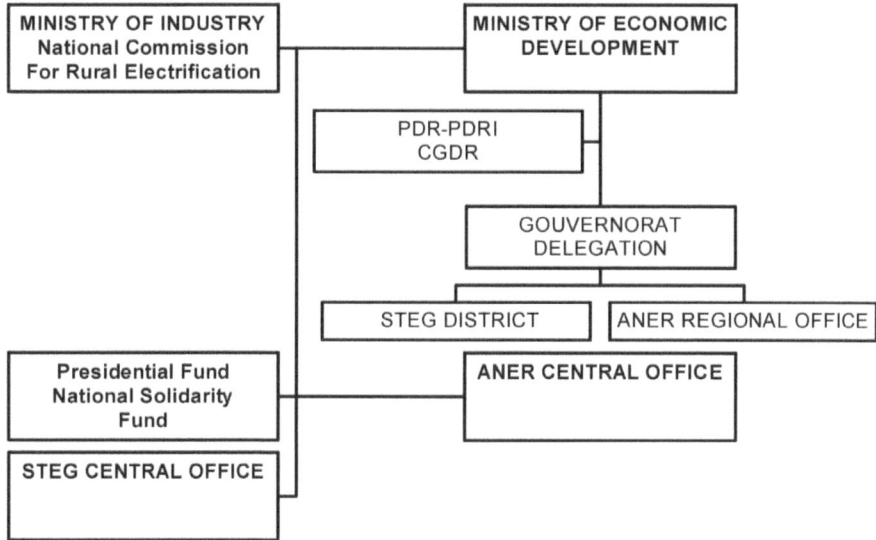

Figure 7-1. *Diagram of Responsibilities for Rural Electrification in Tunisia*

energy. The ANME undertakes PV-based rural electrification projects that aim to connect households remote from the grid. Although much of the ANME's activitiy is centered at its Tunis headquarters, it has offices in the three regions (El Kef, Sidi Bouzid, and Gabès) where it has the most projects.

Agency Coordination

Given that many agencies are involved in rural electrification, considerable efforts are made to ensure their coordination. At policy and implementation levels, coordination is achieved both institutionally and systemically.

The CNER, a coordinating body chaired by the MI's Director of Electricity and Gas, includes representatives of the STEG, the ANME, the MDE, the Ministry of the Environment, the Ministry of the Interior, and the FSN. Through regular meetings, the CNER keeps members informed about rural electrification activities that are carried out throughout the country. It provides a forum in which policymaking, planning, and implementing agencies can exchange views and identify problems.

Such a coordinating body, though useful, is rarely adequate to cope with the entire range of issues that arise during the planning and implementation of a major program. Therefore, in Tunisia, the CNER's work is supplemented by continuous interaction—both horizontal and vertical—among agencies. For example, two-way communication among the MDE, the MI, the STEG, the ANME, and the CNER is continuous when the five-year plans are being drawn up to arrive at rural electrification targets consistent with available financial and technical resources. Similarly, at the regional level, the governorates interact continuously with STEG districts and ANME offices during program execution.

The contents of the five-year plans' rural electrification programs are also developed iteratively through two-way communication between central and regional authorities. In fact, the first estimate of rural electrification projects within the overall rural development budget, established by the MDE, originates at the subregional (delegation) level. Far from being entirely top-down, the process incorporates a substantial amount of bottom-up content, at least at the official level.

Planning and Implementation

The dynamics of this coordination are further illustrated in the planning process, from initial design of the national rural development budget in the five-year plan to the selection of specific rural electrification projects.

The criteria governing these choices are an important aspect of a successful rural electrification program. During most program years, more villages or households wish to receive new electricity service than there are funds available to do so. It is therefore important to ensure that rural electrification planning is open and objective and uses clearly defined criteria to rank villages and households for connection. Clear criteria respond to concerns about social justice or fairness, reduce local political pressure to "jump the queue," and allow for a more rational and economic long-term electrification program. They also greatly facilitate the planning and implementation of rural electrification projects, as they eliminate potential contention among cooperating agencies.

Tunisia has established a project-selection method that is orderly, transparent, and meticulous. Rural electrification plans and targets are made publicly available so that progress can be monitored and assessed. In the STEG program, which accounts for more than 90% of connections, selecting sites for electrification is a two-step process. For the first step, within the framework of the current five-year plan, the MDE identifies the delegations or zones to include in the rural development program. Selection is based on such criteria as income level, unemployment, environmental quality, gender status, expected rate of return from projects, and the costs involved in job creation and improved living conditions.

In the second step, potential rural electrification projects are identified within the delegations and zones selected in the first step. The governorate asks the delegations to list all nonelectrified agglomerations, defined as a minimum of 10 households or adjacent households no further than 200 m apart, built with walls and roofs of permanent materials. Potential sites for electrifying agricultural pumping and water drilling in the delegations are also identified. This list is drawn up in collaboration with the local oumdas.

Next, STEG district offices in each governorate review this list in detail. They make site visits to verify the information provided by the governorates and collect additional data, including lengths of needed medium-voltage (MV) and low-voltage (LV) lines that are available, number of transformers, and number of housing units suitable for electrification. This information is then mapped onto the existing grid.

In this way, a database of economic and technical information is constructed for each STEG district. This information is processed in the STEG's computer-driven economic feasibility model to evaluate the investments per project or *grappe* (several projects served by the same medium-tension line). STEG headquarters then estimates costs, based on STEG unit costs, of electrifying the various households, agglomerations, and pumping and drilling sites. On the basis of this estimate, a table is prepared showing the number of households that can be electrified at various cost levels and estimating the total costs of providing electricity to the number of households at each cost level.

This process permits the STEG to provide the MDE scenarios for electrification. Each scenario gives, for each governorate, the number of beneficiaries, cost of projects, and rates of electrification. Once the rural development objectives of the five-year plan are fixed, these scenarios are used to establish project costs. For example, in the IXth Plan (1997–2001), the MDE fixed the ceiling at 2,200 Tunisian dinars (of which beneficiaries pay TND 200, the STEG pays TND200, and the state pays TND1,800). Thus, all projects that cost less than this ceiling are selected for inclusion in the provisional five-year plan for rural electrification, based on grid extension; however, projects that cost more than TND2,200 may possibly be included in supplementary programs, such as the presidential fund or FSN.

At the regional level, the governorate, in collaboration with the STEG district office, adopts these projects and the funds allocated by the MDE for rural electrification. The CNER then checks for inconsistencies between the adopted governorate rural-electrification projects and those in other programs, such as the PDR, PDRI, PP, and FSN. The governorate program is finally confirmed at the national level in meetings among the MDE, the governorate, the STEG, and the ANME. The five-year plan is the consolidation of the various regional plans.

Regarding the much smaller ANME program, the number of households that could benefit from PV systems is based on census data, results of STEG inquiries, and a 1995 study estimating the role of renewable energy, which concluded that 70,000 households would not be served by the grid. On this basis, the ANME planned in the VIIIth and IXth five-year plans to install 10,000 systems (of which 7,700 have been installed).

The ANME's selection process is largely determined by the advance of the grid-connected system. In the past, because of the time lag between ANME project definition and installation (as long as two to three years), PV projects were sometimes overtaken by arrival of the grid. In other cases, the grid arrived shortly after the PV project had been implemented, thereby duplicating efforts and wasting resources. For example, in 20% of cases, the grid arrived to connect households within three months to a year after they had installed PV equipment.

To avoid such duplication, coordination among the ANME, the STEG, and the governorates is being tightened. The ANME now asks each governorate to provide a list of potential beneficiaries of PV systems. The list is based on rural development needs, present and projected distance from the grid, and household interest. Increasingly, efforts are being made to ensure that projects are situated well beyond the anticipated extension of the grid. The list is therefore checked with the STEG before program implementation.

Achievements and Challenges

Rural electrification experience throughout the world suggests that there is no single institutional structure or process for success. Regardless of the structure adopted, however, certain characteristics are essential. These include clarity of purpose, well-defined roles for all agencies involved, and established procedures that ensure equitable agency coordination. The Tunisian system scores well on all counts.

Nonetheless, Tunisia's rural electrification program has its shortcomings. Coordination has sometimes broken down, as the example of duplicating PV systems and grid extension projects illustrates, resulting in wasted resources. Moreover, although the project-selection process appears admirably clear and transparent, it may be criticized in practice for verging on the mechanical, especially in cases where local costs diverge from the national averages used to estimate total costs of rural electrification. Finally, although the selection process is initiated at the community level, in consultation with the local oumda, this input is considered official rather than at the citizenry level, and could therefore be incomplete.

On balance, however, the successful record of Tunisia's rural electrification program reflects its efficient, well-coordinated processes, as well as its perceived fairness. These factors, in turn, have reinforced a national commitment to improving living conditions for rural residents by making rural electrification an integral part of the country's broader rural development program.

Implementation through the STEG—An Effective Utility

The STEG's long record as an effective, efficient utility has earned it an international reputation as one of the best developing-country power utilities in the world (ESMAP 1990, Hicks et al. 1993). Insulated from unwarranted political influences through its mandate, the STEG has been a key partner in Tunisia's rural development. It is viewed as a model enterprise in the Tunisian government and economy, having attracted the best and brightest Tunisian engineers and economists to implement the nation's rural development mission during the 1970s. Two decades later, the high level of confidence vested in the STEG's technical assessments played an important role in the successful adoption of cutting-edge technology.

Several operational factors have contributed to the STEG's success. They include the encouragement of private-sector participation in the construction phase of rural electrification projects and the promotion of a local supply industry for equipment and material. The STEG also developed a sophisticated computerized inventory management system and rigorous commercial practices, including control of nontechnical losses and effective billing and connection payment practices. Finally, findings on customer satisfaction and quality of service are analyzed.

Mandate and Management Structure

The STEG is a statutory government corporation of a commercial and industrial nature. The STEG, under the MI's supervision, is responsible for the generation, transmission, distribution, import, and export of electricity and natural gas. The utility's 3 departments and 15 directorates report to the chairman and managing director and are responsible for operating the electricity and gas systems and managing the utility.

The Directorate for Electricity and Gas Distribution has primary responsibility for rural electrification through its Department of Program and Budget, logistical directorate, 5 regional directorates, and 34 district branch offices. This directorate is supported directly, however, by the general management, which approves plans and budgets, and by central administrative STEG units, including the Directorate for Finance and Accounts, Directorate for Studies and Planning (which sets tariffs), and the Directorate for Human Resources and Legal Affairs (which trains external contractors).

The STEG's 14-member board of directors includes a chairman and managing director, assistant managing director, nine members representing the state (including a representative of the Ministry of the Environment), two members representing employers, and one financial controller. The STEG also has a cooperative agreement with the Association of Consumer Protection to provide consumer input through regular meetings with STEG headquarters, as well as with field offices and the newly implemented call centers.

The STEG's organizational structure reveals two important reasons for its success in rural electrification. First, the utility has enjoyed the backing of highly professional, experienced administrative units within a large corporation with well-established operating and customer-management procedures. Second, it has benefited from a highly decentralized implementation structure since 1977, when the decision was made to establish district offices in each governorate. Today, in fact, many governorates have more than one STEG district office, which facilitates coordination with rural development planning through the local selection of STEG projects in close cooperation with the regional administration.

Early Computerization and Development of Software Applications

The STEG was the first major Tunisian corporation to computerize operations. This computerization occurred in the early 1970s, the same time that the country's rural electrification drive was launched. By the late 1970s, almost all departments had been computerized, which allowed a sophisticated level of data collection, analysis, and management that contributed greatly to the STEG's ability to monitor and improve its performance in all areas, including rural electrification.

During the mid-1970s, various software applications fundamental to the everyday operations of the STEG were designed and adopted. These applications included such activities as personnel and salary management, billing and collection, and inventory management. During this initial period, the STEG emphasized software development as an operational and business management tool. Engineers and software technicians were recruited to design and put in place these software applications. Software applications were also developed to facilitate the design of rural electrification systems. These applications included the Tanouir software for sizing MV lines and the software for daily account records of LV customers.

Transparent Norms and Guidelines

The STEG's operational norms and guidelines, updated regularly, are used by both STEG technicians and outside contractors to ensure a standardized approach and adherence to contracts. These guidelines illustrate the attention to detail that is paid by the company.

The *Notebook on Specific Administrative Clauses* covers all administrative details, such as costs that can be included in bids, those for which the STEG is responsible, escalating factors for unit costs, terms of payment to contractors, general billing conditions, applicable taxes, penalties, insurance requirements, construction supervision, and project acceptance by the STEG.

The *Notebook on Specific Technical Clauses* includes all general specifications for project construction, such as tolerances when laying out lines, transport and handling procedures for various components, specifications for preparing concrete, installation of line hardware and proper stringing of conductors, and preparation of grounds.

The *Technical Guide to STEG Electricity Distribution* includes all specifications for the design and construction of rural electrification projects and is supplemented by a series of documents detailing Tunisian standards.

The *Global State* (*Mercuriale*) is prepared by the STEG every 12–18 months on the basis of the historical rural electrification costs. These costs are prepared for each assembly that is used in a rural electrification project and include supply costs plus storage fees, overhead, in-country transport from central storage to job site, and installation costs. The *Mercuriale* facilitates the preparation of invoices for construction, equipment, and services rendered by the STEG for its customers, as well as the calculation of project costs, regardless of source of project financing.

The *Tariff Contract* contains unit costs for each task undertaken during project construction as a basis for payment to small enterprises. This document is revised every three years for each zone on the basis of unit costs bid by the large enterprise that is the lowest bidder for large projects in that specific zone (minus the transport costs from central storage to the district, which is the STEG's responsibility for small jobs). Taken together, these guidelines have provided an implementation framework for rural electrification that has reduced costs and raised efficiency considerably through standardization.

Successful Project Implementation and Construction

The STEG's successful implementation and construction of rural electrification projects are based on four major factors:

- the encouragement of private-sector participation during the construction phase,
- development of local industry to supply equipment and material,
- a sophisticated, computerized inventory management system, and
- rigorous commercial practices, including control of nontechnical losses and effective billing and connection payment procedures.

Private-Sector Participation in Construction

Most rural electrification projects are constructed by outside contractors, not the STEG. More often, the STEG's role involves project planning and design, selecting and training contractors, procuring and managing grid supplies, developing detailed standards and guidelines for construction, and monitoring and evaluating completed projects. This approach has succeeded in maintaining low costs and ensuring quality construction, as well as supporting the development of local expertise and enterprises.

Large national enterprises and small local firms alike participate in the construction of rural electrification projects. Those projects whose labor costs exceed TND30,000 require bids. A verification committee, composed of independent evaluators, uses a technically and financially rigorous process to evaluate the bids. For projects whose labor costs are less than TND30,000, the STEG's district office selects small local firms, based on their availability and technical capacity. In 1992, 36% of Tunisia's rural electrification construction was undertaken by large enterprises, 56% by small firms, and the remaining 8% by the STEG itself.

When the drive toward rural electrification first began, the country's few local enterprises lacked the skills needed to construct MV/LV substations and lines. The STEG encouraged these firms to increase their competence by providing trainers from the Sectoral Center for Professional Training (part of the Tunisian Agency for Professional Training). In 1999–2000 for example, this center trained 30 foremen and 63 line workers, who represented firms from throughout Tunisia. This training program has helped to establish a qualified cadre of rural electrification contractors in all regions.

As projects progress, STEG technicians regularly check their adherence to the utility's technical distribution guidelines. The STEG prepares regular project status reports, which are submitted to the regional governments, the MI (which supervises the STEG), and financing organizations. Once projects are completed, a STEG team carries out an inspection to ensure that they conform to the terms of the contract and relevant construction norms. Because the STEG assumes all financial responsibility for subsequent use of the system, these inspections are quite rigorous. The contractor must remedy any inadequacies before payment is made.

Participation of Local Supply Industry

Tunisia's rural electrification program has encouraged the development of national industries to supply its needs. In externally funded projects, local suppliers compete directly with international firms (with a 15% preference given over the lowest international bid), which has pushed local suppliers to improve their product quality and adjust prices to the international market. The bidding process for electricity grid supplies is meticulous. Predefined rules are followed for deadlines, the method of evaluating technical bids independent of price bids, and the method of submitting bids for specialized commissions' approval. These rules guarantee maximum transparency and give suppliers the confidence to make their best offers. Currently, the average share of Tunisian suppliers of grid materials is about 64%.

Now that Tunisia's electrification market is almost saturated, suppliers are turning toward export markets. According to the World Bank (2000a), exporters of electrical machinery are booming, having grown from 1.2% in 1980 to 7.5% in 1997, and are now poised to grow even more. Thus, the STEG's strategy of using local suppliers appears to have not only reduced its own costs but also to have encouraged growth of a national export industry.

Rigorous Commercial Practices

The electricity company also follows rigorous commercial practices in its minimizing of nontechnical losses, billing practices, payment collection, and debt reduction. In Tunisia, nontechnical losses—the financial losses a utility incurs when the power it supplies is consumed but not paid for—are comparable to those of developed-country utilities. In the STEG distribution system, nontechnical losses have been minimized, largely as a result of a customer management improvement program introduced in the 1980s, which reduced losses significantly. For the entire distribution network, in the early 1990s it was estimated that there were only 10.3% technical losses and only 3.1% nontechnical losses, for a total of 13.4% systemwide losses (ESMAP 1990).

In rural areas, fraud and meter tampering are minimal. One major reason is that rural customers have more respect for the electricity utility than do urban consumers. Meters in rural areas have been installed more recently than those in urban areas. Therefore, they are less often damaged so billing problems caused by damaged meters are rare.

The STEG's policy on illegal connections may also be a deterrent. This policy includes frequent, regular monitoring and meter inspection campaigns. Meter readers are rotated regularly among districts. Abnormally low consumption is investigated after generating computerized lists. In addition, bonuses are given for identifying cases of fraud, and strict legal action is taken in such cases. On the technical side, insulated cables are used for networks and supply lines to prevent illegal tapping of power lines by customers.

Customers are automatically billed from two computer centers: one in Tunis and another in Sfax. In the early 1970s, the STEG set up an integrated billing software program whose effectiveness has been proven through rigorous testing. The first customer who requests a connection activates the system. Each customer file is followed closely through connection, cash payment, hookup, and finally metering and billing of consumption. This system allows for daily monitoring of consumption and regular monitoring of installed meters to avoid unaccounted for consumption.

The software used can monitor meter readings and signal any deviation in the bimonthly reading regarding a customer's historic consumption pattern. This system allows the detection of index errors and signals any potential cases of fraud as soon as any unexplained changes in consumption levels occur.

Although low-voltage customers are billed bimonthly, meters are read only every six months in rural areas (compared with every four months in urban areas and every other month for government offices and water pumping). Thus, for rural customers, between each meter reading, two bills are estimated on the basis of the average bimonthly consumption over the past three rolling years. When the meter is read, the actual consumption is calculated, and the amount paid in intermediate bills is deducted. Large customers are metered and billed on a monthly basis. Billing is spread out over time to better divide the handling of customer files and cash flow during the month.

STEG agents deliver statements to customers' business addresses or residences within three to five days; however, this method is expensive. Postal service is also considered unreliable and expensive and faces delivery problems similar to those of the STEG. Both the postal service and the STEG leave bills for more isolated rural customers at the local general store, which serves as an informal post office. This system can result in payment delays and cutting off of service for rural customers.

In most rural areas, customers give top priority to paying their electricity bills. Most unpaid bills originate in the public sector, but, as Table 7-3 shows, payment has improved in recent years. Unpaid bills for LV customers (both rural and urban) represented less than 5% of the STEG's total unpaid bills in 1998.

Payment plans for connection costs are extremely generous because the STEG has learned from experience that rural households can maintain only low payments. When the rural electrification program was first launched, customers had to pay their connection fees over a 10-month period. When even this proved unaffordable for many rural customers, the amount was progressively spread over 40 months in 20 bimonthly payments, and later was extended to 72 months in 36 bimonthly payments, where it remains today. This policy of spreading out payments has greatly reduced the bills of newly connected households; as a result, there are few nonpayments.

Analysis of Customer Service: Problems and Solutions

The STEG has sought technical answers, such as innovative billing practices and the MALT system, to resolve customer service problems. However, little monitoring of customer satisfaction with quality of service has occurred. It is assumed

Table 7-3. *Comparison of the STEG's Unpaid Bills in Tunisia, 1990 and 1998*

	1990		1998	
Unpaid Bills	Tunisian Dinars (million)	%	Tunisian Dinars (million)	%
Total public sector	21.3	81	48.1	79.6
Total private sector	5.0	19	12.3	20.4
LV customers (rural and urban)	3.2	12	2.9	4.8
Total	26.3	100	60.4	100.0

Note: The totals equal the combination of the public and private sectors. LV customers are not included in the totals, as they are part of both the public and private sectors.

that the economic cost of an undistributed kilowatt-hour in a rural area—characterized by low electricity demand—is much less than one in an urban area, that daytime power outages will often go unnoticed by customers, and that economic losses are insignificant.

According to the informal fieldwork carried out for this chapter,[4] power outages, though infrequent, occurred in the villages studied. Some outages were programmed (as part of works in progress), and others were unanticipated (due to natural causes, such as violent weather). Health clinics have complained of not being informed of prolonged outages, which have resulted in spoiled refrigerated vaccines. To protect against such damages, some clinics have had to reduce vaccine inventories or have had to maintain emergency coolers. Communication problems between the STEG and its consumers were also discovered. For example, rural customers have had difficulty contacting the STEG because of out-of-order or inaccessible telephone booths or because they believed that the utility would automatically be informed about the problem.

Voltage fluctuations have damaged domestic appliances and television sets. When regional development authorities and agricultural and agroprocessing customers were interviewed for this study, it was found that voltage fluctuations had damaged electric motors used for water pumping.[5] In the future, such fluctuations could increase as the network expands to include houses located far from MV/LV substations.

Agroprocessing customers in the areas studied were also concerned about lack of access to three-phase power (single-phase power prevails in rural areas). Private silos, usually located near grain fields, require three-phase power because they are fed by electrogenes.[6] Refrigerated collection centers require a power of 15–22 kW and three-phase power. Rural development authorities also mentioned projects that companies are prepared to invest in and that are located in areas where water is available; however, power is limited to the single-phase grid.

Over the past two years, STEG has launched a high-priority effort to improve how customer problems are resolved. Customer service representatives are employed in branch offices to handle customer billing problems and complaints. Moreover, pilot call centers have been set up in certain districts to handle customer inquiries. Additional monitoring of customer needs and service levels is needed in rural areas, which perhaps could lead to educational campaigns for customers and to alternative approaches by the utility.

Financial Sustainability from Grants, Loans, and Revenues

Unlike many developing countries, Tunisia has implemented its rural electrification program without undue stress on government or implementing-agency finances. Four major sources of financing and subsidies have contributed to this achievement. First, during much of the period of rapid rural electrification, Tunisia's economy grew at a fast pace (4–5%), thereby generating adequate budgetary support. Second, decline of investment in electricity generation during the 1980s released funds for rural electrification. Third, rural consumption represented only

4% of total consumption, which minimized the effects of subsidies on operating costs. Fourth, Tunisia had access to loans and grants from a wide range of international donors and agencies.

Rural electrification typically has involved both high capital costs and some type of subsidies. As the grid is extended into new areas, there typically is some type of capital subsidy for system expansion. Once in place, some type of cross-subsidy also is needed to offset the prohibitive costs of providing electricity service to remote communities. These aspects of the program are explained in detail in this section.

Financing Grid Expansion

Grid expansion was achieved through effectively mobilizing the STEG, beneficiaries, and state resources. Although each contributed, the state bore the largest share, either through domestic budgetary resources or borrowing from various international organizations. Since 1977, a formula has been used to define the rural electrification contributions of each of the three funding sources. This formula is similar to some recent output-based aid schemes that have provided incentives for private firms to serve rural consumers. In the case of Tunisia, these incentives have been provided to the public electricity company and have worked quite well.

Beneficiaries are also required to participate in the cost of connections. This participation is fixed at a level of TND200, calculated so that electricity costs less than expenditures on alternative energy sources (candles, kerosene, or batteries). Because initial connection charges are often a barrier to low-income rural families, the STEG spreads them out as 36 bimonthly payments. In some regions, beneficiaries have agreed to contribute more than the required TND200 to expedite household connection (for example, Bizerta's level is TND273, Nabeul's is TND400–600, and Sfax's is TND400).

With regard to the STEG's contribution to grid expansion, a cost ceiling per average connection has been established. This simple, workable formula sets a limit on STEG financial participation and provides incentives to undertake economically justified investments. From the Vth Plan (1977–1981) through the VIth (1982–1986), the STEG contributed up to TND100 per household connection and TND250 for agricultural pumping, thereby providing an additional incentive for the more immediately economically productive activity. However, since 1989, the STEG's participation in household connections increased to TND200, reflecting higher costs and a special national effort to improve the quality of rural life.

For each project, an average cost of electrification is calculated in terms of an upper and lower limit. The lower limit equals the maximum STEG connection contribution and the beneficiary's contribution (each gives TND200 per domestic connection). Thus, projects costing less than TND400 are considered feasible and are financed by the STEG. However, for those projects that cost more than TND400, the state provides a subsidy to the company equal to the additional costs incurred. For such projects, a maximum or ceiling is defined every five years in the Economic and Social Development Plan. For the IXth Plan, this ceiling was set at TND2,200. Those projects that lie between the lower (TND400) and upper (TND2,200) limits are cofinanced by the state under such programs as the PRD and the PDRI. Projects costing more than the maximum (TND2,200) can still draw on special funds available for this purpose (PP, FSN, or a voluntary citizens' rural development fund).

The state, through its various programs, assumes the balance of investment costs not covered by the STEG or beneficiaries. The state's contribution now accounts for up to 85% of total project connection costs, compared to 45% in the program's early years. The practical nature of this subsidy is that by contributing lower subsidies in earlier years, this public utility was encouraged to build a system that would provide them with greater revenues in the early years of the program. As the system expanded, this base revenue resulted in less financial burden on the company.

Sustainable Financial and Tariff Strategies

For long-term sustainability, a rural electrification program must establish a system of tariffs and charges that are self-financing and do not depend on increasingly large subsidies from state revenues. In this respect, Tunisia's tariff policy has avoided many of the pitfalls encountered in other developing countries. The STEG prices power close to its long-term marginal cost and makes considerable efforts to keep rates in line with the cost of providing electricity.[7]

The tariff structure, negotiated between the STEG and the MI, reflects the differing costs in providing electricity supplies to broad customer groups (Table 7-4). Thus, tariffs are lower for high-voltage (HV) industrial customers with high consumption levels and higher for LV customers, which typically are households with low levels of consumption. On the other hand, differences in costs of delivering energy based on location are not reflected in current tariffs. Thus, tariffs are established on a national basis without taking into account, for example, the considerable cost differences in supplying rural and urban households. In this regard, rural household tariffs benefit from a significant cross-subsidy because each new connection costs significantly more than the STEG bills.

According to a STEG-requested tariff study conducted in 1996, HV and MV tariffs, on average, reflect marginal costs of supply. However, LV tariffs were about 10% lower than their long-term marginal costs of supply, despite their being generally higher than HV and MV tariffs.

Table 7-4. *Average Price of Electricity (Excluding Taxes) in Tunisia, by Consumer Group, 1994–1999 (millimes/kWh)*

| Voltage Group | Year | | | | | |
	1994	1995	1996	1997	1998	1999
High	42.3	43.9	43.4	43.7	44.0	44.2
Medium	56.7	58.5	58.7	58.7	58.5	58.6
Low	74.0	76.2	77.1	76.9	76.9	76.7
Average price	61.1	63.1	63.7	64.0	64.1	64.5

Note: 1,000 millimes = TND1.

A second characteristic of the tariff structure is the distinction between peak and off-peak usage in all electricity markets (Table 7-5). In many cases, peak-hour tariffs are almost twice as high as off-peak tariffs.

The low-voltage supply, of which rural users account for 11%, has various tariffs designed to promote social equity and rural development. For example, a low lifeline tariff applies to consumers who use less than 50 kWh per month. These consumers pay 63 millimes (1,000 millimes = TND1) per kWh for the first *tranche* (level), which rises to 90 millimes per kWh for consumption of more than 50 kWh per month. The progressive nature of these tariffs encourages consumers to manage

Table 7-5. *Electricity Tariffs in Tunisia (Excluding Taxes), 2001*

Voltage Level	Tariff	Fixed Charges[a] Subscription (mill./customer/ month)	Power (mill./ kWh/ month)	Energy Price (mill./kWh)[a,b] Day	Peak	Evening	Night
High-tension	4 times a day	—	2,500	42	82	63	29
	3 times a day	—	2,500	44	80	NA	30
	Back up	—	1,000	53	95	68	31
Medium-tension	Uniform	—	300[c]	65			
	Time of day	—	3,000	50	94		
	Water pumping	—	3,000	51	93		
	Agricultural use	—	—	50	Out		
	Pumping for irrigation	—	—	50	Out		
	Back up	—	1,500	63	102		
Low-tension	Economic tranched (1 and 2 kVA)	—	100[c]	63			
	Normal tranche (>2 VA)	—	100[c]	90			
	Public lighting	—	200[c]	77			
	Water heating	400	—	66	Out	66	
	Heating and cooling	300	—	98			
	Irrigation						
	Uniform	300	100[c]	61			
	Time of day	700	—	45	Out	NA	35

Note: mill. = millimes. NA = not applicable.

[a] A value-added tax is applied at the following rates: 18% on all fixed charges and on the energy price (taxes excluded) for all uses except domestic and irrigation; 10% on the energy price (taxes excluded) for domestic and irrigation uses.

[b] A municipal tax is applied at the rate of 3 millimes per kWh.

[c] Millimes per kVA per month.

[d] Below 50 kWh per month; above this, the normal tranche applies.

their consumption to reduce moving into the next higher tranche. Public lighting, which ensures greater public security, benefits from a special tariff.

STEG tariffs are also designed to promote rural development, especially agriculture. Thus, irrigation benefits from the lowest tariffs (Table 7-5). A low off-peak tariff (35 millimes per kWh, compared to 45 millimes per kWh) encourages farmers to irrigate at night. Since the early days of rural electrification, tariff policies have particularly encouraged two activities: oil pressing and milling and grinding. Until 1978, each activity benefited from its own tariff, which was substantially lower than the average low-tension tariff. Between 1979 and 1993, the two tariffs were combined into one that was still lower than the average. In 1994, however, in an effort to simplify, this tariff was aligned with the average low-tension tariff. These advantageous agricultural tariffs are part of a broader program to stimulate rural development, which also includes low-interest loans and subsidies to such projects as irrigation, storage centers for agricultural products, milk-collection centers, and rural industries (including repair shops, bakeries, hairdressers, and weaving sheds).

Unlike tariffs in many developing countries, Tunisia's tariffs are frequently increased to preserve the utility's financial balance. Since 1992, five increases have occurred (7% in 1992, 3% in 1993, 5.9% in 1994, 4.6% in 2000, and 2.4% in 2001), which have yielded an average tariff increase of more than 4% a year. However, this increase is substantially less than the 4.6% cost-of-living increase and therefore represents, in real terms, a decline in overall tariff levels. Tariffs for domestic consumers, including rural consumers, have declined more sharply than the average (by about 16% over the past five years). From 1991 to 2001, the price of the lifeline segment (less than 50 kWh) rose only 6 millimes per kWh, whereas tariffs for consumption above 50 kWh rose 20 millimes per kWh (Table 7-6).

Although the STEG does not provide sectoral accounts, it is believed that over the past decade, the gap between electricity-sector costs and prices has been modest. During the mid-1990s, the costs of supplying electricity may have been somewhat higher than revenues; however, as fuel prices fell in subsequent years, STEG costs and sales revenues probably became aligned.

Overall, the STEG's finances are healthy, with only moderate debt. However, in its accounts, STEG does not distinguish between net profitability of its electricity and gas activities. In the mid-1990s, it is probable that deficits in the overall electricity account were compensated for by gas profits. Although the electricity sector

Table 7-6. *Trends in Low-Voltage Household Tariffs in Tunisia (millimes/kWh)*

	Year										
Tariff Level	1991	1992	1993	1994	1995	1996	1997	1998	1999	2000	2001
<50 kWh	57	59	59	61	61	61	61	61	61	62	63
>50kWh	70	76	79	83	83	83	83	83	83	87	90

Source: World Bank 2005. Rural Electrification in Tunisia: National Commitment, Efficient Implementation and Sound Finances. ESMAP Paper No. 307/05. Washington, DC: World Bank.

appears to have been in balance in the late 1990s, costs have subsequently risen. The price of oil, which accounts for a substantial share of generation capacity, increased sharply, and the cost of connecting households remote from the grid continues to rise. At the same time, tariff increases have not kept pace with inflation.

Financing electricity deficits through surpluses in the gas account may be practicable for a limited period, particularly when gas prices are high. However, the process is vulnerable to changing conditions in the gas market, masking the true financial position of the electricity sector. This type of financing distorts electricity-sector planning and adds to the political difficulty of raising tariffs.

Adoption of MALT—Cost-Cutting Technical Innovations

At the outset of Tunisia's rural electrification program, it was clear that the only way to meet the program's ambitious goals would be to keep investment costs to a minimum. Early on, vigorous efforts were made to cut costs. In addition to the STEG's pursuit of efficient operational and commercial practices, the utility's engineers have continuously developed and adapted technical innovations to Tunisian conditions, thereby reducing the costs of both implementation and maintenance.[8]

Although it is not possible to determine what proportion of Tunisia's rural electrification program has resulted from these cost-cutting innovations, the ADB loan targets from 1979 to 1989 were exceeded by a large percentage for all three rural electrification loans (Table 7-7). In all ADB loans over this period, the length of 30-kV and LV lines, number of substations, and, most importantly, number of new connections far surpassed specified targets. In 1979, a portion of the 72%

Table 7-7. *Targets and Achievements of ADB Rural Electrification Loans to Tunisia, 1979–1989*

Major ADB Loans	30-kV Lines (km)	Substations	Low-Voltage Lines (km)	New Connections
		1979		
Target	500	175	280	17,400
Achievement	910	574	1,375	29,900
% difference	54	330	391	72
		1982		
Target	860	616	605	16,110
Achievement	1,293	1,114	1,531	28,640
% difference	50	81	150	78
		1989		
Target	2,810	2,800	3,900	61,000
Achievement	3,715	3,976	6,590	92,557
% difference	32	42	69	52

new connections was due to devaluation of the Tunisian dollar, but the additional 78% new connections in 1982 and 52% in 1989 were made possible by ongoing, successful reductions in costs. Thus, the cost reductions enabled the STEG to provide rural electricity service to a greater number of consumers.

Commitment to Customized Solutions

One important reason for these cost reductions was Tunisia's early adoption, in the mid-1970s, of a low-cost, three-phase/single-phase distribution system, known as MALT. Unlike most African countries and many other developing countries, Tunisia chose not to adopt the technical standards it had inherited from Europe, which included a three-phase, LV distribution system, which is suited to densely populated areas and heavy loads. Many developing countries that adopted this system, following the advice of European utilities, ended up with a high-cost-per-kilometer distribution infrastructure that was poorly suited to their scattered settlements and low demand levels.

Tunisia's decision to adapt the lower cost, three-phase/single-phase distribution technology used in North America and Australia to its unique environment is arguably the single most important reason for the country's later success in rural electrification (Box 7-1). Wider use of single-phase distribution not only reduced costs dramatically, enabling electrification of far more households within the same budget, but it also fostered in the STEG a unique esprit de corps that grew out of this courageous technical decision. Though much criticized at the outset, it was later proven correct and supported by the political establishment. Moreover, the

---

Box 7-1. Adopting the MALT System: Important Technical Decisions

The three-phase/single-phase MALT distribution system adopted in Tunisia consists of major arteries of overhead lines in three-phase, 30-kV, line-to-line voltage, with four conductors (three phases and one neutral wire) and secondary, single-phase, 17.32-kV, line-to-neutral voltage rural distribution overhead lines (one phase and one neutral wire).

If heavy loads are to be fed, then three-phase lines with four conductors are used. Fuse cutouts protect MV lines. Single-phase transformers give a secondary, phase-to-neutral voltage of 230 V (single-phase, LV lines), which is used by most rural customers. The distribution system is composed of robust materials and equipment that is easy to use and maintain.

When Tunisia adopted the MALT system, it made a second important technical decision: opting for a relatively high, single-phase 17.32-kV voltage, rather than the weak 3 or 5 kV of the North American model. The higher voltage was selected for the single-phase rural electrification overhead lines because of the long distances between villages and the nearest three-phase artery and to provide for future growth over the 30-year lifetime of the lines.

STEG gained confidence through solving numerous technical and related problems involved in setting up the new system. As a result, the utility was motivated continuously to develop and implement vigorous cost-cutting efforts and innovative technical approaches over the following decades.

Steps Toward MALT: Technical and Economic Decisionmaking

When Tunisia's need to accelerate rural electrification became evident in the early 1970s, the STEG undertook a technical audit to assess existing distribution methods, of which there were only two: the North American approach (characterized by widespread use of lines, combined with a three-phase backbone) and the European approach (with extended three-phase lines throughout the service zone). This audit indicated that the predominant European three-phase system was not well adapted to Tunisia's ambitious program of low-cost rural electrification. Given the features of Tunisia's targeted population—low rural incomes, dispersed households, and consumption limited to lighting and basic appliances (mainly refrigerators and television sets)—it was clear that the cost of rural electrification could not be financed solely through tariffs and that limited resources should be invested wisely. This information led the technical auditors to recommend considering a new means of distribution—single-phase lines.

The adoption of the new distribution arrangement certainly was not done without controversy. According to one Tunisian engineer who participated in the program, "Never had a technical recommendation raised as many debates and exchanges of points of view in STEG" (Essebaa 1994). The environment at that time was hostile to the changeover, according to a later ADB report (ADB 1995a), with opposition from both system operators and European partners. However, a technical study for the master plan for distribution confirmed the audit's recommendations. To avoid pitting the European and North American systems against each other, the Tunisians called the new three-phase/single-phase distribution system "mise à la terre," referring to MALT's grounding of the neutral wire.

Having established technical confidence, the decision to change over became an economic question. Thus, economic studies were carried out in several stages during 1974 and 1975. First, a comparative study of distribution systems was carried out in seven typical villages, with positive results for the MALT system, which resulted in 30% savings. Next, the STEG developed a computerized model—an innovation at that time—capable of comparing systems costs in 300 villages randomly chosen from those selected for the Vth Development Plan (1977–1981). STEG staff gathered basic field data on electricity consumption, length of MV and LV lines needed, and estimated future numbers of customers (five years after electrification) for specific end-uses (such as lighting and pumping). Technical assumptions were made about installed power and voltage drops. After gathering the most realistic prices of electrical equipment, these assumptions were used to design and cost-out different scenarios to provide a range of results for both distribution systems.

Results of the model, using data from the 300 randomly selected villages, highly favored the MALT system (Figure 7-2), which projected savings of 18–24% overall.

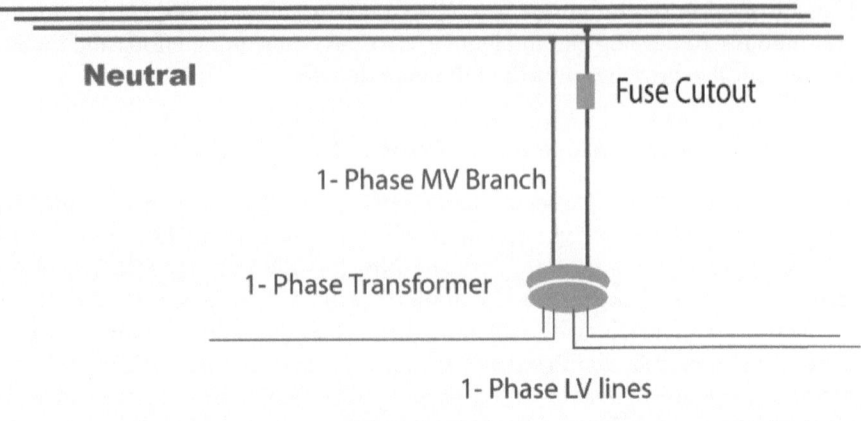

Rural Area

Figure 7-2. *Diagram of the MALT Configuration in Tunisia*

As Table 7-8 shows, the largest savings was at the MV level. In January 1976, soon after these results were made known, the decision was made to switch to the MALT system.

Rapid System Conversion and Resolution of Technical Problems

Once this decision was made, the changeover occurred rapidly, testifying to the STEG's analytical, planning, and logistical abilities. To the extent possible, existing equipment and materials were kept and integrated with the new system to save costs.

The system conversion consisted of two major steps. The first step involved a changeover from the existing 4,000 km of 30-kV grid, which consisted of installing neutral point coils in HV/30-kV substations, laying the fourth neutral wire on the main 30-kV feeder lines, and replacing the constant time protections (relays and current transformers) with reciprocal time protections in HV/MV and ring main

Table 7-8. *Estimated Savings of the MALT System in Tunisia, Compared to the Three-Phase Distribution System, 1975*

Network Level	Cost Reduction (%)
MV network	30–40
MV/LV substations	15–20
LV network	5–10
Overall	18–24

Source: Essebaa 1994.

unit substations. The second step included planning, designing, and monitoring the installation of new construction (lines and single- and three-phase substations) in the MALT system. Both steps posed important questions of technical adaptation, organization, implementation capacity, and customer relations, given the repeated interruptions in supply, which inevitably occurred during the changeover. Box 7-2 gives examples of the types of obstacles that the STEG encountered during the conversion and how it overcame them.

Hydro-Quebec engineers provided technical advice on the three-phase/single-phase system, and short-term technical visits to Canada were organized for district chiefs and system operators and engineers beginning in 1976. However, planning of the new system and resolution of the problems encountered throughout the course of switching to the new system were entirely the work of STEG staff.

The changeover, which was completed in 1980, laid the foundation for launching a vast program of rural electrification in single-phase overhead branch lines. The length of single-phase lines rose from none in 1976 to almost 19,000 km in 2000. As the five-year plans were implemented, the number of kilometers of single-phase lines increased more rapidly than the number of kilometers of three-phase lines, and the single-phase investment grew increasingly dominant. Today, single-phase lines account for 51% of the total network, compared to only 16% in 1981. Similarly, the number of single-phase substations has risen from none in 1976 to more than 22,000 in 2000. Single-phase substations dedicated to rural electrification now account for 70% of all STEG substations.

Technical Solutions for Large-Motor Productive Uses

As the MALT system has advanced and proven its reliability and safety, criticism has diminished, but some negative points are still raised. Single-phase lines present no difficulties for household uses such as refrigerators or color televisions or small motors such as electric pumps or manual tools. However, adaptations and conversions must be made to service large motors above 7.5 horsepower, agroindustrial loads, and deep-borehole irrigation loads. This is a potential problem for large scale industrial development in remote areas, where rural customers must bear the additional costs.[9]

Most industrial development occurs in industrial zones and incorporated villages, which are supplied with three-phase, 30-kV lines. Outside these areas, conversion from single-phase to three-phase lines can be made later, if justified by the load. However, in practice, this has seldom happened in Tunisia.

Technical solutions—widely marketed and practiced in North America—consist of special, more expensive motors, which, for 100-hp loads, can cost an additional US$2,000–15,000. These costs are minor compared to the costs of installing three-phase lines, and it has been recommended to use single-phase lines even in areas with high loads from agroindustry and pumping (Hicks et al. 1993). In Tunisia, however, early experience with single-phase, 7.5-horse power electric pumps was unsatisfactory, which created suspicion among consumers that single-phase lines were somehow inferior. The STEG developed several solutions for rural customers

Box 7-2. STEG's Switch to the MALT System: Typical Obstacles and Solutions for Changing from the 30-kV Network

Obstacle	Solution
Laying a neutral, fourth wire had not been foreseen during construction of the European three-phase network; thus, difficulties were encountered in installing the neutral wire on existing poles while maintaining the required height above ground.	For each type of crossing, considerable imagination and numerous trials and attempts were required to place the fourth wire accurately.
The neutral wire was attached to an LV spool insulator that was later judged inadequate, especially where excessively long spans between poles caused the wires to break.	The LV spool insulator was later replaced by a suspension insulator.
Wires snapped in some spans where the neutral wire had been incorrectly placed, with flashovers occurring between the neutral wire and one of the three live wires.	These anomalies were quickly corrected without significant damages.
The existing, fully saturated current transformers were not well adapted to MALT.	These transformers were replaced with higher performance current transformers.
Difficulties were encountered that were linked to necessary power cuts to replace and adjust protections.	The tripping–reclosing cycles and the automates associated with the new protections were studied, identified, wired, and tested in the laboratory before installation. Field interventions were reduced to installation and connection of a fully equipped panel, which was wired and tested in the laboratory.
Taking the resistant earth protection out of service created much apprehension.	With more experience, it was demonstrated that the resistant earth protection was not indispensable.
The new three-phase (Ynyn) transformers created problems of tank overheating in cases of outages in one live wire.	These transformers were replaced by four-column, magnetic transformers.
The first fuse cutouts and the cabin substation crossing insulators were not suited to the humid climate of the coastal zones or the salinity of Chott El Jérid.	Technical specifications were modified to reinforce insulation of equipment installed in these geographical areas.
Disturbances were encountered in local telephone lines running along long-distance electric lines.	Capacity disturbances were resolved by using filters on the telephone lines; inductive disturbances were eliminated by improving line groundings.

who owned large motors. Still, these solutions are not widely practiced. In all cases, however, the customer must bear the extra costs.

It is difficult to determine to what extent unavailability of three-phase power lines has prevented establishment of productive, large-motor uses in more remote rural areas. In Tunisia today, it is not uncommon for prosperous retirees to return to their rural homes to establish economic activities. Two such examples were encountered in this study's informal rural appraisal: a vineyard and winery under construction had only single-phase connections and would incur considerable costs to purchase motors for both large-scale irrigation and pressing; and a proposed ostrich-raising project would require numerous electric heaters.

Continuing Tradition of Cost-Cutting Technical Innovation

The successful adoption of the MALT system fostered the STEG's aggressive approach to cost-cutting, technical innovation. Throughout the 1980s and 1990s, technical and economic studies and pilot projects were undertaken to further reduce costs of distribution systems. These projects resulted in a number of changes and cost savings, of which examples are given below (roughly in order of importance of cost savings). Though the savings of any one innovation may be relatively modest, the cumulative effect is considerable, testifying to the importance of STEG's culture of continuous improvement.

The STEG has continued to reduce distribution-system costs through further innovations. These include the single-wire earth-return (SWER) and the MALT 4.16-kV, single-phase line. SWER, a variation on the MALT system, was introduced in 1990. It has only one live wire and no neutral wire. Instead, the return current passes through at a grounding point at the end of the line (the MV/LV substations). The technique allows an additional cost reduction of 26–30%, compared to single-phase MALT (according to a 1996 study on village cases). SWER was introduced with a number of precautions because of the potential risks of the returning current to humans and animals if lines are not carefully installed and monitored. In Tunisia, SWER is used in more remote areas, where loads are usually low; hence, the voltage is low and less dangerous. Nonetheless, given the increasing loads in remote areas, effective grounding will need to be carefully monitored in the future.

By late 1996, the feeder lines implemented using SWER as pilot projects supplied 425 villages through 1,148 MV/LV substations. District chiefs have the freedom to decide whether to use SWER in specific rural electrification projects.

The MALT 4.16-kV, single-phase line can reduce the costs of electrifying rural villages in which houses are widely dispersed. As the rate of rural electrification increases, the number of locations with groups of houses decreases, the remaining households without electricity are more scattered, and the average cost per customer increases. This technique also is suitable only for the relatively few projects at the end of the network, where no further extensions will occur. Hence, the gains are relatively small and usually unjustified by the increased management needs of introducing another level of voltage and range of network materials. Nonetheless, district chiefs have the option of choosing to use the 4.16-kV, single-phase line for projects with widely scattered households at the end of the grid.

Thus, the attempts to reduce the costs of rural electrification in Tunisia have been widespread. The company executing the rural electrification program could have taken the conservative approach of overbuilding the systems, as has occurred in many others countries. However, this route would have dramatically increased the investments needed to complete the program.

Photovoltaics: Complementary Strategy for Isolated Users

Tunisia's national PV program underscores the country's commitment to provide at least a minimal level of electrification service to even its most remote rural households, which otherwise would remain unconnected. Interest in PV developed during the early 1980s, based on environmental and social grounds. Several demonstration projects were followed by pilot dissemination projects, which showed that the technology could contribute to meeting the basic electricity needs of isolated rural households and that individual PV systems were better adapted to isolated households than centralized systems, biogas, or grid extension. Major projects included the Gessellschaft für Technische Zusammenarbeit (GTZ)-funded project in Kef and a state-financed rural schools program.[10]

Currently, 7,750 households (about 1% of total electrified rural households), 200 schools, and a few clinics and forest and border posts have PV installations. The grid and PV programs are complementary, and in some cases, PV has become an interesting alternative to the grid. At a connection cost of TND1,900 per household, PV compares favorably with grid-connection ceiling costs of TND1,500–2,200, or even TND2,500 for FSN projects.

The ANME rather than the STEG is the implementing agency for the PV program. Although the ANME has principal responsibility for Tunisia's renewable energy policy and promotion, the Ministry of Economy, in 1993, designated the ANME to play the lead role in PV rural electrification. Since then, the ANME's implementation role has continued. Though the roles of the STEG and the ANME differ, their work is closely coordinated by the CNER, under the aegis of the General Directorate of Energy, which includes representatives of the Ministries of the Interior, Economic Development, and the Environment; FSN; and the STEG and the ANME.

In Tunisia, electricity is viewed as a minimum public service to which every household has a social right. More than 90% of the country's PV rural electrification program is subsidized. Beneficiaries are required to pay TND100 per system, with TND200 financed by the regional government, and the remaining TND1,600 financed by state sources. Currently, consideration is being given to increasing the amount that beneficiaries pay because 20% of system costs today would equal TND500 (TND100 represented 20% in 1990). The largest funding sources are PV-module exporting countries, which have provided supplier credits for some 50% of the PV systems installed to date. The World Bank has provided loan credits for another 25% of installed systems. National development funds, NGOs, and beneficiaries have contributed the remainder. Clearly, the success of national PV rural electrification depends heavily on the availability of credits and subsidies.

Tunisia's PV rural electrification program has sought to meet user needs in several ways. First, system sizes have been increased, initially from 50 W to 70 W, with the present standard now at 100 W, in recognition of greater power needs and less insulation during winter. This equipment feeds a continuous 12-volt current, which can power three light bulbs, one black-and-white television, and one radio-cassette player. Still, surveys have shown that the daily consumption level—up to six hours per day for lighting and television viewing and three hours for radio-cassette player use—is 300 Wh. Households regularly overload their systems, sometimes leading to regulator-induced outages to protect the accumulator battery. To avoid such outages, users connect their televisions directly to the battery, resulting in further damage.

Thus, rural electrification in Tunisia will reach even the most remote households in the next 10 years. Because the cost of serving them with grid electricity would be extremely expensive, the policymakers in the country decided that it would be best to provide electricity to remote populations with PV systems.

Lessons in Integrated Rural Development and Social Equity

Tunisia's achievement of 100% urban and 88% rural electrification is remarkable, all the more so because the country's definition of rural electrification is restricted to connections made outside incorporated areas. Compared to rural populations in other developing countries with high rates of electrification, Tunisia's rural population—although only 35% of the total population—is highly dispersed and isolated, with long distances between small groups of often scattered houses. This characteristic, combined with the social commitment to connecting all households, has highly influenced the rural electrification program's costs and choice of institutional setup, distribution system, and technology.

National Commitment

Tunisia's rural electrification achievement has been motivated by continuing national commitment as part of a broader, integrated rural development program that has emphasized social equality. Since its independence from France in 1956, the country has been at the vanguard in promoting human resources development, particularly gender equity. This is evidenced by the PSC, which was promulgated immediately after independence and the IVth Development Plan, implemented in 1972, whose three pillars were basic education (for girls and boys), improved health services (with an emphasis on family planning), and rural electrification (whose socioeconomic criteria included gender equity).

Integrated Rural Development Context

Regional planning processes and successive five-year plans, which have tightly incorporated rural electrification into broader integrated rural development objectives, have produced synergistic effects. Indeed, growth in rural electrification and national socioeconomic indicators are strongly correlated. Informal surveys in several rural areas attest to the multiple benefits of rural electrification—as perceived by rural householders, especially women—in education, health and family planning, economic opportunities, and enhanced security (Chaib et al. 2001). Integrating gender equity into the socioeconomic criteria for rural electrification has been an important factor in state support for, and subsequent success of, rural electrification.

Effective Institutional Approach

Regardless of the structure or process that a country adopts for rural electrification, certain principles are essential to success. These principles include well-defined, coordinated roles for all agencies concerned and established procedures that ensure agency cooperation that is perceived as being fair. The Tunisian system scores well on both counts. All agencies that participate in Tunisia's rural electrification program have well-defined roles. Equally important, policymaking and implementation agencies at both regional and national levels collaborate closely. Agency cooperation is facilitated through a project-selection process that is meticulous, orderly, and transparent. Through this process, concerns about social justice are addressed, thereby reducing political pressure in identifying projects, allowing for a more rational and economic long-term program.

Utility's Success: Transparency and Innovation

The STEG's effectiveness and efficiency have earned it both political and popular support. Much of the utility's success can be attributed to a clear mandate and a management structure that combines the benefits of centralized planning and design with decentralized operations. Published norms, guidelines, and standard contracts contribute to operational transparency.

The STEG has demonstrated a high-level capacity for adapting technology to meet Tunisia's clearly-defined rural electrification objectives. Early on, the utility computerized its management systems and developed customized software applications, including a sophisticated inventory management system. Introduction of the MALT three-phase/single-phase distribution system has dramatically demonstrated the STEG's high level of innovative technical expertise. Indeed, the utility's switch to the MALT system has been the single largest change introduced into the Tunisian program, permitting rapid expansion of rural electrification. In addition, the MALT system has provided a high level of service by reducing the rate and duration of outages.

The STEG's implementation of commercial practices (including control of nontechnical losses, billing, and collection practices) has been outstanding. Despite difficulties of delivering bills to isolated communities and their limited means of payment, rural consumers have an excellent payment record. Success factors include a customer management improvement program that has focused on sound meter-reading policies and practices, development of an integrated billing software program, and spreading out connection-cost payments. Successful construction and implementation of rural electrification projects owes much to encouraging private-sector participation in construction and promoting local industry efforts to supply equipment and materials.

Robust Financial Arrangements: Effective Tariff Policy

Tariffs broadly reflect the varying costs of supplying high-, medium-, and low-voltage customers. All markets distinguish between off-peak and peak usage to encourage more efficient capacity use. LV supply, of which rural users account for 11%, has various tariffs designed to promote social equity and rural development. These include a lifeline tariff for those who consume less than 50 kWh per month, subsidized public lighting, and low tariffs for irrigation. Such tariffs benefit from a significant, yet apparently manageable, cross-subsidy. Although the STEG does not publish detailed power-sector finances, it is believed that over the past decade there has been only a modest gap between electricity-sector costs and prices.

Complementary PV Strategy

Tunisia's high-profile PV program—with its goal of providing a minimum 100-W level of electricity service to all households by 2010—reflects a commitment to including even the most remote rural areas in national development. The program features the high-quality technical support and robust finances that have characterized the country's rapidly expanding grid program. Success factors have included close institutional coordination with the STEG; careful selection and adaptation of equipment; strong domestic and international donor support; and an emphasis on user needs, maintenance, and after-sales support.

Conclusion

Africa has the lowest rates of rural electrification in the world. In most countries, the rates are 10% or less. But the question also arises as to whether the considerable technical expertise of the STEG and its related partners can be shared with other rural electrification programs in Africa.

In this regard, the MALT system has attracted the attention of various African countries. Both Senegal and Mali have sent their technicians to the STEG for training or to obtain information that they can potentially apply in their countries.

At the request of Madagascar's Ministry of Energy, the STEG carried out a study for a pilot project in that country, and technicians from Madagascar's utility have participated in STEG training courses. Indeed, encouraging the STEG's technical assistance to rural electrification programs throughout Africa and to other developing countries worldwide—through both bilateral and multilateral programs—is an interesting option.

As total rural electrification rapidly approaches, Tunisia still faces many new challenges. Changes toward democratizing Tunisian society may create pressures for greater consumer participation in sectoral decisionmaking and the need for better communication between the STEG and its customers. In theory, Tunisia's project-selection process is transparent and minimizes political pressure, but in practice, it may be criticized for verging on the mechanical. This is especially so in cases where local costs diverge from the national average used to estimate the total costs of rural electrification. Finally, as electricity is provided to the last 12% of the rural population without it, the respective roles of the institutions providing PV systems and grid electrification will require clarification and improved coordination.

Despite these challenges, the program in Tunisia, with its emphasis on coordinating access to electricity with rural development, has been quite a success. The solutions to problems that it faced throughout the past 30 years may not provide a precise blueprint for other countries, but they certainly can provide useful insights. The success of the program in Tunisia may be a beacon for other countries that want their rural populations to have the modern benefits of electricity but are having difficulty figuring out how to approach the problem.

The enabling political and socioeconomic conditions that have contributed to the STEG's success may not be replicated in other countries. Nonetheless, the STEG can provide useful lessons—even in unpromising situations—from its experience in adaptive technology, robust finances, and transparent project selection.

Acknowledgments

The authors are particularly grateful to Raouf Maalej, Anis Besbes, and Radhouane Masmoudi of the STEG Distribution Directorate and to the STEG District Chiefs and offices for their valuable contributions and comments. The authors would also like to thank Neiji Amaimia and Amor Ounalli of the ANER for their analysis of Tunisia's photovoltaic experience. Thanks also go to N. Hammami for his comments and to the ANER office in Kef, particularly M. Njaimi, for field assistance. Thanks also extend to Fundani Matondo of the African Development Bank, who, as Project Manager of the Rural Electrification Program for Tunisia, has also provided all the reports related to the projects financed by the AfDB. Finally, the authors thank Gerald Foley and Robert van der Plast of the World Bank for reviewing drafts of this chapter, and Allen Inversin of the National Rural Electric Cooperative Association (NRECA) for his useful insights and comments.

Notes

1. This section draws on ADB 1995a, 1995b, 1999; Berthier 2000; Republic of Tunisia 1999; STEG 1980-1999, 2000; and Essebaa 1994.

2. This section draws on Bouraoui 2001; *International Herald Tribune* 2001; Nordic Consulting Group 1997; and World Bank 2000a, 2000b, 2001.

3. This section draws on Hicks et al. 1993; FAO 1993; STEG 2000; and ADB 1999.

4. Chaib, Aissa, and Ounalli 2001.

5. One disadvantage of electric water pumping for agricultural use is that it rapidly exhausts the water table. For this reason, it must be carefully monitored. The water authority sets maximum outputs, which farmers cannot vary according to their needs; therefore, a reservoir must be constructed. Some farmers and water associations also complained about bimonthly billing (they would prefer monthly bills), the inconvenience of peak-load management periods, power-factor penalties, and taxes on electricity bills.

6. An electrogene is an electric generator that is usually found in a remote area and is not connected to the main transmission grid.

7. This section draws on STEG 1980-1999, 2000.

8. This and the following two sections draw on Belkhiria 1996, 2001; STEG 1996; Essebaa 1994; ADB 1995a; and Masmoudi 1997.

9. This section draws on NRECA 1999.

10. This section draws on AME and GTZ 1999; ANME 1996, 1998, 1999; Amaimia and Ounalli 2001; and ESMAP 1996.

CHAPTER 8

Rural Electricity Subsidies and the Private Sector in Chile

Joseph Andrew McAllister and Daniel B. Waddle

C HILE'S LONG HISTORY OF RURAL ELECTRIFICATION began in the late 1930s, when rural electric cooperatives (RECs) were formed to support agricultural development in the fertile lands surrounding the country's regional capitals. Low population density and the rather narrow valleys in which agricultural projects supported themselves largely limited the size of the cooperatives. In the early 1980s, the country's public electricity distribution companies that served regional district capitals were divided and sold to private investors. The cooperatives remained as distribution companies at this time. Both the private distribution companies and the cooperatives have nonexclusive rights to serve customers. This means that in theory the companies could compete for each other's customers, but in practice this happens for only larger commercial or industrial clients.

The privatization program established a new ownership structure, but it did not provide incentives for private nor cooperative service providers to expand service to less populated areas. Expansion of service coverage proceeded gradually, so that by 1990 rural coverage remained below 50%, with no formal government initiative designed specifically to increase access to electric service in rural areas.

In 1994, to encourage private companies to increase rural coverage, the government of Chile (GOC) initiated a concerted effort to increase rural electrification from 50% to 75% by the year 2000. The administration of President Eduardo Frei established both regional and national goals for rural coverage and concurrently attempted to rationalize the use of government subsidies to achieve them. The result was the establishment of Chile's rural electrification program (Programa de Electrificación Rural [PER]). The program has been quite successful in meeting the challenges of rural electrification through private-sector companies and existing cooperatives. The presence of a competitive environment, combined with the willingness of many mature distribution utilities to participate, has been important to the success of the program.

Several factors contributed to the success of the new government program to expand access to electricity. First, the rural electrification program had reliable infrastructure for generation and transmission services at its initiation. Also, the distribution companies were already mature, so the program was based on well-established construction practices, experienced construction service providers, and an already existing materials industry. On the government side, the program was implemented through a competent and motivated planning agency with sufficient authority to develop and guide a national electrification policy initiative. The development of policies that were applied fairly and equitably also enhanced program success. In this respect, the government agency had well-defined project-selection methods, and there was a strict adherence to this methodology in project selection. The PER also was aided by a sustained political commitment that allowed it to be implemented over a long period.

By 1999, a year earlier than planned, the PER had achieved 75% rural electric coverage. At the same time, by encouraging management improvements and facilitating stronger negotiation between governments and utilities at the local level, the proportion of state subsidy had declined over time. Building on this success, the new administration of President Ricardo Lagos has extended the program, and now it is well on its way to achieving 90% coverage by 2005.

Background on Chile

Chile's population of more than 15 million is about 86% urban and 14% rural. Almost one-third of the country's residents live in Santiago. Communities in Norte Grande and the southern regions are isolated, separated by vast, unpopulated areas. The country's major ethnic groups include *Mestizo* (persons of mixed Spanish and Native American ancestry) (66%), European (25%), and Native American (7%), as well as various indigenous communities (2%). Spanish is the official language, with limited use of Native American languages. Roman Catholics make up about 77% of the Chilean population.

Chile is one of the world's narrowest countries, extending some 4,270 km along South America's Pacific coast. The country is about twice the size of California. The landscape is dominated by the Andes Mountains, which stretch from the Bolivian plateau in the north to Tierra del Fuego in the south.

The terrain of Chile makes it difficult for providing rural infrastructure. Both electricity and roads are difficult to establish and maintain in such an environment. Of Chile's more than 79,000 km of roads, some 14% are paved. The Pan American Highway runs the length of the country, forming the backbone of the road system. Railroad lines are limited to the northern two-thirds of the country. Many coastal areas rely on water transportation, and the country also has 725 km of navigable inland waterways.

Chile is at a relatively advanced stage of economic development. The level of economic development is reflected in the annual gross domestic product of Chile, which is about US$4,000 per person. Its educational system is quite advanced. The state-operated University of Chile in Santiago is internationally respected. In

1997, the nation's literacy rate was 95%. The number of people living in poverty decreased by almost one-half between 1990 and 1995. As of 1995, 91% of residents had access to safe water. By 1997, life expectancy had risen to 72 years for men and 78 years for women. In 1999, infant mortality had declined to 10 per 1,000 live births. During the 1990s, production and consumption grew about 8% per year (Energy Information Administration 2000).

History of Rural Electrification

Rural electrification programs have a long history in Chile. CGE, the Compañía General de Electricidad, electrified regional capitals in the 1930s. Not long thereafter, rural electric cooperatives were formed to support agricultural development. Even after decades of gradual expansion, the largest of the rural electric cooperatives had less than 25,000 customers.

Steps toward Deregulation in the 1980s

The first step toward deregulation of the electricity sector began in the early 1980s; it involved the breakup of the country's largest state-owned electric utility, known as Chilectra, into several companies. What followed over the next decade was the emergence of broadly diversified, private conglomerates whose core activities centered on electricity generation and distribution (Urrutia 2000). In 1995, one of those conglomerates, Enersis, gained control of Chile's largest electricity generation utility, Endesa, to become the largest electricity-focused holding company in South America (Walker 1998). Before 1995, Endesa's largest business had been Chilectra Metropolitana, the distribution company serving metropolitan Santiago.

In rural areas, the shift toward deregulation was no less fundamental. Thus, in the early 1990s there was a diverse group of retail service providers. This group included some six larger investor-owned utilities (IOUs) and more than two dozen smaller distributors, many of them RECs. Then, as now, the IOUs were generally better organized, better capitalized, and more aggressive in their business outlook than the RECs.

Decentralization of Infrastructure Planning

Concurrent with deregulating the electricity sector was a move to decentralize infrastructure planning, which reinforced the regional character of the distribution utilities. Many administrative and managerial activities, previously under the central government's purview, moved to the regions. Ahead of their time, but in keeping with the now familiar model for healthy decentralization, Chile's regional governments took on the daily duties of infrastructure planning and oversight. The central government's role became one of defining overarching policy frameworks, allocating funding across regions, and developing planning tools for regional application.

The central government largely succeeded in encouraging professional sector-specific planning and in building regional administrative capacity functions within the context of each elected administration's broadly defined political goals. Some friction—both cultural and decisionmaking—exists between the central and regional governments. In addition, each region has its unique political landscape, which influences regional investment priorities. President Eduardo Frei's administration, which came to power in 1994, promoted market competition, based on the idea that fair competition allows consumers to obtain quality services at low prices (Jadresic 2000). Importantly, regional competition among electric utilities provided added impetus for the regional governments to "get it right."

Shift toward Consolidation

During the late 1990s, consolidation proceeded rapidly at all levels of Chile's electricity sector. Currently, there is some concern that concentration in the generation sector has become dangerously high. Every four years, the national energy commission (CNE) calculates node prices using a BAT methodology. Node prices are used to determine maximum tariffs at the distribution level for each area in the interconnected grid. These calculations are extremely sensitive, and CNE makes a great effort to maintain propriety and impartiality in gathering and validating generation and transmission cost data from all participating utilities, as well as in documenting power flow models used and their parameters.

In addition, international ownership of the generation and transmission systems has increased; Chile's largest generators were recently purchased by multinational firms. As of late 2000, Endesa-Spain was the principal owner of Enersis, which owns Chilectra, the largest generator in Chile. Chilquinta, the country's third largest generator, is owned by Sempra Energy and PSEG Global. Transelec, the country's principal transmission utility, was purchased in 2000 by Hydro-Quebec, a large Canadian utility.

Concentration at the distribution level has been slower, but fewer owners controlling an increasing share of distribution service territory has resulted in some level of concern. Over the past decade, the number of electricity distributors has decreased by one-third, and many smaller utilities have been absorbed by their more powerful competitors. Chilquinta, Sociedad Austral de Electricidad (SAESA), and Emelectric have been the most active regional consolidators. In the process, the RECs have been hit especially hard; many have been purchased by their IOU competitors.

By 1994, the CNE, the principal government entity responsible for setting energy-sector policy, had decided that the cooperative-based model had run its course and had turned its attention to the IOUs (CNE 1994b). A few years later, somewhat ironically, CNE staff softened their attitude considerably toward the cooperatives because they had gained a practical appreciation of the challenges involved in promoting private investment in rural electricity. However, the CNE's subsequent efforts to include the RECs in regional rural electrification activities did little to stem the slow demise of the cooperatives.

During the mid-1990s, the larger RECs restructured their operations to achieve greater efficiency. In some cases, they created holding companies, within which their distribution companies became for-profit corporations distinct from their other businesses. By the late 1990s, many RECs had been absorbed by their IOU competitors. In Maule, Chilquinta (now owned by Public Service Enterprise Group of New Jersey) took over the Curicó, LuzAgro, and Parral RECs, and Emelectric absorbed the Talca cooperative (now an Emelectric subsidiary called EMETAL); in Los Lagos, SAESA purchased the Osorno and Paillaco cooperatives (Castillo Antezana 1999). Similar tendencies prevailed in other regions.

The CNE's performance-based regulatory strategy has had the expected effect of market consolidation. Regulation monitors technical performance standards and sets price caps for various tariff categories. If distribution service providers are unable to operate sustainably under these circumstances, then CNE expects that changes will be made within the service provider. From the commission's perspective, the consumer's interest has a much higher priority than supporting low-functioning distribution utilities.

This approach seems to have worked well during the initial phase of regulatory reform, but there is increasing concern that consolidation may go too far. In recent years, mergers and buy-outs of the larger distribution companies have been rumored, which could reduce rather than engender competition. However, with the PER continuing to achieve its goals with reasonable connection costs, it appears that the electrification strategy will not be threatened by further consolidation.

Early Role of the Cooperatives: Rural Services Provider

Chile's rural electrification efforts began in the late 1930s, when, in similar fashion to the RECs in the United States, Chile's cooperatives were small and focused on providing power to energize agricultural loads. Unlike the U.S. model, however, Chile's RECs were not supported by an agency charged with providing ongoing financial resources for strengthening and expanding electricity service. As a result, expansion beyond the original service territories was slow, and cooperatives remained relatively small.

In the early stages of rural electrification, particularly in the southern regions, the RECs played a central role. In many mid-size cities, as well as in most rural areas, the cooperative framework provided the organizational structure needed to pool resources from interested sectors of society in order to provide large capital investments for constructing and managing the nascent electric system. Most RECs also provided ancillary rural services and products, which allowed them to maintain reasonably low electricity rates.

Typically, along with electricity an REC offered credit for domestic appliances, agroprocessing equipment, and seed, fertilizer, and farm supplies. The repayment for this equipment could be made part of a consumer's monthly bill. The REC credit terms were often more favorable than those of local commercial banks, and they also offered flexible monthly or quarterly repayment periods. It also sold electrical machinery, appliances, stereos, televisions, vehicles, and even toys on credit to those members wishing to purchase such items. In addition, it provided to communities water and irrigation services, as well as health and educational support. These services and products were sold to members using the cooperative principle of demand aggregation to secure lower prices for them.

During the country's first push toward industrialization of agriculture, cooperatives provided these important credit services in response to the needs of their members. For example, in Region VII, LuzAgro, a cooperative based in the town of Linares, provided electricity to power the refrigeration and processing equipment for the burgeoning fruit industry. The producers were members of the cooperative, and LuzAgro was able to supply the equipment as well. Similarly, in southern Chile, many cooperatives grew with the success of the dairy industry, including Llanquihue Electric Cooperative (CRELL) in the Lake District of Region X and SOCOEPA, based in the town of Paillaco in Region VII. (CRELL and SOCOEPA continue to function as cooperatives.) In the early period of Chile's rural development, diversification of services was healthy for the cooperatives and advantageous for their members.

Effects of Privatization on Rural Cooperatives

Chile was one of the first countries in Latin America to make privatization of the electricity sector a priority. The electricity law of 1982 defined the terms under which the newly privatized sector would operate. It stipulated that, to use public property legitimately in its operations, an electric utility had to become a concessionaire. In the late 1980s, regulators and other national and regional officials increasingly led cooperative management to believe that, if the RECs did not apply to become concessionaires, their access to public rights of way would be threatened. By 1990, most of the major cooperatives had become concessionaires. As a concessionaire, a distribution utility became subject to the regulated tariffs promulgated by the CNE. It required concessionaires to submit detailed financial and operational information to regulators for the calculation of these tariffs.

CRELL did not apply for concessionaire status because it did not rely heavily on roadside rights of way for the passage of distribution lines and the construction of substations. As a result, CRELL has not faced the fiercely competitive environment experienced by other cooperatives and retains some freedom in its tariff and business structure. Planners at Llanquihue (Region X) believed the costs of applying to become a concessionaire outweighed the benefits; for example, although the utility was responsible for serving users along the corridor of concession in perpetuity, its rights were nonexclusive.

To minimize subsidies for nonelectric services and thereby minimize the cost of electricity service to all consumers, CNE also required that distribution utilities follow strict rules regarding the separation of energy and nonenergy business transactions. CNE did not allow cross-subsidies between business units to ensure that all distribution utilities would report costs of electric distribution operations without mixing other business costs and income in financial reports submitted to CNE for ratemaking or other purposes.

The transparent accounting that resulted revealed many inefficiencies in the cooperative electric system operations. Downward price pressure exerted by the

CNE as a means of providing more reasonable energy prices to consumers, combined with the exclusion of nonelectric system income, resulted in financial losses for a significant number of cooperatives. By 1999, 6 of the 13 cooperatives had been sold to IOUs, and others continue to experience difficulties.

During the late 1980s and early 1990s, the larger cooperatives formed holding companies, the shares of which were held by the cooperative membership. The holding companies, in turn, established a number of wholly owned subsidiaries, which included a series of business units that had previously operated under the single cooperative banner but were now required to sustain themselves as independent corporate entities. Most cooperatives formed companies that separated the electricity from the nonelectricity businesses. These cooperatives included an electricity distribution unit to manage all electric energy commercialization, a construction unit to manage construction of new infrastructure, and a commercial enterprise unit to sell agricultural goods, appliances, vehicles, and other items of consumer interest.

Comparative Advantages of the Private Distribution Companies

The IOUs that were formed after privatization of the state electricity distribution system have aggressively expanded service in direct competition with the CGE and many RECs. CGE has not participated substantially in the PER, and neighboring IOUs, including SAESA and Emelectric, have aggressively pursued PER funding. They have far outperformed the RECs, even though the cooperatives would appear to have been better positioned to develop project portfolios and to expand their service coverage.

At first glance, rural electrification markets may not have appeared to represent a significant opportunity for the IOUs. It seems that the principal reason for their engagement was to establish themselves strategically to take advantage of future load growth. Over the past five years, load growth in Chile has averaged 8%, but most of it has come from urban areas. Even so, if a company can add infrastructure at a fraction of its actual cost even though load density might be low at the outset, this situation could prove to be an effective strategy once the economy in the new service territory begins to grow more rapidly.

The IOUs also had long-established sister corporations that provided such services as engineering, procurement, and construction of new projects for the distribution companies. The nature of the subsidies designed for PER (discussed in the next section) meant that participants could profit from project construction and electricity sales. In addition, the companies could take advantage of a fiscal credit equivalent to the full project value, whereas because of the subsidies provided under PER, the cost to them was only a small percentage of actual cost. In short, these short-term benefits made the projects attractive to the IOUs.

The cooperatives could also avail themselves of these same benefits, so it is somewhat surprising that the IOUs have been far more successful in competing for PER program funds. The RECs were much smaller than the IOUs, and they had far fewer institutional resources to enable them to compete effectively. They also lacked experience in competitive markets. By contrast, the IOUs were well managed and willing to take reasonable risks to open new and attractive markets.

They had more engineers, financial analysts, and financial resources available to develop projects. In most cases, the quality of their proposals and the price competitiveness far exceeded those of the RECs. Their larger size enabled them to lower their stated subsidy requirements, in effect, underbidding the cooperatives to win the most contested projects. Their large size also helped them to recover funds lost in such competitions by obtaining higher subsidies on projects for which there was no competition. For embattled cooperative directors, one of the ironies of competition is that the IOUs knowingly charge the state higher subsidies where they have a monopoly and use their size to make lower subsidy demands in cases where competition is intense, effectively creating their own cross-subsidy. Finally, the IOUs understood how to lobby the political system, an area in which the RECs were relatively inexperienced.

In summary, the IOUs have been far more significant PER players than have the RECs. As discussed in the next section, IOU participation is a principal reason for the PER's success, which is predicated on competition. Had the IOUs not shown an interest in the PER, the cost of construction per customer would likely have been much higher and the impact less. However, in fairness to the cooperatives, it should be noted that the level of PER investment increased dramatically at the very time that they were forced to restructure themselves and focus on cost-cutting across all operational fronts. The end result was that almost half of the RECs did not survive, and of those that did, only a few were subsequently involved in expanding service through the PER.

PER Origins, Design, and Implementation

In 1994, when Eduardo Frei was elected president of Chile, the general feeling among the populace was that the country's emergence as a strong democracy and the recent gains based on market-opening economic reforms required increased emphasis on equitable distribution of wealth. This feeling had largely been ignored by the previous military regime. In response to public sentiment, one of the Frei administration's first actions was to establish an aggressive rural infrastructure investment program aimed at poverty alleviation, and at its centerpiece was the rural electrification program.

The focus of the PER and complementary rural infrastructure programs was expansion of rural *residential* service. As a consequence, PER's goals and indicators of success were linked directly to residential connection targets. Although impact on local economies and business development was intended, at the time of the formation of the program it was of secondary importance. PER's policy also aimed to complement other development programs by emphasizing the electrification of schools, rural health clinics, and churches as an integral part of the projects it funded.

Origins: Rationale for Government Involvement

PER architects were confident about dedicating resources to the program. Electrification was viewed as a relatively good public investment. A subsidy mechanism already existed for electrification, meaning it was possible to restructure the existing program rather than having to initiate a new one. The expansion of electricity to rural areas was viewed as having both social and developmental benefits, helping to integrate them into the rest of the country. It is an interesting twist that Chile's development indicators were designed in such a way that having electricity increased a family's score on the poverty index scale, thus providing the government an efficient pathway for monitoring the effect on development. Finally, a successful rural electrification program was viewed as having potential political benefits.

The Frei administration established a six-year goal of increasing rural electrification rates to 75% by the year 2000. This goal would require serving some 120,000 households, many located in remote areas. As Figure 8-1 shows, most households without electricity were located in a small number of regions. More than 80% of the households without electricity in Chile were in the southern regions, where mountainous terrain and severe weather cycles significantly increase construction and operating costs.

The focus on poverty alleviation is another important key to understanding the nature of the GOC's commitment to rural electrification. From the beginning, it focused explicitly on assisting projects that the private sector would not undertake on its own. Thus, the PER focused on projects with negative financial rates of

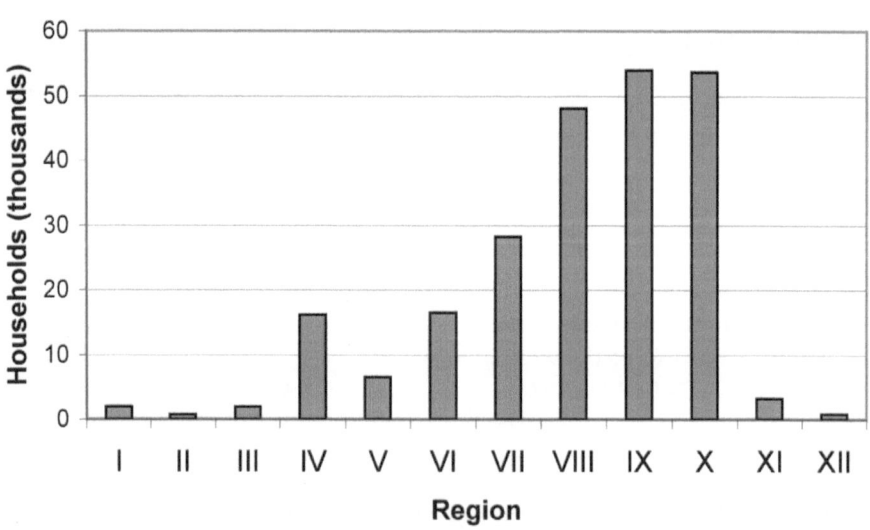

Figure 8-1. *Rural Households without Electricity by Region, Chile, 1992*

Note: Regions I-XII are presented in north-to-south order; metropolitan- Santiago falls between Regions V and VI.

Source: CNE 1994b.

return considered unacceptable to private and cooperative distribution utilities. The key to a successful PER, then, was structuring public subsidies in an efficient way while causing as little market distortion as possible.

Many economic development models have not definitively established a relationship between access to electricity service and poverty alleviation. However, national demographic surveys have established at the least a correlation between poverty and access to basic public services such as potable water, primary education, primary health care, and access to electricity service. This correlation means that without access to these services, rural families are considered to be living in poverty.

In Chile, this development model has been accepted. From its very beginning, PER officials acknowledged the links between rural electrification, productivity, and economic development, but they chose to focus on poverty alleviation rather than quantify the real economic gains achieved through rural electrification. Such gains might include a minimum increase in rural income, new business starts, or specific increases in productivity. Setting goals based exclusively on the number of connections made simplified the need to measure specific effects of electrification but prevented the true economic value of the program from being understood.

Design: Subsidy Allocation Methodology

The centerpiece of the funding process is the evaluation methodology developed by the Ministry of Planning and Coordination (MIDEPLAN). To be eligible for PER and National Fund for Regional Development (FNDR) funds, a given lifespan of a project's economic benefits must exceed total costs, but the lifespan of a project's financial benefits must be less than total investment and operating costs (CNE 1994a). Thus, a project with a sufficiently positive financial internal rate of return would not be considered for subsidy; in theory, the market should dictate that an interested utility would execute such a project on its own. However, in rural Chile, private utilities on their own initiative have only rarely financed such projects because of the exceedingly high cost of construction relative to projected energy sales.

The MIDEPLAN project evaluation methodology consists of a three-part process. First, an economic analysis is performed of the project. This analysis is based on an analysis using demographic data derived from the most recent census performed in the project area. For both economic and financial present-value calculations, demand per capita is assumed to grow at a latent rate of 1.5% per year, and population growth is included per historical rates in the project area. Second, a financial analysis of the project cost and financial rate of return is performed. Finally, given the results of the project's financial rate of return and the economic rate of return previously calculated, a maximum amount is set for the project subsidy.

The first phase of screening of each project is its economic evaluation (or "social" analysis to use the Chilean terminology), which is taken as indicative of the project's net economic value as a function of existing demographic patterns and conditions of the project site. Two demand points are estimated, each characterized by a consumption Q and a unit price P (Figure 8-2). This figure

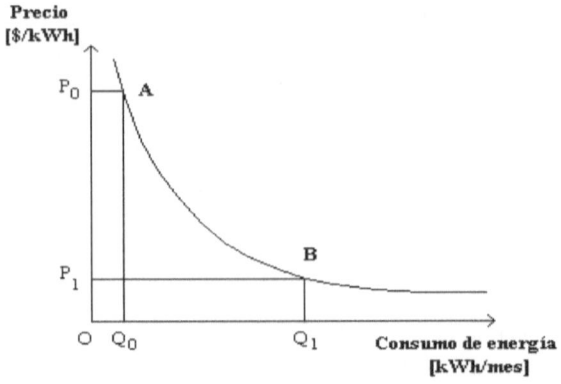

Figure 8-2. *Conceptual Demand Curve in Chile*

Source: Ministry of Planning and Coordination of Chile 1996.

appears in the Mideplan–CNE methodology document (Ministry of Planning and Coordination of Chile 1996), and a variation of this technique is well accepted and is commonly used to assess the project benefits of rural electrification. The existing situation—the baseline situation without electrification—is given as Q_0, P_0 (Point A). Q_0 is estimated through surveys that identify the quantities and forms of energy currently in use that would be displaced by electrification, in kilowatt-hours. For example, kerosene lanterns or wick lamps may be present, candles are quite common, dry cells provide power for radios, and automotive batteries are commonly used for televisions. Simple energy equivalencies are used to convert each of these into kWh units. P_0 is the price currently paid for such services. The area under the demand curve between the preproject situation and the postproject projection represents the gross economic (social) benefit to be created by the project. The net benefit for the project is equal to subtraction of the costs of the project (approximately equivalent to P_1, B, Q_1, O in Figure 8-2) from the total area under the demand curve (generally approximated as P_0, A, B, Q_1, O).

In a typical, relatively poor rural area, perhaps 3–5 kWh are consumed each month for these basic energy services, at a price of about US$4 (1,600 Chilean pesos, Ch$) per kWh equivalent. In practice, these surveys are not done systematically for each project but are estimated using census data and other periodic broad rural surveys conducted by the Chilean government.

The future situation (with the project implemented) is represented by Q_1, P_1 (Point B). The quantity of energy expected to be consumed if the project were executed is Q_1, and P_1 is the price per kilowatt-hour of electricity that is an estimate of the approximate cost of service. The quantity is estimated for the project site from actual consumption data gathered from similar projects already executed in areas demographically similar to the potential project site. Again, for line extension projects, this estimate now rarely requires field surveys but is made by taking advantage of the extensive electrification database present in each region. In many rural areas of Chile, apart from the far South, average monthly consumption of the lowest socioeconomic stratum is around 25–30 kWh per month per family. For renewable-based systems, for example, photovoltaics (PV), basic service might provide around 7 kWh/month. P_1 is given by the regulated tariff for the particular end-user group; for rural residential users, the two-part "BT1" rate, or the first tariff category of low voltage service, applies. Given these two demand points, Q_0, P_0, and Q_1, P_1, a demand curve (Figure 8-2) is constructed that represents the consumer "willingness to pay" for electricity services in a rural household.

For an entire project, families are classified by socioeconomic status; the gross social benefit is calculated for each group and then summed for all users. As a final step, the economic (social) net present value is calculated over 30 years, usually using an economic discount rate of 12% and applied to net project revenues that include the value of initial investment in year zero, annual operating costs, and annual project revenues. If the ENPV is positive, the project passes this first screen and goes into the regional electrification database as one entry in the overall project portfolio. If not, then the project is rejected as a candidate for funding.

In parallel with the economic analysis, a financial analysis is carried out to determine if the project will be financially attractive for the distribution company after taking into consideration the subsidy of the capital costs. This is a traditional assessment of the project's financial net present value (FNPV) to 30 years. Again, costs include both initial investment costs and annual operating costs. Logically, for the financial analysis to come out with a positive FNPV, net annual cash flow generated from user payments must cover depreciation and interest payments and all operating and maintenance costs. For grid extension projects, given probable demand characteristics, determining cash flow for any user requires a simple calculation. With newer technologies for which real-world experience is more limited, maintenance requirements and particularly their costs under different management schemes constitute unknowns that can also increase long-term risk, or at least the *perception* of risk on the part of the project implementer.

The FNPV and financial internal rate of return are calculated for each project. Finally, the PER and FNDR subsidy is set just above the financial break-even point at a 12% discount rate, to allow a reasonable rate of return to the implementing utility for its commitment to manage the project over its 30-year lifetime. Thus, the project must be economically viable for the country and financially attractive for the utility to qualify for the subsidy from the government.

Implementation: CNE's Role in Funding PER and Supporting Decentralization

The design of the PER, and the ultimate responsibility for achieving its electrification goals, fell to the Chilean national energy commission (CNE), which was then part of the Ministry of Energy. In 1999 the CNE was placed within the Ministry of the Economy, which represented something of a demotion. This change was seen as a sign that the program had succeeded in setting up an effective rules-based program, reducing the need for high-level political support.

At the same time, the CNE recognized that the important decisions influencing electrification were made at the regional level, planned and negotiated among the regional governments, municipalities, and utilities. Furthermore, the CNE did not control infrastructure funds; these funds flowed from the national treasury through the FNDR, located in the Ministry of the Interior. The CNE's primary role, then, was to allocate funds proportionally to regional governments according to electrification indices and the willingness to participate in the program. CNE also provided a coherent set of goals for electrification and a plan for achieving them. Combined with CNE's leadership, and backed by the Frei administration's political will, the regional governments followed their design and executed rational electrification programs within their respective political jurisdictions. The CNE in fact directly controlled only a small portion of total PER program funding, but it played an important and strategic role in regional program design, evaluation, and education and created high visibility for electrification activities at the national level. In short, the PER "is composed of a group of complementary actions which tend to strengthen the decentralized rural electrification process, accelerating the rhythm of project implementation, improving program efficiency and increasing coverage above and beyond current growth rates" (CNE 1995).

The CNE initiated a variety of activities in the regions to encourage planning, professionalism, and expansion of regional electrification programs. Specific activities included hiring or seconding consultants for extended periods in five regions; sponsoring studies to do such things as examine demand patterns and determine reasonable unit construction costs; organize interregional conferences and exchanges; and spearhead international cooperation efforts aimed at pilot programs for new technologies. These and other CNE support activities in large measure have stimulated each region to meet the proposed electrification goals. A 1995 CNE publication acknowledges the PER's dependence on regional partners. In sum, the CNE's leadership, more than its funding, was critical for stimulating each region to achieve the aggressive objectives proposed by the Frei administration.

Thus, in 1994, as part of the Frei initiative, infrastructure funds loaned by the Inter-American Development Bank (IDB) to the GOC were increasingly designated for electrification in the regions, in both absolute terms and as compared to those available for other infrastructure. Then, in response to the PER's initial success, in 1995 an additional pool of funds was created by allocating Chilean national treasury funds, apart from the original IDB line of credit. These additional funds were channeled through the same FNDR/MIDEPLAN mechanism, and from the local perspective they are largely identical to the FNDR/IDB funds. The only exception here is that the CNE and regional governments possess some added discretion to promote the use of the additional funds for individual projects that fall outside the strict IDB guidelines (for instance, in the case of pilot projects that would test a new technology, or those that use innovative contractual arrangements, which might be ineligible for IDB funding). The PER's main challenge has been to create the conditions in each region under which efficient use of the increased funding stream can occur. With this in mind, we next describe a few of the salient characteristics of the PER.

Decisions for infrastructure investment are made largely at the regional level. The gross amounts of infrastructure funds given to each region are determined in accordance with the projected need in that region, as well as some basic measure of the regional government's demonstrated ability to use the funds cost-effectively. During each year's funding cycle, staff of the regional government, Secretaría Regional Ministerial de Planificación (SERPLAC), develop a program portfolio, and make recommendations based on their analysis. The Consejo Regional (CORE), a regional board of elected officials, reviews, modifies, and approves the annual investment plan. The national government does not normally exercise direct, official influence in these decisions.

The development of a regional portfolio of electrification projects is an ongoing task that begins with interactions among communities, municipalities, and utilities to identify and quantify the need for services. Individual projects emerge from sustained interaction among rural municipalities and one or more local electricity-distribution utilities. A municipality, represented perhaps by a mayor, requests that the utility elaborate a feasibility study for the electrification of an area conduct. Some projects may be feasible without a subsidy; the utility may decide to implement these on its own. More commonly, however, the project is determined to require a subsidy. At this point, the municipality presents the preliminary project outline to the regional SERPLAC, making sure that the project enters the queue for the upcoming funding cycle. SER-PLAC electrification staff then evaluate the subsidy requirements of each project application, ensuring that all necessary studies and other documentation are complete. The economic evaluation verifies whether the project is minimally viable. If viable, the project becomes part of the regional portfolio. The project is then compared to all others in the portfolio for financial viability. Some projects are funded during the first cycle after application, whereas others must wait several years.

Community participation is emphasized as an important aspect of the PER. The demand for electrification in most cases begins from the expression of need from an organized community, and municipal-level political processes reflect community priorities. Moreover, given that the FNDR subsidy can only be authorized once for each family, community members have a broad incentive to make sure the most effective solution is selected. To a large extent, this sort of community participation is not new to the PER because the funding process for electrification before the PER rewarded community initiative with a greater likelihood of project approval, funding, and construction.

Additionally, during the course of each project's definition and approval, the community (i.e., the project beneficiaries) participates in some important decisions. The most important is the monetary contribution of the individual users to the initial cost of the project. In general, the user must pay for, at a minimum, the corresponding residential service drop and interior wiring installation, which typically amount to around 5% of the total project cost. Depending on each user's ability to pay, this cost can be financed by the utility, to be recovered over time through a periodic electric bill. In projects for which funds are so limited that economic viability (and therefore subsidy approval) is in doubt, the user or municipality may make additional contributions to increase the possibility of approval by the SERPLAC. Relative contributions to a project are typically negotiated among the utility, the municipality, and the future users. However, once a project has taken shape, the utility supplants the municipality as the primary driving force.

Implications of the Subsidy Funds

At the regional level, the government receives project proposals from electric utilities and cooperatives, which compete for limited grants. In general, projects that provide new electric connections with the lowest FNDR subsidy per connection are funded progressively, until the subsidy funds for that year are exhausted. The funding strategy is thus designed to reward utilities requesting lower subsidies on a per-user basis, even if each utility proposes a unique set of projects. The successful utility is awarded more of its proposed projects and is paid the subsidy for each on initiation of construction. In its simplest form, competition for FNDR subsidies occurs implicitly between rival utilities in a cycle of strategy, action, and reaction.

Competition for Subsidies

The increased scale and public profile of the PER prompted a number of SER-PLACs to take fuller advantage of the existence of competitive conditions in their regions. Several SERPLACS standardized unit construction costs in an effort to extend subsidies. Most of the southern SERPLACS also used competitive negotiation to lower subsidies for specific projects presented by more than one utility. In one of the southern regions, the CNE consultant and SERPLAC officials asked competing utilities to submit multiyear plans for specific groups of larger projects to be constructed in multiple phases.

The high stakes and more explicitly competitive playing field ended up saving the region around US$2 million over three years. Finally, several regional governments constructed geographic information systems (GISs) and typically began the task of data entry with electrification data from the regional PER. Almost immediately, GIS-based analysis better enabled regional planners to track progress toward meeting electrification goals and provided more complete bases for the demands the SERPLACs made of the utilities. Over the longer term, GIS systems have improved planning in other sectors as well, by permitting the integration of a broad array of physical, demographic, and social services data to create a powerful tool for analysis and evaluation.

Caps on Subsidies

The state explicitly limits the FNDR subsidy to a maximum of a project's initial capital cost. Therefore, each potential project must generate sufficient cash flow to cover operation and maintenance costs. If a project cannot generate sufficient cash flow to cover operation and maintenance costs, for example because of low projected demand or low customer densities, then it will not be approved for a subsidy. Although such a high maximum subsidy may not seem limiting, the result does in fact reduce the potential for gross unit cost excesses of some programs in other countries and represents a clear limit on the willingness of the state to pay the private sector for the provision of social services, such as rural electricity. Moreover, allocating capital subsidies provides a strong incentive for utilities to extend distribution lines sooner rather than later. Once lines are built and consumers are provided with electricity service, the utility assumes responsibility for all maintenance and operating costs.

Lack of Technical Assistance for Productive Uses

The majority of rural distribution companies have placed little or no emphasis on load promotion in their newly constructed lines. Several reasons may be behind this. First, the continued existence of PER funds through the FNDR, and the competitive nature of their allocation, has meant that distributors have an incentive to focus their time and effort on system expansion, including short-term planning of construction-related activities. This emphasis is natural, given the fact that the PER was designed with temporal goals and a predetermined lifetime—clearly these funds must be won now, rather than later. In the competitive allocation environment for these funds, focusing on new construction is strengthened even further because each utility must make a case to its regional SERPLAC, CORE, and municipalities to gather support for its annual array of project proposals. In short, load densification can wait.

In the past, as well, distribution utilities have not concentrated on load intensification to lower costs per kilowatt-hour. Instead, their view was that as long as profits could be made from the construction program, load growth could wait. The potential for productive load growth appears at least on the surface to be rather low in rural Chile. Rural residents in Chile have not traditionally engaged in a great deal of economically productive activities but rather seem to fit traditional definitions of rural poor.

Overcoming Problems in the Subsidy Program

During the early years of the PER, the most competent of the regional PER managers uncovered several ways in which participating utilities sought to overestimate the subsidies they should receive. Coordination between PER regional coordinators allowed several actions to be taken to address the effect of PER investments. These changes in program implementation policy included establishing standardized unit construction costs for PER projects, direct negotiations with the utilities, use of portfolio-based competitive bidding, and exerting executive influence through the regional governors (*intendentes*) on the utilities when necessary. An important factor in addressing these issues was the provision of electrification consultants by the CNE to several of the regional governments.

The following problems and their subsequent solutions were experienced during the first years of the PER. One problem was that when a distributor had little competition for a viable project, the utility tended to overestimate construction costs. This tactic increased the stated capital cost of the project, resulting in an increase in the subsidy provided by PER. To address this problem, several regional governments collaborated to establish reasonable standardized unit construction costs rather than accepting individual project estimates, the unit costs of which might vary dramatically from utility to utility and from region to region.

In the early years of the PER, utilities overstated subsidy levels by underestimating load growth. This tactic resulted in understating projected annual revenues for the life of the project. The decreased nominal financial feasibility meant that they could claim a higher subsidy to reach the break-even point over 30 years. To address this problem, CNE and participating regional governments quantified consumption levels and load growth for newly electrified communities, documenting patterns and applying these findings for future projects.

Another technique used in an effort to inflate subsidy levels involved omitting a significant percentage of the potential users in the feasibility planning, resulting again in underestimating consumption levels and load growth. In Chile, most potential consumers adopt grid electric service when it becomes available. The solution in one of the southern regions was to require that utilities include all residences in the project corridor in the project profile. This approach has reduced underestimation of connection levels, consumption, and load growth, resulting in lowering subsidies required to reach financial viability for these projects.

Finally, the success of the PER highlighted the need for advanced planning tools at the regional level. Whereas the majority of projects proposed by the municipalities and utilities made sense from a system point of view, some did not. As coverage increased, regional governments sought to take a more proactive role in negotiating not only the costs of some projects, but also their optimization from a planning perspective. In particular, using newly developed GIS systems, several regional governments are now able to make sure that the annual portfolio of approved electrification projects fits into its broader development plans for the region.

Not all PER utility participants used these measures to increase program effects resulting from PER subsidies. However, the problems described herein were deemed to be sufficiently widespread that actions were taken to mitigate them. As indicated, most participating utilities have engaged in PER projects to position themselves for strategic load growth rather than to cash in on short-term financial returns. In any case, the CNE made efforts to correct problems that arose during early program implementation.

The Success of the Subsidy Program

For the government, the primary means of measuring PER success was to monitor the cumulative number of newly connected households in each regional program. To facilitate achieving 75% coverage by the end of the year 2000, the program was designed to proportionately allocate a greater percentage of funds to those regions with the highest number of unelectrified households. The PER subsidy was considered by the government as one means of addressing poverty alleviation and equity through increasing access to rural electric service. Thus, funds were allocated so as to reach the largest number of rural poor households, rather than allocating equal funding levels to each region of the country.

Over a seven-year period (1992–1999), rural electrification investments nationwide totaled US$211 million, of which US$133 million (63%) was provided by the FNDR. Of the maximum annual allocation of US$43 million, reached in 1998, FNDR subsidies accounted for more than US$26 million, or 60% of the investments for that year (Figure 8-3).

Over the same seven-year period, the maximum number of annual connections was also achieved in 1998 (CNE 1999), although the change from the previous year was relatively modest because of rising per-connection costs. The rural electric service coverage reached the 75% goal in 1999 (Figure 8-4). For the program, costs per consumer increased over time as coverage expanded, population densities decreased, and the so-called "marginal new connection" required more effort to reach (Figure 8-5). During this period, almost 120,000 households nationwide were provided with electricity.

Although the cost per connection increased, the proportion of subsidy funds declined over the life of the program. The per-connection subsidy amounts grew more slowly than the total cost per new connection over time. This means that the growth of nonstate contributions to new electrification projects outpaced increases in total cost. In 1992 and 1993, the FNDR funds made up 70% of the project cost. After the PER began in 1994, this value declined in the last two years of the program to a low of 61%, at the same time that the cost per connection rose

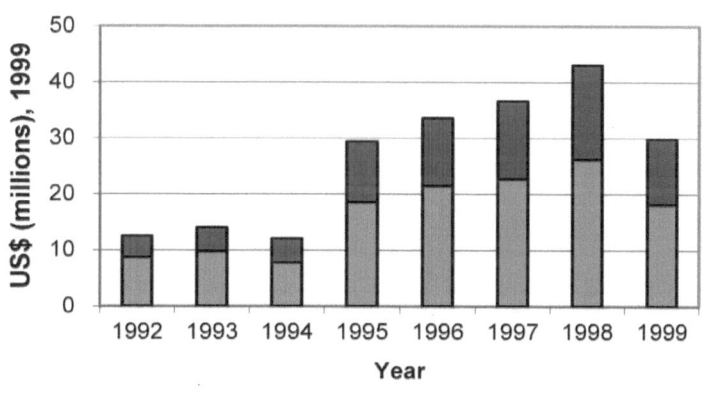

□ FNDR Subsidy Funds ■ Utilities, Municipalities, and Users

Figure 8-3. *Total PER Investment in Chile, 1992–1999*

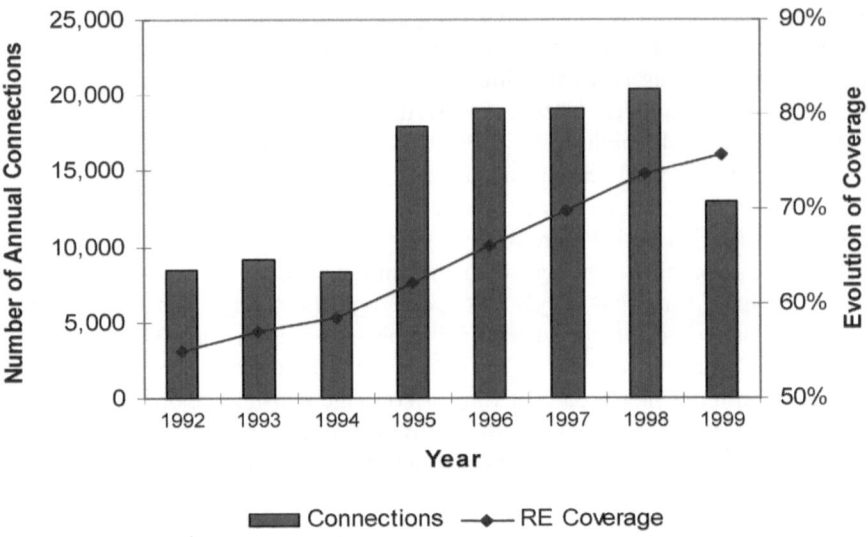

Figure 8-4. *National PER Results in Chile, 1992–1999*

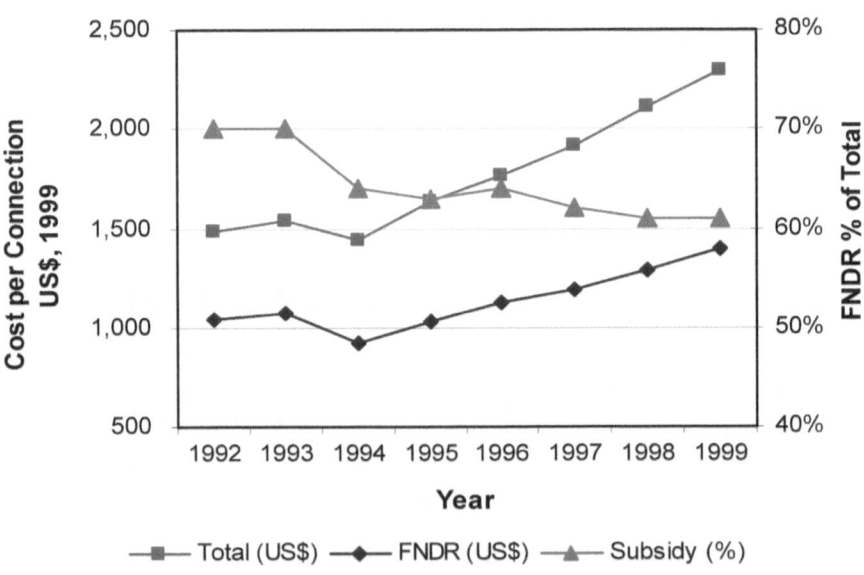

Figure 8-5. *PER Connections and Costs in Chile, 1992–1999*

55%, from $US1,480 to US$2,295. Maturation of regional program staff, combined with a competitive environment engendered by the presence of multiple service providers, resulted in increased efficiency in the use of PER funds over time, thereby contributing to program success.

The Case of El Maule

The case of El Maule provides a concrete example of how the PER program worked in one region. Historical data, made available by SERPLAC officials for the purposes of this study, revealed a number of trends that attest to the challenges PER faced as electrification coverage increased. Well before 1992, the southern regional governments, including that of El Maule, financed all infrastructure projects using the same fundamental approach. National funds flowed to the regions according to a formula that combined current needs for each type of infrastructure with past records of efficiency in using PER/FNDR funds. The regions that received the greatest share of funds tended to have larger investment requirements and a proven ability to use the funds efficiently. This was true not only for rural electrification but also for road construction, water, health services, and other basic infrastructure projects.

The total cost of PER projects increased substantially from 1992 to 1996, followed by a decline of almost 50% in subsequent years. In the early years, program funds were well distributed among service providers. For example, in 1993, Emelectric received US$502,000 in PER contributions, compared to the US$493,000 awarded to the Talca Cooperative that same year. In later years, the distribution utilities experienced a significant consolidation, with Emelectric garnering a significantly higher percentage of total PER funding.

The percentage of total cost covered by FNDR subsidies declined significantly from a peak of 87% in 1995 to below 77% in 1999, and the subsidies PER awarded various service providers tended to converge. Applying the competitive strategy, as well as setting standards for unit construction costs and subsidies, resulted in reducing subsidies overall and normalizing subsidies across service providers. These tendencies reflect the adoption of regulatory strategies by the CNE and the regional government.

From 1992 through 2000, total cost per user increased steadily, from about US$1,000 in 1992 to more than US$2,000 in the most recent project cycle (Figure 8-6). Given that the remaining populations without electricity are found in increasingly more remote locations, some rise in cost in real terms is to be expected; however, an increase of 130% seems somewhat high. Accordingly, the per-connection subsidy rose 110% over the same period, from US$870 to US$1,830. In effect, CNE and regional government efforts to reduce subsidies as a percentage of total cost were only partly able to offset per-connection cost increases.

Over time, the average cost per connection among the utilities converged. This trend was confirmed by conversations with utility officials, who expressed that, as the SERPLAC and regional governments were pressured to reach electrification

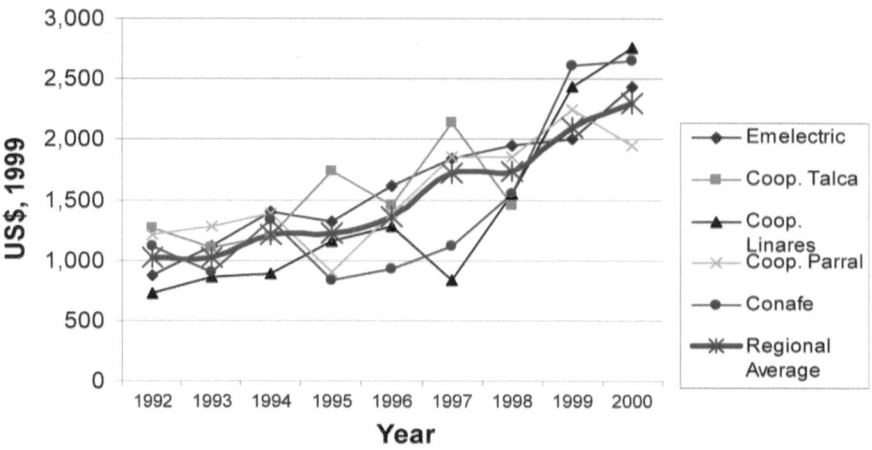

Figure 8-6. *PER Project Cost per Connection, by Distribution Utility in Chile, 1992–2000*

targets and gained more experience managing larger project portfolios during the PER's early years, negotiations between the state and individual utilities intensified. Increasingly, the SERPLAC held all utilities in its region to similar cost standards, and, at the same time, pressured them to contribute a greater portion of project costs. As a result, the utilities tended toward uniformity in per-connection costs, and the percent contribution of FNDR decreased slightly over time. The most growth-oriented utilities accepted slightly lower subsidies than they might have initially proposed to win a greater share of available subsidy funds.

Emelectric, an active IOU in El Maule and neighboring regions to the north and south, garnered an increasing share of projects and subsidies. Given that the IOU's construction costs are just as high as other utilities and that its technical designs are similar, what accounts for Emelectric's increasing success over time? Compared to the RECs with which it competes, Emelectric covers a much larger area, and its consumer base is many times greater. CGE, the other major IOU in Region VII, supplies urban areas, including the regional capital city of Talca, and thus does not compete with Emelectric for PER subsidy funds. CGE has had only one PER-subsidized project in Region VII; executed in 1998, it benefited 17 rural families. Another reason for Emelectric's success may be its ability to scale up to absorb increased subsidy levels by mobilizing a large group of subcontractors, planning designers, and procurement staff to accommodate ever-larger cash flows. Finally, compared to other utilities in Region VII, Emelectric has been relatively more effective in communicating with and lobbying the regional government during negotiations.

In summary, the developments described above introduced an element of competition that prompted the most solvent utilities to reduce the amounts of subsidy solicited. In El Maule, negotiation has been managed with each utility individually, and yearly portfolio data have been assembled, primarily as a tool for oversight.

Renewable Energy Options and the PER

The seminal PER policy document of 1994 stated that, where possible, electrification with the interconnected grid would be the preferred rural electrification technology. This statement means that if grid extension is feasible under the MIDEPLAN methodology, this option will be funded even if its per-user cost is higher than alternatives sources, including remote power systems and renewable energy. Apart from this stipulation, the MIDEPLAN method ensured that the most cost-effective projects—those that provide electricity to new users at the lowest subsidy per user—were executed in any given year.

In 1995, however, PER planners began to require more rigorous consideration of alternative technologies in regions with the largest deficits. Thus, the MIDEPLAN evaluation method was expanded to include financial and economic evaluation methodology for proposed renewable energy projects. Over the next several years, a number of electrification projects using renewable energy technologies were implemented with PER/FNDR funding (McAllister 1999). Observations and suggestions based on these experiences are highlighted below.

In the early years, projects that used solar photovoltaics and small wind were initially problematic and, in some cases, led to a loss of consumer confidence. For example, in one region a project used low-quality inverters in PV systems. This quality caused initial user dissatisfaction, which in turn, was reinforced by the implementing utility's slow response. In Region IX, a wind project experienced premature battery failure, which required local utility staff to make many visits to diagnose and repair the problem, as well as increased reliance on the backup generator with its high running costs. In both examples, technical and managerial shortcomings combined to compound problems that reduced quality of service, increased costs, taught new users not to expect high system availability, and eroded users' willingness to pay.

Traditional utilities, versed as they are in grid-based systems, are unlikely to embrace alternative technological options unless there is a clear means of earning a profit by using technological diversity. Thus, although utilities are unlikely to renounce renewable energy technologies explicitly, they are also unlikely to pursue alternative technology projects unless pressured by the CNE or the regional government. Where utilities are involved in renewable energy projects, subcontracting arrangements with specialized firms are likely to emerge as a likely option to reduce costs and to manage more effectively the remote project sites.

In contrast to grid-based electrification schemes, tariffs for remote power systems are unregulated. In theory, utilities can use specific market information to set tariffs. Renewable energy installations, particularly those using wind energy, generally need to be designed with detailed attention to the particulars of each site. To achieve cash flow commensurate with cost of service, each project becomes a unique case that presents challenges for negotiating tariffs with the municipalities and regional governments. For larger projects with multiple sites, cost aggregation may be advantageous.

Fundamentally, the MIDEPLAN methods focus on energy consumption, rather than any equivalency measure for service provision. This arrangement means that stand-alone systems using efficient appliances do not receive as positive an evaluation as do inefficient end uses typically connected to the grid. In this respect, the methodology does not provide a level playing field for renewable energy projects. The MIDEPLAN method also does not account for the positive environmental attributes of renewable energy technologies. These attributes are difficult to internalize at the project level, but rural end users should not be expected to accept lower levels of service (or subsidy) for the broader public good.

The foregoing leads to the conclusion that the PER program methodology does not take into account all factors that could result in higher levels of subsidies for remote communities that will not qualify for traditional grid extension projects. There has been some reluctance on the part of PER planners to correct these distortions in the methodology because renewable energy projects have proven to be institutionally difficult to support, due to the lack of enthusiasm on the part of participating PER concessionaires and the lack of response from renewable energy service providers. Today, as the percentage of households without access to electric service becomes smaller, off-grid, renewable energy solutions should play an increasing role in PER funding. There is little indication that this will, in fact, transpire, however.

Lessons from the Chilean Experience

The Chilean experience in rural electrification is unique for a number of reasons. First, the country has a professional public sector with a tradition of consistency and relative impartiality. Moreover, Chile has relied on private firms for providing many basic services and infrastructure far longer than have many other developing countries, and public–private interactions are relatively well demarcated. In addition, the strength of Chile's economy during the 1990s—which included the period coinciding with the Asian economic crisis to which Chile was intimately linked—has provided a stable foundation for providing subsidies to fund poverty alleviation and infrastructure development, including electrification.

Chile's rural electrification model may have substantial relevance to programs in other countries. However, it can only be replicated if a competitive environment exists or can be created for subsidy funding. Chile's model has particular relevance to programs designed to develop and apply consistent mechanisms for prioritizing projects based on appropriate methods. The decentralized nature of infrastructure decisionmaking, another important aspect of Chile's program, is also relevant in countries that are anxious to move some administrative oversight to local governments. The following sections summarize some of the important characteristics of the Chilean rural electrification program and the degree to which these characteristics contributed to program success.

Sustained Political and Financial Support: Rural infrastructure programs require long time frames in which to mature, and their benefits cannot be realized within a single presidential term. The success of Chile's PER hinged on a high level of multiyear financial and political commitment through various national agencies, including the CNE and MIDEPLAN, as well as strong regional government buy-in.

Basis for Competition: Competition depends on the presence of a well-organized, well-trained, and experienced set of actors that can provide services to electrification programs. The group of private and cooperative utilities in Chile had this critical experience. They are one of the principal reasons that the PER reached the level of success reported.

Focus on Regions without Extensive Coverage: The PER's priorities were established and maintained to focus on regions with the highest number of households without electricity. Thus, most funds were allocated to the southern regions. Also critical was the use of objective criteria for system planning and resource distribution within each region.

Objective Selection Methods: Although the PER had clear political motivations, its project portfolios were assembled and approved based on economic and financial criteria rather than political processes. Objective methods were used to determine levels and applications of program subsidies. The fact that subsidy percentages decreased over time illustrates that the subsidies were applied in a clear way.

Effective Management and Coordination: PER managers modified program characteristics over time to compensate for changing conditions, which increased positive program results. Continued involvement of a well-trained, motivated program coordination group was of paramount importance to program success. Coordinated by the CNE, the program coordination group consisted of professionals from several agencies.

The Clearly Defined Role of the Central Stakeholder: The need for a central stakeholder's ongoing involvement may not always be recognized. Countries considering adapting Chile's model to their own situation should carefully examine CNE's critical role within the PER. Within the Frei administration, CNE provided ongoing political support, which may have been its greatest contribution to the PER. Without the central agency's continuous support, PER funding may well have terminated when Frei's presidential term ended. But the program's success created the momentum needed to carry it over from one administration to the next. CNE was able to translate this political support into successful implementation through regional governments, with little CNE-controlled financing behind it. This is a powerful testament to the quality of CNE's staff and the clarity of its goals as defined within the PER.

Need to Establish Construction Standards before Implementation: In Chile, the distribution utilities have lacked incentive to review construction standards to optimize costs because PER subsidy contributions cover most construction costs. Because the utilities usually hire construction contractors affiliated with the utility holding company, the more the system costs, the greater the margin yielded by the project. Construction standards should be developed by independent parties and reviewed by participating utilities to ensure that neither group compromises use of program funds and that, after construction, operating costs are no higher than necessary.

Combining the Capital Cost Subsidy Approach and Project Financial Viability: Limiting the subsidy to a maximum of a project's initial capital costs has encouraged utilities to construct new lines. However, once a line is built, the requirement that a project be financially viable encourages the utility to maintain high service levels. In addition, no matter how high the maximum, an initial subsidy approach represents a limit of the state's willingness to pay the private sector to provide rural electricity and other social services. Thus, the Chilean subsidy approach encourages the distribution utilities to both expand and maintain service long after the payment of the subsidy.

Decline in the Role of the Cooperatives: Chile's RECs enjoyed almost six decades of providing rural communities ancillary services. They developed productive-use support programs for agricultural activities supported by credit and equipment programs offered through central and branch offices. They provided health clinics, educational support, seed and fertilizer, and water and irrigation services. Ironically, these very advantages, which later distinguished them from their IOU competitors, were negated in 1997, when it was ordered that all nonenergy services be unbundled and separated from cooperative utility operations. Some RECs have responded by becoming more like their IOU competitors, abandoning many of the social services developed by their membership for general membership. Others have moved away from electricity distribution as a core activity. The cooperative SOCOEPA, for example, now derives less than 30% of its income from electricity distribution. Although, on the surface, it made sense to separate dissimilar lines of business, the effect has been that many rural communities now have less access to essential services.

Lack of Promotion of Productive Uses: Lack of emphasis on and lack of progress in promoting productive uses of electricity have limited the PER's economic impact. If promoting productive uses is a program priority, then the rural electrification program must mandate it, and funding must be provided to finance productive use loans. Without such a specific program focus, distribution utilities will not make it a priority.

Inefficiency of Duplicate Distribution Lines: Each distributor owns the lines it uses to transport energy from the nearest purchase node to the end user; in only a few special cases does one distributor pay another for the use of its lines. Consequently, in many areas, two (or even three) sets of primary distribution lines run in parallel to serve customers in a single area. Lack of geographically exclusive concessions in newly electrified areas can lead to such cases. Currently, CNE is discussing the possibility of establishing a multicarrier service model for electricity, which would allow an individual user to choose his or her supplier, ostensibly including a toll charged by the company that owns the lines locally. Such a system would, in theory, obviate the need for constructing duplicate lines.

Limits of the Poverty Alleviation Model: Chile's national census, taken every 10 years, uses the socioeconomic characterization, as measured by MIDEPLAN, to gauge the level of socioeconomic development for each family in the country. Because the socioeconomic characterization allocates development points based on a family's access to an electricity connection, poverty reduction, by definition, is viewed as following mechanistically from electrification. Setting the bar intentionally low has made PER measurement of achievement simple; on the other hand, the true economic value of the program is not accurately nor fully measured.

Conclusion

The Chilean rural electrification program is often referred to as an example for which subsidies have been used successfully as incentives for the private sector to promote rural electrification. Given the high costs associated with the remaining areas without electricity in remote Chile, without the unique subsidy program many households in rural Chile would still be waiting for electricity. However, the Chilean program was fairly sophisticated in its approach to providing this subsidy, and there was extensive planning, evaluation, and local involvement in the process.

From all respects, PER has been and continues to be a highly successful rural electrification program. The special conditions that exist in Chile should not be minimized, however, when reviewing the applicability of this approach for use in other countries. The goal of creating a competitive environment for subsidy funds is important if not essential to maximize program effects, but such competition depends on the presence of an electricity sector that is mature and only requires a few incentives to push their coverage further into rural areas. The private and cooperative utilities in Chile were already fairly mature when the program was implemented. This certainly was one of the critical reasons why the program reached such a high level of success.

Acknowledgments

The authors thank Irene Righetti Simón and Xenia Corvalan Latapia, of the Region VII SERPLAC office in Talca, who provided invaluable data on the electrification program in that region. Also in Region VII, Medardo Navarro of Compañía General de Electricidad and Guido Zamora of Emelectric provided interesting viewpoints on the value of the PER, as well as access to several projects for inspection. Javier Castillo of the Region X Intendencia facilitated understanding of the administrative evolution of the PER in that region. Orientation about the PER and its broad goals and directions was provided by Rubén Muñoz B. and others before him at the National Energy Commission. Finally, Nancy Whittle of MIDEPLAN in Santiago provided some historical understanding of the development of the electrification projects evaluation methodology.

CHAPTER 9

National Support for Decentralized Electricity Growth in Rural China

Xiangjun Yao and Douglas F. Barnes

O VER THE PAST HALF CENTURY, China has witnessed tremendous growth in the development and delivery of rural electricity services. The initial challenge of rural electrification was tremendous, especially in light of the size of China's population, with most people living in rural areas. The success in the development of rural electrification can be attributed to several major factors in a program fairly unique in the developing world. They include the development of local and regional power companies that always had a high degree of independence, even though they were public companies. These local distribution companies were supported by state investments in the construction of both local and regional grids for distribution, and they also had assistance from the country's power industry. In other countries, rural electrification extended from central companies. In China, however, many local companies with state support were established and then over time spread out and were integrated into the national grid program.

The program went through several significant phases during the past 50 years. The first stage involved a commitment of rural communities to gain the benefit of electricity on their own. Through concerted and sometimes misdirected efforts, they built local systems themselves. This step was followed by a government program to support agriculture, irrigation, and small industries. Although somewhat successful, the lack of focus on local production of electricity and benefits for rural households led to a pilot program in 1979 called the "100 counties" program. During this stage, it was thought that rural electrification and small hydroelectric power (SHP) would improve the quality of life and rural productivity. The last stage is marked by a shift from a planning to a market approach to rural electrification, although this market approach had significant government support and involvement, as is typical of many Chinese programs. During this stage, local entrepreneurs and business leaders became involved in power production through small hydroelectric and other forms of power development. Finally, China adopted a market-oriented approach to rural electrification much sooner than it did for other industries and began developing complementary ways to have local generation in the context of a national grid. Today, the State Power Corporation (SPC) functions more like a commercial company. Rural areas have benefited from increased power supply as grid-based rural electrification has expanded, and surplus power has been sold to rural areas.

In parallel with other business activities, rural power companies were influenced by changes taking place in China during this time, moving away from authoritarian management systems to ones that are more oriented toward service. China's most successful rural electrification experiences have centered on the development of decentralized power systems, especially small hydroelectric power, which have exploited indigenous water resources, adapted inexpensive technologies and materials, and capitalized on local expertise. Other factors important for the success of SHP were the national support for capital investments, favorable policies for taxation and pricing, flexible methods for raising project funds, and the protection of self-supply regions.

Background on China

With approximately 1.3 billion people, China has the largest population in the world, and geographically is the world's fourth largest country. As in most countries of this size, there is a vast difference among the various parts of the country. China's topography ranges from mountainous deserts and plateaus in the west to plains, deltas, and hills in the east. The country's highest point is near the border with Nepal, the same range as that of Mount Everest.

The ethnic group of Han Chinese make up about 92% of China's population. The remaining 8% are made up of Chuang, Hui, Uigur, Yi, Miao, Mangchu, Tibetan, Mongol, Ruyi, Korean, and other ethnic groups. Mandarin is the official language. China does not have a dominant religion, but has a long tradition of Buddhist, Taoist, and Confucian beliefs as well as Christian and Muslim influences.

China's history dates back some 5,000 years. The country was first united under the Qin Dynasty (221–207 B.C.), which standardized the writing system and constructed the Great Wall. Buddhism flourished under the Tang Dynasty (618–908 A.D.), often regarded as the Renaissance of Chinese history. Following multiple conflicts, which characterized the first half of the twentieth century, Mao Zedong led Chinese communists to establish the People's Republic of China in 1949.

China has a large power supply system. In 1996, China had an installed capacity of 232 gigawatts (GW), with 13 power grids, 5 of which were wholly operated by regional administrative groups. For electricity generation, about 83% was thermal and 17% hydroelectric power. Coal-fired power plants made up more than 90% of thermal generation, and oil-fired plants accounted for the remainder, although this number has been declining in recent years. The largest electricity consumer was the industrial sector (74.1%), followed by households (10.7%), services (9.1%), and agriculture (6.1%). Households in rural areas used mostly biomass and coal for cooking and heating.

In 1978, China began a transition from a Soviet-style, centrally planned economy to a more decentralized, market-oriented system, with increased foreign trade and investment. One result of this transition was that by 1997, there were 270 television sets, 56 telephone main lines, and 6 personal computers per 1,000 people. As of 1995, 81.3% of the population was literate (89.9% men and 72.7% women). In 2001, life expectancy was 69.8 years for men and 73.6 years for women. China is one of the fastest growing economies in the world.

History of Rural Electrification: Local Beginnings to National Integration

The amazing accomplishment of providing electricity to more than 1 billion Chinese people started over 50 years ago. During this period, the program changed significantly from its modest early beginnings. This evolution can be divided into five stages, and each of these stages has been unique (Table 9-1). The problems faced in each stage are in some sense a reflection of the changing economy and industrial structure of the country (Cheng 2001). Also, these stages are somewhat dependent on the growth of electricity consumption (for a categorization scheme based on electricity consumption, see Pan et al. 2006). Rural electrification has not always been the central focus of the government's investment activities in China, but it nonetheless has benefited both directly and indirectly from them.

In 1949, the government encouraged villagers to initiate their own programs. As a consequence, this first stage (1949–1957) focused mainly on lighting and food processing, with a vast number of differences in the various regions of the country. The second stage (1958–1978) marked the formal beginning of state-supported rural electrification, with an expanded focus on flood prevention, irrigation, and rural industry. The third stage (1979–1987) is characterized by central government involvement in special rural electrification promotional efforts. Reflecting trends in China's overall economy, the fourth stage (1988–1997) marked a shift away from state control to decentralized, market-oriented development of the rural

Table 9-1. *Evolution of Rural Power Consumption in China, 1949–Present*

RE Stage	Fiscal Years	Rural Power Consumed (% of total)	No. of Villages with Access by End of Stage	% of Total Villages[a]
I	1949–1957	0.6	—	—
II	1958–1978	0.8–13.1	412,517	61.1
III	1979–1987	13.5–15.3	722,831	78.7
IV	1988–1997	16.3–21.1	726,993	97.7
V	1998–present	21.2–22.0	733,172	98.2

[a]Villages were restructured in 1987, which decreased their total number. As a result, the percentage of villages with access to power increased greatly during Stage IV, even though the number of villages did not change dramatically.

Source: Government of China 2001.

electrification program. Since 1998, the beginning of the fifth stage of the program, the focus has been on the reform of rural power property rights, integration with the power grids, and reform of the rural tariffs to equalize them with those in urban areas.

Stage I (1949–1957)

Attention to rural electrification began in 1949, right at the time communist China was founded. Rural communities lacked access to grid-supplied power, which was being directed to nearby cities and suburbs. However, villagers understood the advantages of electrification and decided to invest in power stations, relying on local energy sources in the form of coal, diesel, and hydroelectric power. These sources of energy were used to produce electricity through small power stations. Rural communities played an important role in raising funds, managing construction of such small-scale stations, and securing institutional technical assistance to construct them properly. However, during this period only a few small power stations provided rural villagers with electricity, and they used it mainly for lighting and food processing. By 1957, rural energy consumption had reached only 100 million kWh, just 0.6% of the nation's total power consumption.

Due to limited finances for extending the central grid to vast rural areas, the government mainly supported large-scale hydroelectric power projects. By contrast, rural communities raised funds for constructing SHP stations. It was determined that stand-alone power generators (such as locally developed SHP stations) were more economical than the centralized grid for reaching mountainous and other remote regions.

Stage II (1958–1978)

In 1958, the government became more involved in policies for rural electrification, and it started by targeting irrigation and flood prevention. By the early 1960s the state shifted the focus of rural electrification programs more toward the productive uses of electricity. The idea was to reinforce the progress being made in other programs supporting agricultural mechanization and water conservation. In addition, rural electrification was necessary for the development of small industries, including steel production. By 1978, rural power consumption had climbed sharply to 51 billion kWh (Table 9-2).

To accomplish the reorientation, the central government organized on-site workshops to disseminate successful experiences in the development of SHP. At the time, there was a shortage in raw materials for such electrification schemes, so the planning sector developed a quota scheme giving priority to the sector. Provincial and local groups—major players in rural electrification promotion—also played their part by setting aside a portion of the materials they received, such as steel and wire, for the development of SHP. In addition, local manufacturers that produced goods necessary for the development of rural electrification were given both financial and technical support.

Table 9-2. *Timeline of RE Activities and Features in China, by Stage*

RE Activity or Feature	Stage				
	I (1949–1957)	II (1958–1978)	III (1979–1987)	IV (1988–1997)	V (1998–Present)
Central government	Sets up special sector in MWR responsible for SHP development; initiates pilot SHP projects	Formally initiates RE; constructs power stations and expands grid; transfers technology and experience	Creates NPRECP	Continues NPRECP; market focus versus state planning	Focuses on pricing reform and property rights
Local government	Initiates pilot station construction for SHP and other decentralized power sources	Implements state-required power development planning; sets up manufacturing plants	Targets 100 counties	Adds 500 counties to the NPRECP	Adds another 400 counties to the NPRECP
Financing source and type	Mainly local and farmer funds	Limited state funds; more input from local communities	100 million RMB provided annually to each NPRECP county	Shift to shareholder system, with participation of private and other sectors	300 billion RMB committed over a three-year period
Technology used	Locally made; small-scale; no grid connection	SHP	SHP and grid expansion	SHP and grid expansion; higher quality demanded	Reform of rural power network
Local participation	Self-inspired by villagers	Community organized; farmer-built infrastructure	Paid input of farmers	Wealthier farmers share property	Farmer investment in indoor facilities
Other RE features	Few RE villages; periurban only; power mainly for lighting	RE power used to promote small steel industries and rural production, including irrigation	High initial RE growth for TVEs	Agricultural reform; improving living standards; 96% coverage	Falling operating and market price
Approximate village coverage and installed capacity at the end of period	NA 7 MW	61% 3,000 MW	78% 9,700 MW	97% 18,000 MW	99%

Note: MWR, Ministry of Water Resources; SHP, small hydroelectric power; NPRECP, National Primary Rural Electrification County Program; RMB, yuan; TVEs, Township and Village Enterprises; and MW, megawatts.

During the 1960s and 1970s, the central government provided important construction materials, including steel, to demonstration sites, whereas other projects had to compensate for shortages locally. Because of their strong demand for rural power, local communities made use of whatever resources they had (e.g., unused iron and copper collected to replace wiring in manufacturing plants). Small hydroelectric turbines were manufactured locally, which enabled project developers to focus on procuring other needed inputs. In addition, provincial governments funded the training of local technicians in system design and operation.

Stage III (1979–1987)

In 1979, the era of nationwide reform began. This reform was a significant departure from the previous years of government control over the total economy. As part of the process, the government recognized the importance of rural electrification for continued development. The Ministry of Water Resources (MWR) conceived and later managed the National Primary Rural Electrification County Program (NPRECP), whose aim was to demonstrate the benefits of SHP in promoting rural economic development and improving farmers' living standards. Thus, it was conceived that SHP would be a catalyst for rural and town development in China. In 1983, the State Council approved the NPRECP and, two years later, initiated its implementation in 100 counties.

The NPRECP selected counties with power shortages that had promising conditions for rural electrification, including having ample hydroelectric power resources, the demonstrated enthusiasm of local authorities, strong community participation in the program, and well-designed planning for construction. The central government provided interest-free financing, complemented by local government financial support in the form of labor force investment and villager contributions. The result was a dramatic nationwide upsurge in rural electrification.

China's success in rural electrification is closely linked to power sector development and policies. As investments in capacity and grid expansion began to generate surplus power, it became possible to connect more villages to the grid. During 1978–1994, when shortages in energy supply occurred, the state issued investment policies that increased supply, thereby promoting rural electrification indirectly. In 1981, for example, the State Council issued a document stating that bank credit should be used primarily in the energy sector and the light chemical industry. Then, in the mid-1980s, a report was issued that stressed investment in the energy and transportation sectors and provided for a tax to fund important construction projects. Power sector investment led to capacity additions that supplied excess power to rural counties, and additional capital was used for specific rural electrification investments.

Early on, China's government recognized the importance of SHP in alleviating power supply constraints. Starting in the 1950s, it used professional personnel to disseminate technologies through national and provincial meetings, training courses, and workshops. Combined with technology transfer, local governments selected counties with ample hydroelectric power resources and sufficiently knowledgeable and enthusiastic rural communities to receive funding for construction of power generation stations and supply grids. During the 1980s, the typical investment per county was 1 million yuan. Based on lessons from the first demonstration projects, local governments gradually expanded the project's scale, encouraging surrounding counties to construct SHP plants.

Stage IV (1988–1997)

The fourth stage of rural electrification began with the national economic system's shift from centralized planning to a more market-oriented approach to development. This stage, accompanied by the movement away from agricultural communes to smallholder farming, saw living standards improve dramatically. In addition, there was a change in the emphasis of programs involved in rural industry structure. Instead of stressing services related to agriculture, rural industries began the production of goods and services geared toward the overall national economy. Rural electrification was a necessary input for this changing rural economy, so it necessitated an increase in both quantity and quality of electricity supply. As a result, the state decided to continue supporting the NPRECP, adding another 500 counties to the program.

The shift from market planning brought many social and economic changes. New facilities were developed based on both market principles and state-supported township and village enterprises. There arose multiple forms of ownership of such enterprises, including communities, local shareholders, and private-sector investors. The consequence was that central government funding for enterprise development fell from 95% to less than 20%, with contributions directed mainly toward support for rural industry development in the form of technology research, pilot-project demonstrations, training, and national standardization.

Based on successful NPRECP experience, the state formulated a national standard for regions supplied by large-scale power grids to accelerate rural electrification. The NPRECP standards centered mainly on county-level planning, related hydrological codes, design of SHP stations, and economic assessments of projects.

Concurrently, the former Ministry of Power developed the Program on Poverty Alleviation with Power, which included 13 counties without electricity. By late 1997, the total percentage of households with electricity reached 96%. In regions supplied by large-scale power grids, more than 500 counties attained the national county standard for rural electrification, and 332 counties in regions supplied by hydroelectric power met the standard.

The central government issued a law that outlined major requirements for power production and use (Government of China 1995, 1996). Except for those designed specifically to promote SHP, the government developed standards and codes for constructing and rehabilitating rural power networks. These standards included ways to reduce rural electrification costs through codes for designing small transformer stations and the distribution of electricity with lower voltages.

China developed two national rural electrification codes: one for rural areas supplied through the national grid and another for more remote areas that rely on SHP. For example, developers used *Specifications for Formulating a Preliminary Design Report for Small- and Medium-Sized Hydropower Stations* as a technical guide (Fujian

Hydropower Design and Research Institute 1996). The standard code applies to primary counties supplied by the national grid or those whose economies are developed. The other code applies to pilot counties, most of which are located in poor mountainous and other remote regions. The primary county standard is set as a target that pilot counties must achieve.

Stage V (1998–Present)

The current stage of rural electrification focuses on reform of rural power markets. The slogan for this stage is "two reforms, one price." The hallmark of this period is a change in property rights of investors in rural electrification, continued large rural power grid development, and standardization of prices between urban and rural areas. The major activities are to clarify and adjust power system property rights, reduce grid line losses, and improve bulk power sales and distribution efficiency. All activities aim to reduce operating costs and market prices for rural end users. Recently, the state decided to invest, over a three-year period, more than 180 billion yuan in the improvement of rural grid construction. The stated goal of the program is to promote rural economic development in a faster and more sustainable way.

Thus, China's rural electrification program has prospered through a long history of government changes and modifications in policies toward the program. China's rural electrification program is characterized by its beneficial link to the power sector, localized use of grid-based power, and government financing of demonstration projects. Other important features include the exploitation of SHP resources, extensive use of local expertise and materials, development of technical codes and standards based on demonstration projects, and pricing policies that limit the profitability of SHP stations with high production costs. Finally, the program reflects the transition from a centrally planned to a decentralized, market-based economy.

Decentralized Industry Structure with Central Support

Throughout China, both local and the central government have been important for promoting rural electrification. The role of the central government included the setting of policy objectives, support for financing, and technical assistance. Local electricity companies working with county and regional governments adapted these policies into their own rural electrification programs. In this section, we detail how the program has worked from several points of view.

Diverse Decentralized Power Companies and Government Support

The power companies serving rural areas in China are quite diverse in terms of their size. Regional and provincial systems provide energy supplies primarily to large cities and towns. About 700 out of 2,400 rural counties in China are supplied power directly from SPC, which owns and operates regional or provincial networks. In rural areas and smaller townships outside the jurisdiction of these regional systems, electricity supplies are usually the responsibility of decentralized power companies (DPCs). These companies are administered by the government at the township, county, or prefecture level. DPCs are not just in the power distribution business, but also own and operate subtransmission (35-kV) systems, medium- and low-voltage distribution systems, and (in most cases) generation plants. However, most DPCs are interconnected with adjacent large grids, and about 1,000 receive most of their supplies from these grids. The other 800 DPCs generate at least 70% of their energy using SHP units. They rely on the grid to augment their supplies during the dry season and to purchase surplus energy during the rainy season.

The DPCs exist because of the great contribution these small power companies have made in rural China, especially in the early part of the program's history. Beginning in the 1950s, local political units in rural areas were encouraged to develop their own infrastructure, such as electricity, to enhance economic development and improve quality of life. The central government provided advice and partial funding, and for the most part local residents conducted project planning and implementation. Local communities supplied much of the needed equipment, including the local manufacture of turbines and generators.

Originally, the DPCs had their own small power stations. Individual grids were constructed using funds raised mainly from villagers and local communities. Villagers willing to connect to the minigrids paid for the service, both through locally set prices and in many cases through local investment by communities. Before constructing a power system, DPCs usually surveyed potential users' willingness to connect to the grid and predicted power load for daily consumption and agricultural production. However, in cases of obvious power shortages, DPCs sometimes developed projects without preproject surveys. The result was a dramatic increase in the number of people with access to electricity, a trend that continues today, but at a reduced rate as market saturation nears.

Selection Criteria for Guiding System Expansion

Early in the program, village-selection criteria were important for ensuring that safe and reliable power was made available to villages. To guarantee the profitability of DPC operations, village-selection criteria included an assessment of community, industrial, and agricultural production capacities; current consumption levels of other energy resources; and education level (Box 9-1). During the initial stages of planning demonstration projects, state and local governments allocated partial project funding to develop model successes. A team of local authorities and experts carefully completed the selection process. At first, for commercial projects, villages with better economic conditions were chosen on the basis of their ability to pay monthly fees. The rationale was that it was necessary to ensure DPC profitability. However, for some cases in which electricity was geared toward poverty alleviation, other indices were used.

Box 9-1. Typical Village-Selection Criteria

Rural households (%) with access to
- Public well as a water-supply source
- Private well as a water-supply source
- Electric lighting, including
  - Radios, televisions, sewing machines, refrigerators, and electric fans
  - Water pumps for agricultural use

Households (%) that use
- Charcoal as cooking fuel
- Wood as cooking fuel
- Gas as cooking fuel
- Other modern fuels for cooking

Dwellings (%) constructed in the past five years

General health indices:
- Rural population density
- Population birth rate
- Population growth rate
- Ratio of population to local physicians

School enrollment ratios:
- Upper elementary to lower elementary
- Lower secondary to upper elementary
- Upper secondary to lower secondary

Agricultural assessment (%):
- Gross area in agricultural use
- Arable land under rice cultivation
- Arable land under field crop cultivation
- Total area under fruit trees and tree crops

Baseline electrification data:
- Consumption in villages already electrified
- Ratio of electrified households to total households
- Percentage of households already electrified

Overall village characteristics (average size):
- Household
- Village population

Shifting Control from Central to Commercial Management

In the early part of the rural electrification program, almost all rural electrification development activities were managed by state-owned administrative agencies, such as the central Ministry of Power, and provincial and county power bureaus. The reach of these bureaus extended even to the township level. Before adjustment of the economic system, power bureaus at provincial and county levels were in charge of policy formulation, planning, power distribution, and power generation. Unfortunately, investing them with so much power in many cases led to corruption. After the market reforms in the early 1980s, these administrative agencies lost much of their power over local companies. At the central level, the State Economic and Trade Commission is responsible for macro administration of rural electrification, and the SPC and its branches function as commercial companies responsible for their own profitability and long-term development, rather than as government agencies.

This shift in policy highlights changes that have led to increased profitability of the DPCs and an increasing orientation toward customer service. The change is due to several factors. Within companies, employees are now more concerned with the business's success because their incomes are closely tied to company profits. The DPCs are conscientious about completing careful preconstruction studies, lowering production costs, and improving operation and maintenance. Power companies understand that their consumers are now the main source of company revenue.

Technical Standards and Codes

The development of technical standards for rural electrification was important for keeping costs low and making sure that the country does not have incompatible systems. Standards and codes for rural electrification system planning, design, and approval are divided into SHP- and grid-related specifications (Box 9-2).

Box 9-2. Typical RE System Standards and Codes

Small Hydroelectric Power

- Acceptance examination code of primary RE (GB/T15659–95)
- Regulation for development programming of the region's electric power mainly supplied by rural hydroelectric power (SL22–92)
- Standard of Supervisory Control and Data Acquisition function for rural hydroelectric power system (SL/T53–93)
- Code for hydrology calculation of SHP stations (SL77–94)
- Hydroelectric energy design code for SHP projects (SL76–94)
- Specification on compiling RE planning for SHP areas (SL145–95)
- Economic evaluation code for SHP projects (SL16–95)
- Technical code for constructing SHP stations (SL172–96)
- Introductory guidance for calculating loss of SHP (SL173–96)
- Specification on compiling initial design reports of SHP stations (SLT/179–96)

Grid Construction

- RE standard (DL/407–91)
- Construction and acceptance examination code on high-voltage facilities in electrical equipment installation project (GBJ147–90)
- Specification on designing 35- to 110-kV transformer stations (GB50059–92)
- Construction and acceptance examination code of batteries in electrical equipment installation projects (GB50172–92)
- Construction and acceptance examination code of 35-kV and lower voltage power lines in electrical equipment installation projects (GB50173–92)
- Specification for designing low-voltage power distribution systems (GB50054–95)
- Construction and acceptance examination code for electric lighting facilities in electrical equipment installation projects (GB50259–96)
- Specification for designing small-scale, rural transformer stations (DL/T5078–97)
- Specification for designing 35- to 110-kV transformer stations with no person on duty (DL/T5103–99).

Note: GB = state-issued standards; SL = Ministry of Water Resource-issued standards; and DL = State Planning Commission-issued standards.

These standards and codes are based on lessons learned from current and past projects. They are periodically adjusted to promote technical improvement. Regarding the design of rural transformer stations, for example, a 1992 specification on designing 35- to 110-kV transformer stations (GB50059–92) was changed in 1999, based on progress in technical research. The countries are eligible to receive financing and technical assistance once the authorization agencies confirm that counties are adhering to the new standards.

Technical specifications and codes provide guidance to local system designers and developers to ensure quality performance of the system and equipment (Box 9-3). The rural electrification standards and acceptance examination codes help administrative agencies to check the construction results when a project is partially funded by the government.

Even if the grid is connected to the regional grid or is in a decentralized supply area, all standards are valid. Counties follow rules to meet specified standards before being classified as a rural electrification county. Before a final check by a team of experts and officials, a county completes a self-check in representative villages, according to a national standard (Box 9-4).

Three Main Approaches to Electricity Supply

The supply of electricity to rural people in China involves several types of approaches. Most rural electricity is delivered by decentralized power companies, but there are three major types of electricity supply to companies serving rural areas. Each of these types involves a separate set of management issues, but they cover 99% of the 2,400 counties involved in the country's rural electrification program. The state power system serves 716 counties (30%) and directly supplies people in rural areas through provincial power corporations. The DPCs that purchase power from the SPC are called the interconnected counties, and they involve 1,004 counties (42%). Finally, the DPCs that generate most of their own power are called the self-generation counties and serve 652 counties (27%).

State Power System

Based on the national economic objectives of five-year planning periods, the SPC has made power development a priority. Over the past few years, 716 counties that are directly supplied power by the SPC have been the central government's focus of economic development. Grid expansion has also received significant attention, resulting in a well-developed network and enormous power consumption. The large quantity of power consumed by these counties—54.5% of the total power used at or below the county level—has prompted state and local governments to invest in grid extension, leading to rapid rural electrification.

The SPC, through its provincial branch company, operates county power corporations. The provincial power corporation nominates managers of county power corporations, whom the SPC then approves. The county corporation's profits are

Box 9-3. Supply and Demand: Model for SHP Development Planning

In 1995, the MWR issued *Specification on Compiling RE Planning for SHP Areas*, a model document for DPCs to use in their SHP development planning (Government of China 1995). The document contains chapters on general rules, load planning, power-resources planning, balancing of power-generation capacity, grid planning, research planning, personnel training, management, economic evaluation, and feasibility analysis.

Plan Contents

Each development plan is based on a county's unique situation and includes the following:

- baseline study (including availability of SHP resources for further development, existing distribution grid, and power generation sources);
- analysis of present problems;
- forecasted power load, based on population growth and economic development;
- planning of power-generation sources, based on a balanced analysis of forecasted power consumption and supply;
- planning of the distribution grid;
- technology to be adopted and technical innovations;
- financial and economic analysis; and
- implementation strategy.

Balancing Seasonal Surplus and Shortage

Because most pilot counties are supplied with SHP, balancing power supply and demand is a problem during periods of surplus (rainy season) and shortage (dry season). Thus, planning must include an optimization study of the entire local grid to ensure that it can meet the increasing demand load. During the rainy season, the price of power is lowered to encourage households and farmers to consume the surplus (e.g., electric cooking, grain drying, and other electricity-powered tasks). When selecting the power load for planning purposes and to minimize investment, some industrial users decide to use power on a seasonal basis (e.g., silicon steel manufacturing plants). In the dry season, load centers are requested not to use power so that residents, basic industry, and commercial sectors have sufficient power. If a gap between supply and consumption remains, hydroelectric power stations equipped with adjustable reservoirs or thermal power plants are used. Moreover, some areas encourage connecting with large grids and then selling surplus power in the rainy season and purchasing it in the dry season.

Source: Guangxi Bureau of Hydropower 1995.

considered part of those of the SPC. Most counties with county power corporations are located in fairly developed regions with a better developed grid, greater power consumption, and relatively few power resources.

Under the SPC system, the rural electrification program's institutional and management structure has six geographically based levels of authority. They include the state, regional, provincial, prefecture, county, and township levels. In Shanxi Province, for example, these six levels correspond to the SPC, North China SPC, Shanxi Provincial Power Company, Linfen Prefecture Power Company, Jiaocheng

Box 9-4. Checklist According to National Rural Electrification Standard (DL/407)

General Rule

Rural electrification standards

- are formulated to establish clear rural electrification objectives in China and support healthy development;
- take account of county–rural electrification characteristics and the appraisal rules for 100 rural electrification demonstration counties using SHP supply, and
- may be adopted for any county, as long as required criteria are satisfied, as follows:

Required Criteria

Access to electricity (%)

- Township: 100
- Village: 100
- Household: more than 95

Guaranty of power supply (%)

- Irrigation: 100
- Domestic use in towns and villages: more than 90

Annual electricity consumption (average kWh/person)

- County resident: more than 300
- Rural resident: more than 160
- Domestic use per rural resident: more than 50

Power grid and source construction

- Rural electrification is a component of local power-development planning and is suitable with rural economic and social development.
- Power-generation sources and grid are structurally sound, efficient, and have low losses.
- Rate of power voltage is more than 90%.
- Rate of power frequency in county grid is more than 95%.

Qualified rate of facility (%)

- Main equipment: 100
- First-grade equipment: more than 70
- Transmission, distribution, and transformer equipment (capacity of 10 (3,6) to 110 kV): 100
- Transmission, distribution, and transformer equipment (first grade): more than 70
- Distribution equipment (380/220 V low voltage): more than 95
- Distribution equipment (first grade): more than 70

Check and appraisal

- The provincial government should arrange for relevant sectors to check rural electrification status annually.
- On approval, the provincial government should report to the ME for rural electrification county certification.

(continued)

> ## Box 9-4. *(continued)*
>
> • The county applying for rural electrification certification must complete a self-appraisal using this standard and file a report with the provincial administrative agency, which, in turn, will verify certification. If the county cannot satisfy a certain standard criterion, it will be asked to improve and attain it within a limited period. If the county still cannot attain the standard, the provincial administrative agency should suggest ways to deal with the unqualified county and alert the certificate-issuing organization to revoke certification.
>
> *Note*: The standard was created by the Rural Electrification Committee of the China Generator Engineering Society and adopted by the National Rural Electrification Pilot County Program. It is carried out mainly in counties supplied by the national grid; its major task is to reform and expand the rural power grid and improve rural electrification management (more than 500 counties participate in the program).
>
> *Source*: China Ministry of Energy 1991.

County Power Company, and Township Power Station, respectively. At the state level, the power management agency has a special Department of Rural Power Development, which is responsible for organizing and managing implementation of reform of rural power systems and grids. (Other rural electrification–related, state-level departments cover hydroelectric power and renewable energy, grid construction, and power sources and distribution.)

The Department of Rural Power Development also supervises and monitors the quality of rural grid innovation projects and cooperates with other relevant sectors to develop policies and measurements of rural grid construction and technical progress. Department success has been based on well-prepared annual planning, regular site visits, workshops, meetings, and coordination with other central government sectors. At the regional, provincial, and prefecture levels, power companies have rural power divisions responsible for the industrial management and implementation of policies, rules, and standards.

Historically, power bureaus at various government levels have managed generation and distribution through administrative agencies. Expansion of the distribution system relied on management expertise and power bureau funding but also included government interaction. Today, with the separation of administrative functions, the SPC focuses more on power distribution. However, as a state-owned enterprise, it also owns, either in whole or in part, some power generation stations. Therefore, expansion of the distribution system is based on economics, focusing on profitable areas. At the same time, the SPC or governments subsidize certain projects that specifically target grid expansion and emphasize poverty alleviation.

Interconnected Counties

Interconnected counties make up almost half of all counties in China and represent about 32.7% of total energy consumed at and below the county level. These counties have played an important role in rural electrification because villagers invest in, construct, and directly own the power distribution networks that target them. They perform well because of independent decisionmaking by local government sectors (DPCs fall within this category). The county power corporations in the interconnected counties purchase large quantities of power from the state grid, which they then sell at a profit to end users. However, they also generate some of their own power through the use of microhydroelectric and power stations. The SPC has little or no ownership share, but it still influences power purchasing negotiations.

Because of their rich energy resources, most interconnected counties own their power generation stations. Typically, they only buy power from the state grid to compensate for the difference between demand and self-supply. During power shortages, the priority customers are industrial, and the counties rarely have been able to purchase the planned amount that they require. As a result, self-generation in county power systems has played an important role for rural electrification because rural power demand always has second priority to industrial demand.

In general, county grid expansion has followed the principle that the investor also manages and owns it. Because direct state investment is limited, local governments, communities, enterprises, and consumers have played a critical role by constructing the county grids. Despite their achievements, these grids have suffered from reliability problems. They lack system planning and have used varying standards and local materials and technology. Resulting problems have included problems in grid system design, higher line losses, unreliable power supply, and inefficient equipment—conditions that have left open the door for considerable technical reform.

The SPC and provincial power corporations have invested in and constructed grids to provide power to county power corporations responsible for grid investment and management to points outside rural substations with 10-kV lines. Farmers and township power administration stations have shared maintenance and management of facilities with voltages lower than 10 kV.

The customers for the county systems must pay for all management, operation, and maintenance expenses. All households are charged fees, typically collected by village committees. The committee appoints one or two technicians to take charge of meter reading and other technical issues. Certain fees, such as interior wiring and metering, are charged to individual households before construction, and fees for line connections and substations are paid for by the village committee. The committee deducts fees collected from collectively owned businesses or farmers from annual income and typically charges poorer households less or provides some of its services for free.

Self-Generation Counties

Most self-generation counties are located in mountainous and other remote areas, where it is economically unfeasible for the state to extend the grid. In such areas, local governments and communities have strongly supported SHP and small thermal power development, having set up independent grids to satisfy local demand. Decentralized power grids are managed by the water conservation sector or are comanaged with county power companies. Because of their special power resources, 12 counties manage themselves independent of the power and water conservancy sectors. Some self-generation counties are interconnected with the state grid to supplement shortages during dry seasons.

Most counties supplied with SHP are located in southwestern and central provinces, where water resources abound (e.g., Yunnan, Guizhou, Guangxi, and Zhejiang). Counties supplied with small thermal power are located mainly in Henan and other economically undeveloped regions rich in coal resources. Locally constructed thermal power grids are becoming less common because of increasing state pressure regarding energy efficiency and emissions.

In the early stages of SHP development, design of distribution systems was based on production rather than consumption. System structure was based on generation capacity of the SHP station because most villagers lacked power supply beyond the grid. Although self-distribution covered a relatively small region, it played an important role by providing farmers power for both domestic consumption and agricultural production (e.g., irrigation and grain milling). As more SHP stations were constructed, several distribution regions were merged to ensure villagers a more reliable power supply. Today, some self-generation counties have one interconnected network that is also connected with state or regional grids to receive power from the large grid in the dry season and sell power in the rainy season.

Like the power sector, the water resources sector takes a hydroelectric power development approach at all levels—from the central government to the township. At the provincial level, a division within the Department of Water Resources is responsible for SHP development in regions with existing SHP plants or with hydroelectric power potential. In such minigrid service counties supplied with SHP, the responsible water conservancy organizations cooperate directly with power utilities because minigrids have the right to sell or purchase power from the grid, according to state regulations.

Most SHP stations have been financed and constructed by local (i.e., county or township) governments and have included village committees. Most work has been completed by township power distribution stations, jointly supervised by county power agencies. Local governments, communities, and farmers—whose inputs include labor, land, and cash—have shared funding of the systems.

Development of locally available resources, such as hydroelectric power, has always been a priority for rural electrification development in China. SHP development began with the building of small water conservation systems to solve rural supply shortages in mountainous areas. From 1949 to 1998, installed capacity increased from 4 MW to more than 22,000 MW (Table 9-3). By the end of 1999, it had increased another 1,500 MW, with more than 43,000 SHP stations producing almost 72 million MWh of electricity. More than 25% of China's rural population relies on SHP-produced electricity.

Table 9-3. *Historical Development of SHP in China, 1949–1998*

Year	Installed Capacity (MW)	Annual Power Generation (MWh)
1949	4	na
1955	7	na
1960	251	na
1965	330	na
1970	1,019	na
1975	3,083	670
1980	6,926	1,270
1985	9,521	2,410
1990	13,180	3,930
1995	16,646	5,540
1998	22,024	7,130

Note: The statistical range of SHP has varied over time: <500 kW (in the 1950s), <3,000 kW (in the 1960s), <12,000 kW (1970–1985), <25,000 kW (1986–1995), and <50,000 kW (1996–on); there are no power-generation data before 1970.

na = not available.

Source: Government of China 1999.

Operational Problems and Overlapping Responsibilities

The natural growth of the rural electrification programs in China led to competition, misunderstandings, and conflicts among various agencies and bureaucracies. As local power systems grew, their role changed from providing power for low levels of electricity use from locally generated sources to buying and selling electricity and serving ever larger customer demand. However, the system, with all its complexity, has resulted in an expansion of electricity to almost 1 billion people.

Development of Distribution Systems

Developing rural distribution systems varied according to power supply source. When power was supplied from large grids, distribution systems were constructed based on the principle of first extending the grid to relatively developed areas with a high load and urgent power demand. By contrast, the distribution grid for decentralized power systems was developed to serve consumers with lower demand. For example, when an SHP station was built, an individual distribution system was first constructed to provide power to surrounding consumers. Over time, after the development of several small distribution systems, they would be merged to form a self-supply network. Finally, to supplement the differences between daily and seasonal power demand fluctuations, self-supply networks eventually were connected to a nearby large grid. In cases of shortages from self-generated power stations, power was purchased from the large grid, thereby creating an integrated distribution system.

Early on, the government promoted rural electrification development by having the local power supply bureau manage the financing of distribution systems. This investment was recovered when power was sold. With this revenue, the power bureau gradually extended distribution systems until all rural consumers were covered. Unlike financing distribution for large grid systems, the financing distribution for decentralized power systems was expected to be fully shared by users during the construction stage, a practice that followed the principle of self-construction and self-management. In fact, because they desired electricity, rural populations contributed significantly to the construction of distribution systems. Not only were they investors, but they also purchased needed facilities when materials were in short supply. More importantly, they worked on civil construction projects and were paid directly for the number of days worked. This type of voluntary contribution reduced total cash demand and made the construction system more feasible.

Township power administrative stations paid close attention to development issues because these stations directly operated distribution systems and worked with rural consumers. Various administrative systems—financing, personnel, technology, and safety—were developed locally to improve services. A nationwide assessment was regularly conducted to choose model stations for demonstration and training so that other businesses could follow by example. The standards ensured that manufacturers produced the highest quality of electrical goods.

Changing the Mix of Investments for Financing Rural Electrification

The means of financing rural electrification in China has changed over the course of the program. Before the 1960s, local communities provided most of the capital and labor for decentralized power systems, and the state mainly financed demonstration projects. For the demonstration projects, the financing coverage included all primary capital, but not labor, investments. Subsequently, county and state government capital contributions increased dramatically, especially during the NPRECP. With further rural electrification development in the mid-1980s, investments of the state governments decreased and mechanisms to promote market development were developed.

Today, there are multiple financing sources, including governments (central, provincial, and local), communities or collectives, and the private sector. The sources include grants from central and provincial governments, loans from policy banks, funds raised by county governments, private-sector capital investment, and villager input in the form of both investments and in-kind contributions. Grant sources include government budgets, special SHP development funds from the MWR, and other targeted funds, including those for general poverty alleviation. Villager inputs include both labor and capital. Since the 1990s, bank loans have accounted for a considerable share of total investment in SHP construction.

After 1997, state grants represented only 5.7% of total financing, whereas locally raised funds accounted for 30%. During the same period, bank loans increased to 45%. The role of banks in financing DPCs will likely grow as SHP development becomes more and more commercial.

Government subsidies have played a critical role in promoting rural electrification, especially for the development of SHP. During the 1970s, relying more on local and central government input for SHP system development was appropriate, based on the percentage of state-established subsidies. For an SHP station with a single turbine with a capacity of less than 500 kW, the state allotted the station 150–200 yuan per kW of total installed capacity. Capital was provided from a specific fund, which accounted for almost 20% of primary station investment. For a station with a single turbine larger than 500 kW, the state grant was 40–60% of total cost, much higher than that for smaller stations. Remaining county-implemented and provincially approved projects were listed as local infrastructure for water conservation or power development planning. These were also partially subsidized or received favorable loans from policy banks.

Another form of government subsidy was a priority supply of low-cost building materials, which was more helpful than direct capital input. Before the 1980s, the state controlled all materials through a quota system. For SHP construction, each province could use turbines and power generation sets for self-installation that were manufactured in the province, and the state allocated important raw materials for making equipment and constructing stations and distribution systems.

By the 1980s, the central government set up a fund to subsidize SHP and rural electrification. The state supported low-interest loans of about 5% with favorable financing terms of approximately 10 years through policy banks. Commercial banks charged interest rates of 15% or higher with six-month or even shorter repayment periods, so such loans were not feasible for the long repayment schedules necessary for rural electrification projects. Additionally, local governments could set up a rural SHP development fund through charges incorporated into the local tariffs. The mechanism for most provincial subsidies during this period was a contribution of about two cents per kilowatt-hour to such rural electrification funds and thus became a source of subsidies for local government promotion of rural electrification.

Energy Pricing and Cost Recovery

The price of electricity in rural China is based on local costs, but of course is not entirely a commercial price because of the significant subsidies that have been involved in the program. However, the local power companies are expected to be commercially viable after taking into consideration the local and state incentives to provide service to rural people.

The price per kilowatt-hour depends on financing sources, local investment, the technical level of installations, and other administrative factors (National Economic Committee and others 1985). It also varies according to the commercial situation of the DPC, including the cost of generating and distributing power. This situation includes the financial gains or losses generated through sale and purchase of electricity from the large grids. But generally, the price varies according to the cost of electricity at the generating station, which includes depreciation of fixed assets, operational cost, and commercial loan interest. Currently, the tariff for rural residents includes the following:

- a state-issued tariff for the 10-kV power supply system (0.30 yuan per kWh),
- a state construction fund (0.02 yuan per kWh),
- funds raised for the Three Gorges project (0.007 yuan per kWh),
- the loss of subtransformers (7%),
- losses in low-voltage lines and lines connected to households beyond subtransformers (18%),
- adjustment of power coefficient of subtransformers above 100 kVA (0.02 yuan per kWh), and
- electrician wages (0.02 yuan per kWh).

Therefore, rural residents pay almost double the tariffs paid by people who live in urban areas. Even though the state is working toward a uniform tariff, it may take several years to complete this work.

Provincial power bureaus adjust the SPC-approved tariff annually or occasionally. Officially, a province adopts a single price, with an allowable scope for change (± 5%), to account for differences in power sources and distribution systems. To adjust the tariff, the county first estimates the purchase or production cost of self-owned generators; then, based on the status of the existing distribution system (including line loss and operational expenses), it determines the tariff. Finally, the power bureau (on behalf of the county government), submits a detailed written report on the estimated tariff to the provincial power bureau for formal approval.

Connection policies encourage rural residents to connect to the grid. Village communities must organize the connection, and installers must be certified electricians. Households pay only for connection materials (i.e., lines and a meter, which cost about 300 yuan). All other expenses are shared by all households. In addition, there are some cross-subsidies in the tariffs. The DPCs have effectively subsidized many customers, including sectors that consider energy critical (e.g., agricultural product processing). Many such customers are companies that provide employment and other social benefits to the exclusion of commercial practices, such as payment of market-based energy prices. Central and local governments sustained this situation by creating a difficult operating environment for many SHP companies.

System Planning and Load Promotion

System planning and estimates based on load growth have been important to the development of rural electrification in China. The rational estimate of power load has been important to ensuring a sufficient, reliable power supply, as well as avoiding over- or underinvestment in systems. Based on experience in more than 100 counties, a system was developed using quantitative measures to calculate the planning for additional capacity.

To expand the existing grid or to build a new power distribution network, major development stages have included baseline surveys, planning for construction of new power generation stations and distribution networks, optimizing technologies for power generation, fundraising, and implementation and construction of all planned activities.

Through renovation of the existing grid and lowering expenditures, users pay less as a result of a lower power price; this in turn encourages an increased power load. In addition, different prices for peak consumption encourage consumers to use more power. In places with sufficient self-generated power, favorable power pricing for domestic cooking has been successfully adopted to promote power demand.

Planning, Bill Collection, and Technical Issues

When public electricity supplies were first developed in rural areas, little attention was paid to operational efficiency. Most local equipment manufacturers could not meet the required tolerance standards of the specialized factories that had traditionally produced this equipment. This meant that the energy efficiency of equipment was below internationally accepted standards, resulting in inefficient practices. Hydroelectric resources were abundant in many regions, but site development planning often failed to maximize the energy potential of a given location. Moreover, expansion of infrastructure was given priority over the operational efficiencies of the new systems. This planning included such issues as keeping distributional losses low and ways to improve bill collection. Unfortunately, many of these approaches became accepted practices and are reflected in current power system planning.

However, changing market conditions are forcing reevaluation of traditional approaches to DPC administration, planning, and operation. As indicated, drastically reduced state funding for system expansion means that other funding sources must be found. This demand means that DPCs will have to self-finance a significant proportion of new projects, and these will undergo more scrutiny. The tariff structures and levels under which the companies have traditionally operated do not provide for investment in sector expansion. The Bureau of Water Resources in consultation with the regional grid (the responsible provincial agency) sets the tariff that DPCs with SHP systems can charge to the regional grid. Unfortunately, availability of commercial loans is limited, in part because the DPCs have not been operated as autonomous, financially independent enterprises.

Customer Billing and Payments

When extending distribution lines to villagers, meters are installed in all households at the same time. To ensure accuracy, meters are installed just beyond the internal-line connection inside the household yard. In each village, the village committee selects a technician who meets training, educational, and public relations criteria to be in charge of metering and billing. Village technicians submit collected cash to township power stations. The amount charged to each village must coincide with the gross consumption metered. Employees from other township power stations monitor the main meter.

When village technicians meter and collect bills, township power stations are monitored to ensure that the money received matches the village metering. Because only one technician manages the charges for each household, corruption can occur. Thus, several safeguards have been put in place. First, each household's monthly bill is publicly posted in the village. Second, a metering team at the township level is set up to replace village technicians; team employees rotate and are responsible for villages other than their home ones. Third, the main bill submitted to the township power station lists household names and payments; soon after rural residents pay their bills, they are sent a receipt.

Residents in most rural villages have a repayment rate of about 95%, in part because Chinese custom frowns on carrying a debt. If a household does not pay its bill, either the village committee pays (a type of social welfare) or it is evenly divided among all households. Village committees pay the bills of the poorest households (e.g., elderly couples without children or members with disabilities) from collective finances or charges remitted from power stations. However, the default payment rate of rural businesses is much higher than that of rural residents, especially in areas where rural industry develops rapidly and competition bankrupts unprofitable enterprises. For example, one town in Zhejiang Province estimates that bill defaults have accumulated to more than 300,000 yuan; half of the bills cannot be collected because the enterprises have closed down.

Today there are some new approaches to customer billing. Rural residents or businesses can purchase power through local banks, reducing the workload of township power companies.

Problems of Technical and Nontechnical Losses

New hydroelectric sources for expanding generating capacity are becoming more difficult to locate and more expensive to develop. The most favorable sites for economic generation of hydroelectric power have been fully exploited in many rural areas. The few project development sites available are significantly more expensive today than in the past. Central government policies restrict the use of oil for power generation and coal (which can only be used in areas where it is indigenous). Combined with the difficulty in securing funding for system expansion, these considerations have increased the need to emphasize more efficient use of human and natural resources by optimizing the technical efficiency and reducing the financial requirements for new investments and restructuring the institutional framework of the DPCs so that they operate as commercial, financially viable enterprises.

Many of the DPCs have been poorly planned and have expanded their systems in a haphazard way. To some extent, some of the high prices found in rural areas are due to both technical and nontechnical losses because they affect the average tariff charged to rural households.

Township power stations are responsible for distribution system maintenance. Village electricians do most of the maintenance work in a village. If requested, township technicians may assist, in which case they are paid up on submitting receipts for their expenses to the township power stations.

Various options are being implemented to try to improve both the technical and nontechnical losses of the DPCs. To optimize the economic operation of substations, various options should be considered: parallel operation of multiple sets of transformers, selection of highest capacity transformers, and adjustment or stoppage of transformers when load is light to reduce losses. In addition, regular system maintenance can ensure effective DPC operations.

Nontechnical losses result mainly from internal management of DPCs because the operational costs of administrative employees and technicians (e.g., overemployment or low efficiency) are a component in calculating tariff costs. In addition, technicians' questionable or unfair practices (e.g., exempting village household leaders and their relatives from paying monthly bills) increase the average tariff for other households. There are some potential solutions to these problems. One solution is for township power stations to hire all technicians, none of whom can work in their home villages. An even more effective solution is to announce monthly bills publicly.

Hydropower Development: Lessons from Anji County, Zhejiang Province

The process of rural electrification has local origins with national support. The county of Anji in Zhejiang Province illustrates the importance of both local initiative and national policies on shaping their rural electrification program. In this section, we explore how this process worked for Anji County.

In Zhejiang Province, rural electrification followed a path familiar to the rest of China. RE evolved both by expanding the centralized national power system, including regional systems, and by developing decentralized power systems, including SHP stations and small thermal power. Hydropower resources abound in the southeastern subtropics of the province. Taking advantage of this resource during the early stages of rural electrification in the province, virtually the only source of electricity for local villages and towns was from isolated SHP stations. The next stage involved the development of an expanding network to provide electricity to rural people near the villages and towns. Finally, with the economic growth of Zhejiang and the resulting increased demand for electricity, it became increasingly difficult for SHP stations to meet all local needs for power. In response, the conventional power grid was extended from urban to rural areas in a way that complemented the SHP network.

Electricity Development and Its Effects in Anji County

Anji County has played a leading role in Zhejiang's SHP investment for almost 50 years. It has a population of 447,200, with 264 administrative villages and 19 administrative townships. It also is one of the 15 demonstration counties that participated in the second phase of the NPRECP between 1991 and 1996 in Zhejiang Province. Over this period, SHP investment grew as a result of government support, promotion of favorable policies, and the introduction of innovative financial and ownership sharing arrangements. The county also adopted new technologies and was active in renovating and updating older power plants. In late 1994, Anji was chosen as a national model county for rural electrification development. Five years later, the number of SHP stations totaled 111, with an installed capacity of 39 MW, representing 74% of exploitable capacity in the county.

The county benefited greatly from the program for the development of rural electrification, which in turn supported the development of its agriculture, township and village enterprises, and local industries. In this respect, SHP played a major role in rural electrification, enhancing local economic and social development by providing an inexpensive, secure source of electricity. Local production of both agriculture and small industry grew at a rapid rate during the period. From the early 1980s through 1999, the output value of agriculture and industry increased close to seven-fold, from 176.9 million to 1.24 billion RMB. Likewise, the county's per capita net income grew from 963 to 3,756 RMB. From 1989 to 1999, the output value of township and village enterprises climbed from 222 million to 4.65 billion RMB.

Not only did the local generation of electricity through SHP benefit economic growth in the county, but it also provided a source of revenue for the county as well. In 1999, about 118 million kWh of power generated from SHP stations was sold to the grid, netting more than 30 million RMB after taxes and expenses. In the poorer mountainous areas, SHP became the main economic resource for towns that generated and sold power to the grid. The proceeds were used to support local development in the form of building schools and administrative buildings and also to improve transportation and communication systems.

The social effect of rural electrification in the county also should not be overlooked. Rural electrification provides children more time to study during evening hours, facilitates televised educational programs, and encourages the use of electric tools to improve household productivity. For rural households, a secure power supply also boosts the standing of women by encouraging their participation in cash-earning domestic industries, such as sewing.

SHP Development Stages

The development of SHP in Anji County can be divided into several stages, which parallel those reviewed at the national level. Over time, the parties making major investments in SHP shifted from the state and county institutions to social institutions, enterprises, and farmers. This shift to multiple owners resulted from the development of a relatively clear definition of ownership and an increased emphasis on profitability.

1949–Late 1960s. During the early stages of rural electrification, the development of 10 SHP stations resulted in the replacement of oil lamps by electric lights. During this stage, the government offered technical support, but it did not invest financially or institute preferred policies. Instead, it was up to farmers and other residents in the county to provide labor and raise funds for the construction of the systems.

Late 1960s–Late 1970s. During this 10-year period, Xin'an Jiang power station was installed and interconnected to other regional electricity grids, and the county began to exploit SHP more efficiently as a result of better planning. In 1979, installed capacity exceeded 10 GW. These developments helped to free laborers from heavy manual work, improved the environment, and promoted rural economic development. The government role at this time involved the implementation of improved policies for investment, lending, taxation, electricity pricing, and connection to the outside grid. These policies accelerated SHP development throughout Anji County.

1976–1990. During this stage, there was a major SHP development in Zhangcun with the construction of Baishawu reservoir and power station. Overall, this development boosted the installed capacity in the county by 40 kW. In 1979, through the construction of Shuangyutang reservoir and two power stations, an additional capacity of 910 kW was added to the counties' power resources. Overall, during this time, 13 SHP stations were installed, with a capacity of 1,458 kW and annual power generation of 2.7 million kWh. By late 1985, 70 SHP stations had been completed, with an installed capacity of 16 MW and a generating capacity of 63 GWh. All of the stations were built with the support of state funds, community investment, and farmers' labor input.

Early 1990s–Present. In the early 1990s, a joint venture cooperation system was instituted using local funds and complemented by favorable investment policies. In 1994, Dragon Mountain station, with an installed capacity of 820 kW, was constructed using joint shares from the forestry sector and local farmers. Later, four additional stations were built, supported by a joint venture of the forestry, water conservancy, and power supply sectors. By the end of this period, installed capacity had reached 4,000 kW and power generation had reached 11.2 million kWh. Thus, in Anji County, the development of its hydroelectric power potential for generating electricity was an essential component of its growth, providing both power for local industries and revenue from the sale of electricity to the regional power grids (Table 9-4).

Pricing Policies

Before 1984, most hydroelectric power stations were fairly large-scale state-constructed stations. This type of hydroelectric power was priced quite low, and this pricing discouraged investment in SHP. At the time, this type of hydroelectric power was sold to the grid at about 0.04–0.05 yuan per kWh, compared to 0.10–0.20 yuan per kWh for conventional power. However, after 1984 there was a major policy shift in pricing SHP. During implementation of the NPRECP in Zhejiang Province, the price that the grid was required to pay the generators for SHP more than doubled, and even higher prices were possible for

Table 9-4. Investment in SHP Station Construction in China (Zhangcun, Anji County)

Station	Installed Capacity (kW)	Start-Up Year	Ownership	Total Investment	Government Subsidy	Community (including labor)	Public Sector	Collective Sector	Private Sector
						Investments by Source (1,000 yuan)			
Baima	73	1970	Community	200	20	150			
Baisha	100	1971	Community	120	20	180			
Changtan	40	1971	Community	60	20	100			
Furang	30	1972	Community	40	20	40			
Shmen	160	1974	Community	112	15	25			
Gao'er	40	1977	Community	33	20	92			
Huangtianwan	75	1979	Community	152.5	15	18			
First project, Shuangyutang	500	1981	Shareholder	2,450	40	112.5		460	800
Second project, Shuangyutang	410	1985	Shareholder	410	1,190			82	75
First grade, Dragon Mountain	820	1995	Shareholder	1,800	253			500	1,300
Shanshui	400	1996	Shareholder	1,500				1,000	500
Langcun	75	1997	Shareholder	258				150	
Shimenkan	125	1997	Shareholder	450		108			450
Mafeng'an	1,130	1998	Shareholder	3,900				2,000	1,900
Longkou	1,320	1999	Shareholder	3,500			960	1,640	900
Total	5,298			14,985.5	1,613	825.5	960	5,832	5,925

Source: Yao 2000.

new generators. This pricing adjustment stimulated SHP development and, after 1994, the SHP price was set for each generator (Table 9-5).

In addition to the price of SHP sold to the grid, the market price of power sold to end users has also varied by development stage. Before 1990, the provincial government asked that each power enterprise adopt a principle of "no profit and no loss" when selling SHP to the grid and distributing it to end users. Therefore, the market price of power was determined using this principle. During the electricity supply shortage of 1991–1992, various levels of government encouraged local residents to raise as much funding as possible to construct new power stations. To encourage local development of SHP, the government decided to provide a higher price for power generated by self-funded stations. The result was a price of 0.25 yuan per kWh for power controlled and adjusted by the state and a price of 0.38 yuan per kWh for power from self-funded stations. This

Table 9-5. *Evolution of SHP Pricing Adjustments, Zhejiang Province in China*

Adjustment Year	Classification of SHP Stations	Price of Power Sold to Grid (yuan/kWh)
Before 1974	All SHP stations	0.294
1975	Counties supplied directly by state-owned power stations	0.042
	Counties that purchased power from the national grid in a batch	0.05
1984 (starting of first NPRECP counties)	Existing stations	0.05
	Newly constructed stations	0.10
1990 (starting of second NPRECP counties)	Existing stations	0.10
	Newly constructed stations	0.20
1994	Stations operational before 1984	0.23
	Stations operational during 1986–1990	0.29
	Stations operational during 1991–1993	0.39
	Stations operational after 1994	0.45
1996	Stations operational before 1990	0.32
	Stations operational during 1991–1993	0.41
	Stations operational after 1994	0.45
1998	Stations operational before 1990	0.34
	Others	No charge
2000	Remit 0.02 yuan/kWh as SHP development fund to all stations	
	Stations operational before 1990	0.33
	Stations operational during 1991–1993	0.39
	Stations operational after 1994	0.45

Source: Yao 2000.

pricing policy effectively boosted investment for SHP construction from diverse sources, especially the private sector.

Favorable Tax Policy, Capital Subsidies, and Ownership Rights

To encourage SHP development, the state issued a preferential tax policy for SHP compared to industry. In poor and remote areas, the provincial government allocates a certain portion of its annual budget to subsidize the construction of new stations or reconstruction of older ones. The subsidy reduces the amount of construction capital necessary for the development of SHP and thus makes it more attractive to investors.

The local and private ownership of SHP generators also was a hallmark of the program. Historically, most SHP stations in Zhejiang had been under state or collective ownership. Even under this situation, however, government sectors can intervene in station operation by reassigning managers, resulting in potential corruption.

With preferential policies on power pricing, purchase quotas, and taxation in place, the state encouraged local authorities to implement property rights reform for SHP. Because Zhejiang is one of China's most developed provinces, property rights reform of SHP stations started quite early.

In 1998, the provincial Department of Water Conservancy issued an official document to enforce implementation of the reform. The reform of SHP station property rights in Zhejiang has the following components. First it linked property rights to the restructuring of station ownership (e.g., shareholder, cooperative, and multiple forms of ownership). Second, it transferred ownership of smaller scale, state- or collective-owned stations with poor profitability to any purchaser to reclaim national capital and avoid loss of state-owned assets. Third, it adopted a long-term lease system for stations equipped with reservoirs used mainly for flood prevention and irrigation to separate ownership from management. The owner of the reservoir still owns the station and receives profits from the station operation; however, the use rights are transferred to station employees or other enterprises to improve station profitability and maintain smooth operation of the reservoirs.

Finally, under the leadership of the Ministry of Water Conservancy, the reforms transformed hydroelectric power investment companies from commune or county owned systems to a blend of both public and private ownership depending on the local situation.

In the past, under the planned economic system, stations were state-owned and profits went to the water conservancy sector. The station operation involved no risk because, even when a shortage of electricity supply or a loss occurred, station profitability was usually ensured through preferential policies or government subsidies. By contrast, under the new market-oriented system after the property rights reform, the government allocates fewer subsidies but uses more incentive policies.

Furthermore, obligations between administrative agencies and power enterprises were clearly defined. The generating stations were no longer part of the water conservancy sector but rather belonged to the companies' boards of directors or owners, who were responsible for making profits. This setup means that the

Ministry of Water and Power can focus more on general principles of development at the macro level and providing technical assistance and supervision needed by power generation enterprises. For example, if an individual station has a conflict with the power grid, it has little influence by itself because of its small size. Thus, some counties have set up SHP industry associations to negotiate with the grid on the price and sale of power. Working out the details of the ownership patterns, the pricing policies, the subsidies, and other matters was no small accomplishment. The measures not only benefited SHP development in Zhejiang but also provided valuable lessons to other provinces throughout China.

Lessons from Decentralized Electricity Development

A question can be raised as to whether the unique experience of China's road to successful rural electrification can be duplicated in other countries. The answer to this question is not straightforward. If the implication is that others could wholly adopt the aspects of China's approach to rural electrification, the answer is probably no. The Chinese approach to rural electrification, as is apparent in other countries as well, is embedded in the political process. However, many aspects of the Chinese program can be replicated or adapted to the situation in other countries.

The rural electrification sector has allowed for multiple forms of public and private investors. Moreover, policies favorable for investment—including pricing, taxation, and financing—have been essential to rural electrification success in China, especially during initial program stages. China's planning-oriented economic system, combined with state and community ownership, has been important in promoting rural electrification development. This is in contrast to most of the country's other sectors, which have involved a significant degree of state ownership.

Regional Focus of Grid-Based Rural Electrification Development

Given the range of economic differences between regions, China used a step-by-step strategy to develop grid-based rural electrification appropriate to different regions of the country. In relatively poor regions, the focus was on improving rural residents' access to power; in more developed regions with limited, low-quality supply, a greater emphasis was placed on improving electricity supply. Finally, in rural regions with sufficient supply, the program sought to improve service levels to those of urban areas. Thus, the program provided important areas of support to local electricity systems based on their overall needs.

The success of SHP development is an example of this strategy. Since the 1960s, three principles have guided SHP development, as follows:

- Self-construction—Local governments and populations are encouraged to use local water resources, technology, and materials to build SHP stations under a unified system of planning. Counties raise capital investment, and, in some areas, local producers manufacture some components of SHP generation equipment.

- Self-management—Investors own and manage SHP stations; this situation can avoid administrative interference in economic laws and can protect local communities' enthusiasm for developing SHP.
- Self-use—Electricity produced by SHP stations should be used locally if stations have their own power-supply regions and can form integrated markets. This requirement includes SHP generation, power supply, and electricity consumption. The conventional grid is not allowed to compete in these markets.

The central government policies encouraged rather than discouraged local companies to participate in rural electrification development. The process involved state-funded construction of pilot projects, but local governments and property owners are the primary service providers and must take responsibility for development planning, power pricing, station and grid management, and facility manufacturing.

Multisector Coordination of Planning and Standards

Rural electrification planning is closely linked to local economic and agricultural development planning. The level of success of rural electrification is associated with the status of the local economy, but it is a precondition for promoting rural economic development, including agricultural production. Rural electrification planning for local development involves assessing energy options, coordinating the development of the grid and other power sources with other development plans, focusing on improved economic profitability, and ensuring that the environment is protected.

According to each county's situation, a development plan is created that includes a baseline study. The analysis includes evaluation of current problems, forecasting of power load (based on population growth and economic development), and planning of the distribution grid. This planning generally results in a fundraising strategy, an annual work plan, and measures to ensure successful implementation.

Appropriate Low Cost System Design to Reduce Costs

Since the 1990s, standardization of technologies has been stressed. In grid construction, new technologies and standards have been strongly recommended to ensure a safe, reliable, and economic power supply. Well-planned distribution lines and substations will both reduce costs and improve electricity service. The construction of the local grids is based on previous experience involving transformer distribution and the optimal length of electricity lines. In developed regions or areas with longer distribution lines, increased substation allocation was suggested. In rural villages, the layout of low-voltage distribution systems must take into account village planning and conform to the low-voltage power regulations.

The system design issues involving SHP technologies include many relevant factors. Government-funded research and development has led to ways to reduce overall costs and improve efficiency and reliability. Moreover, at the outset of each project, qualified experts are hired to develop a detailed design and conduct a feasibility study. By investing in optimal design and careful planning, operational benefits are ensured over the long term.

Training of Electricity Company Staff

The SPC requires systematic training of employees involved in rural electrification. Individuals must obtain certification approved by the Ministry of Labor before being hired by electricity companies. Once hired, technicians receive orientation training and must take courses regularly to update their skills. Companies and bureaus pay for training, with the exception of national rural electrification workshops and seminars, which the SPC sponsors on an occasional basis. At such workshops, staff from institutions at all levels share experiences, discuss new policies, and learn about the latest grid-based rural electrification technologies. Moreover, certain companies have established their own universities and training centers.

Training also has been vital to SHP success, and it is conducted at national, provincial, and county levels. State funds are used to train local technicians in SHP design, construction, installation, maintenance, operation, and administration. Twelve universities in China have special SHP departments. Most students who enroll in SHP courses later become trainers, thus transferring their knowledge to local technicians. DPCs and counties supplied by large grids are trained in power distribution and customer service. Training takes many forms—from in-class lectures given by expert technicians to study tours and on-site information sharing. The state emphasizes the development of training manuals, whose content is developed by either the water resource or power sector. For the most part, trainees are subsidized.

Financial Viability and Development of a Favorable Climate for Investments

The ownership status of China's electricity companies is complex, with a mix of state, county, and local control over the companies. The policies developed for investors have the overriding principle that the companies should be profitable to maintain and expand service to their customers. From 1958 to 1978, an important ownership principle for rural electrification was developed and accepted: those who invested could own, manage, and benefit from their investments, even though state ownership had dominated in the past. Adoption of this principle encouraged multiple owners to be involved in rural electrification development, provided the impetus for the development of shareholding, and later limited privatization of rural electrification.

The grid distribution system typically involves a combination of state and community ownership. The SPC and the county power bureaus are involved in certain types of ownership for rural electrification systems. At the local level, the county power bureau usually invests in and constructs the high-voltage distribution grid and then charges users an added-value price, which is used to repay the loan. Although all paying users own the system in theory, it is generally agreed in practice that the county power bureau owns the high-voltage grid. For the low-voltage distribution system, the villagers own the system because they directly contribute primary capital to it.

The main investors in SHP systems are recognized as their owners by national regulations. When the central or provincial government is the investor, the system is state-owned. When the township government or village committee is the major investor, the system is collectively owned. Similarly, when private enterprises invest, systems are privately owned.

Local supply bureaus that purchase power from the SPC grid have been allowed to make a profit from the local sales for their own sustained development. Similarly, power sector enterprises can make profits from local SHP stations and local power grids to use for reinvestment in further expansion and development. Profit distribution in power supply sectors varies by county. In those counties directly supplied by the SPC, power supply and generation profits are combined. In consultation with local branch agencies, the SPC invests in rural grid construction, using profits mainly from the power generation sector, as well as the supply system; consequently, grid expansion and improved power quality stimulate the use of rural power and enhance the development of generators.

The price at which an SHP station sells power to the grid affects each station's capital accumulation and development capacity, which, in turn, affects domestic and foreign investor confidence. Even during the electricity shortage, the state made it clear that the amount of power produced by SHP stations could be priced locally, according to market demand. In addition, the price of SHP is determined by the generation cost, plus tax and profits.

To encourage rural electrification expansion in counties that buy power from the grid, the state recommended tax incentives, including a "free income tax" policy or charging an income tax and then returning it to the power sector. Given that China's production tax rate is about 33%, the "free income tax" policy benefits rural electrification development remarkably. The final decision on such incentives was made by local authorities because income tax is directed to local finances. In economically developed counties whose local governments emphasize rural electrification development, this measure has been successfully implemented.

Types of Investments and Subsidies in RE

The investments for rural electrification in China involve subsidized loans with long repayment periods. This system is similar to the type of investment terms other countries have used to fund rural electrification programs. As an example, in 1998 for the rural grid construction and reform program covering more than 2,000 counties, the state provided loans of about 180 billion yuan, with annual interest rates of 5–8% (a little less than other types of loans) and with a favorable repayment period of 10–20 years. The next year, the State Council announced that individuals or sectors legally responsible for projects on rural grid reform could negotiate repayment periods with banks up to 20 years. This type of loan is essential for rural electrification because of the high costs of equipment that last for 20 years and sometimes longer. The repayment period for rural electrification is much longer than that for agricultural projects, which is usually less than five years.

To encourage investment in rural grid construction, state banks allocated loans to developers, based on an analysis of the construction projects and capacity to repay. Mortgages were not required for the power bureaus because the banks were owned and managed by the government. However, commercial banks offered power sector developers mortgage loans, thereby ensuring successful operation of banks and helping developers make careful investment decisions.

In addition, low-interest loans are available for SHP development. To encourage towns and villages to construct SHP systems, a low interest rate of 3.6% has been adopted. The loan repayment period for the purchase of equipment was extended from the normal 3–5 years to 10 years for the developers. Currently, money-lending institutions are the main funding source for local SHP development. The Agriculture Bank of China, the Industry and Commerce Bank of China, and the Construction Bank of China have special loans for rural SHP development with extended payment periods and low interest rates that can be deducted from business profits.

Conclusion

Today, most rural villages in China have access to electricity. This vast achievement was accomplished with a mix of incentives for local involvement through an evolving process of national support. Although one is struck with the accomplishments of the local electricity companies in rural China, the role of supporting agencies in local and state administrative agencies should not be underestimated. The role of the state was conceived as providing assistance and support for the overall process of rural electrification by creating local incentives for providing electricity service.

The result was a unique blend of local accomplishments made possible by a complex system of central support. The old question of whether electricity growth supports development or is a result of development did not bother the Chinese. China's rural electrification development strategy evolved to support electricity development in ways that were consistent with local conditions. The effort can be characterized by a step-by-step trial and error process in solving the challenge of providing electricity to more than 1 billion people.

Acknowledgment

The authors thank Norma Adams for editing and making substantial contributions to the original version of this chapter.

CHAPTER 10

The New Deal for Electricity in the United States, 1930–1950

Paul Wolman

T HE WORLD HAS LONG SEEN the United States as a developed country, but its situation 65 or 70 years ago was not nearly so monolithic. In the mid-1930s, rural areas in the United States lagged significantly behind metropolitan centers in development. U.S. agriculture suffered from a long-term depression. Electrification of rural households was barely 10%, compared with almost 85% of urban households. In these and other ways, the unevenness of U.S development in the early twentieth century resembles that of many of today's developing countries. One might therefore expect the U.S. experience of rural electrification to contain some lessons for today's rural electrification programs.

Large-scale energy corporations dominate the U.S. market and U.S. energy development activities abroad and have done so for generations. It therefore may be surprising for some to recognize that large corporate utilities did not lead rural electrification in the United States. Instead, rural electrification grew from locally run nonprofit cooperatives advised and financed by a federal agency, the Rural Electrification Administration (REA). Another federal agency, the Tennessee Valley Authority (TVA), built hydroelectric and flood control projects in the lower Midwest and South. It provided low-cost power for more than a quarter of the early cooperatives (Brown 1980, 73). With federal assistance, rural electric cooperatives grew within the initial 20 years of the U.S. rural electrification program's founding to serve more than 4 million of the nation's poorest and most dispersed people. The program's part in this growth is especially impressive, considering the federal REA's close-to-market-rate financing and the local cooperatives' economical costs of line construction, self-sustaining fiscal status, and rapid amortization of loans for construction of distribution networks and home wiring.

Overall, rural electrification in the United States increased during this period from about 13% to 94%. That is a stunning achievement. Electricity production and distribution by the federal–cooperative partnership was only a small fraction of the overall U.S. market then, and it remains so now. Was the U.S. rural electrification program essential in electrifying rural America? Or was it merely an artifact of Depression-era deficit-spending programs that slightly accelerated what would have occurred soon anyway? Those are difficult notions to prove or disprove definitively. Still, it is clear that even with electrification, development was slow in many areas of the rural South and Southwest. They remain part of the "other America" that Michael Harrington (1962) described in the 1960s—lagging in health care, nutrition, education, and income. But another question may be more valuable to explore in the present context. Does the formative experience of U.S. rural electrification, with its emphasis on federal–cooperative partnership, bear valuable lessons for developing countries? A review of the growth of U.S. rural electrification and its political, economic, organizational, geographic, and social dimensions may provide some answers.

The Early U.S. Electrical Industry

The distinctiveness of the federal–cooperative partnership for rural electrification that emerged in the 1930s is better understood in the context of the industry's evolution as a system dominated by large, integrated corporate utilities. Some historians have seen the industry as molded by technological innovation and economies of scale (Hughes 1983, *464–465*). Yet social, geographic, and political factors were clearly at work as well. Indeed, the same factors that hastened electrification in U.S. industries and cities also retarded electrification in households, in the countryside, and on the farm.

Origins of Grid-Based Electrification

Almost from the start, the electric power industry in both Europe and the United States tended toward central-station, grid-based systems. Investment bankers such as J. P. Morgan and entrepreneur-innovators such as George Westinghouse focused in particular on the building of regional electrical networks in the United States (U.S. Department of Energy 1996). Crucial technological advances in alternating current (AC) also enabled the transmission of higher voltage and lower current power over longer distances—essential in the geographically extensive United States.

The earliest U.S. urban grid was based at Thomas Edison's Pearl Street direct current (DC) generation station, serving lower Manhattan. However, AC systems soon dominated. AC conserved on expensive copper wire and incurred lower resistance losses (Friedlander 1996, *38*). AC technologies also gave industrial, commercial, and municipal consumers the flexibility to vary the voltage and application of power—for lighting, heat, unit-driven motors of various horsepower ratings, streetcar and railway traction, and so on (Hughes 1983, *86–87*). Westinghouse facilitated the rapid dominance of AC by acquiring crucial transformer patents from the French inventor Lucien Gaulard and his British business partner, John D. Gibbs. Westinghouse also used Nicola Tesla's patents for multiple-phase AC generators as the basis for his 1896 Niagara Falls project, which initiated large-scale generation and transmission of AC power (U.S. Department of Energy 1996). Other innovations between the 1880s and 1903, such as steam and gas turbines and AC metering, supported large-scale generation and use of electricity (Hyman 1995).

The exploitation of highly productive oil and natural gas fields in Texas in 1901 and in Oklahoma in 1905 kept the costs of generating electricity comparatively low in the early twentieth century. These developments also made petroleum products readily available and substitutable for industrial and commercial heat, light, and motor power (Yergin 1991, *82–87*), thus holding back the rise of electricity prices. Production for World War I rapidly pushed the use of electricity to almost 50% of industry's energy consumption (DuBoff 1979, *53*; Nye 1990, *261*). Increased demand from wartime and peacetime industry and from urban commercial and residential consumers led to the emergence of regional, turbine-based power systems. Depending on their locations, these systems often had access to more than one fuel (e.g., coal, oil, gas, or hydroelectric power), which helped them to inaugurate flexible load dispatching, or, as it was then called, *rationalization*. For example, when hydropower generation became impossible, as when droughts affected river sources or freezing curtailed operations at high-head plants, utilities could use coal or fuel oil to generate electricity in a complementary fashion (Hughes 1983, *367–368*).

A regional system based on low-cost turbine generation and AC power had another advantage in serving the densely populated but widely separated urban and production centers of the United States. Utilities could spend rapidly on distribution networks in and near the cities. Urban-centered regionalism also had a cost for the United States, however. It left rural areas, except those immediately around cities and large towns, mostly unserved. In contrast, most Western European countries, whose rural areas were less extensive, substantially outpaced the United States in rural electrification before 1930. Compare, for example, the rural electrification rates at the time of Czechoslovakia (70%); France and Denmark (50%); and Finland (40%) with that of the United States (10.4%). Germany, New Zealand, and Sweden also outpaced the United States (U.S. Census Bureau 1976, 827, series 110; Brown 1980).

Another distinctive feature of the early electrification period was that household electrification trailed significantly behind industrial and commercial electrification because of the comparatively high investments required for local distribution systems, home wiring, and appliances. In 1907, only 8% of all U.S. residences had electric service. The gradual dispersion of electrification into the household sector in the United States took place according to the social ladder of energy consumption that later became typical in many developing countries. That is, the wealthier households became the earliest adopters of successively more "modern" fuels (World Bank 1996). Thus, high-income, urban households, better able to afford wiring, piping, and appliances, were the first to adopt electric lighting, complementing it with more common fuels and newer appliances, such as natural gas for hot water and cooking and fuel oil for heat. Middle-income households adopted electricity for lighting at lower rates and continued to use coal for cooking and space heating. Poor urban and rural consumers remained dependent on wood, coal, and kerosene (Rose 1984). This broad distribution of energy sources remained typical for several decades in the household sector.

In the United States of the early twentieth century, the spread of household electrification was also slowed by the fact that household technologies were then just evolving. Even in the cities, appliances such as hot water heaters, refrigerators, washing machines, irons, and vacuum cleaners did not gain true mass markets until the 1920s and 1930s (DuBoff 1979, *53*; Nye 1990, *261*; Friedlander 1996, *3–4, 66*). Household appliances reached most rural areas even later.

Structure of the Electrical Industry

Apart from factors of geography, resources, technology, and demography, the way the electrical industry evolved suggests why electrification of rural areas did not become a priority in the United States. In the late nineteenth century, U.S. municipalities granted private utilities extensive, and sometimes overlapping, distribution franchises in efforts to extend service and encourage price competition. From such struggles, large corporations typically emerged to dominate the nation's major cities. For example, in Chicago, Samuel Insull, head of Commonwealth Edison, bought out 20 competitors by 1907, innovated the use of large, high-capacity steam turbines to cut generation costs, and gained market share by increasing the utility's load factor through off-peak sales to industrial consumers (Anderson 1980; Hirsh 2002). After all, why should industries maintain their own steam plants and facilities for exhausting waste heat? An electricity consumer needed to pay only for electricity metered and used. Was it not then tempting simply to purchase the consolidated utilities' distributed power (Hughes 1983)? Many consumers agreed in Chicago. In Detroit and New York, too, the electric utility business underwent substantial consolidation, and, overall, vertically integrated private utilities controlled about 40% of the market early in the twentieth century.

The electrical industry, among many others, was undergoing a wave of consolidations, as corporations rapidly engrossed proprietary, unintegrated, and smaller-scale enterprises. One historian called the 1882–1907 period one of "regulation by private competition" (Hellman 1972, *8–9*). Other interpretations have emphasized technology (particularly the efficiency of capacity utilization) as the driving force for industrial development and consolidation (Hughes 1983; Chandler 1990; U.S. Department of Energy 1996). Overall, the period of corporate reorganization is seen most comprehensively as an epochal shift permeating social relations, laws, institutions, values, and policies as well as economic alignments and technical development (Sklar 1988, *14–20*).

Despite the corporate consolidations, however, some enterprises, especially industries, retained an interest in maintaining their own generation plants and in conserving the option of selling excess power directly to other consumers or back into the grid. In particular, heavy electricity users—such as traction companies, electric railroads, and smelters—were motivated to continue self-generation and resisted corporate utility blandishments to buy central-station power. So did commercial and large residential buildings that generated steam for district heating with electricity as a by-product (McGuire and Granovetter 1998). These groups lacked coordination, but they still controlled 52% of generation in the early twentieth century. Industrial generation of electricity actually outpaced that of the utilities at a national level until 1912 (U.S. Census Bureau 1976, 820, series 36, 40).

Another obstacle to complete corporate domination of the early electric utility industry was the fact that municipalities engaged in nonprofit power generation and franchising of private generation. Although municipalities controlled only about 8% of gross generation, public utilities still represented 27% of the facilities in operation before World War I (U.S. Department of Commerce 1915). More-over, these facilities did not peak in number—at more than 2,500—until 1922 (Hirsh 2002). An important reason that municipalities continued to found and operate nonprofit utilities was their desire to ensure stable electricity prices and reliable public service. Responding to the increasing rates of consolidated utility providers, cities would generate their own power, usually on a limited scale, and would license new franchises in an effort to limit prices through competition. Supported in these efforts by the League of American Municipalities, the National Municipal League, and the American Federation of Labor, cities would also issue decrees against price-fixing. However, large utilities often bought out new fran-chises or drove them into receivership, and here municipal decrees often had little effect. Moreover, franchising was often accompanied by flagrant public corruption. In Chicago, for example, the "Gray Wolves," a group of city councilmen, became notorious for selling their votes to prospective franchisees. Other impediments to efficient public service were frequent and costly lawsuits over franchise rights and municipalities' inability to control prices and practices as utilities spread beyond city limits (two leaders in nonprofit rural electrification, Morris Cooke and David Lilienthal, had prior experience in utility litigation in Philadelphia and Chicago, respectively; see, e.g., Cooke 1948).

The tumult of consolidations, public corruption, and inefficiencies of competi-tion in urban electricity markets raised some vital questions about the relation-ship of public and private entities and nonprofit and for-profit utilities and about the nature of regulation in the sector. State-based investigations in New York and Wisconsin soon proposed regulation that led to the establishment of public service commissions in many states (Anderson 1980).

The New York initiatives led to the first state public service commission in the United States, founded in 1905 under Governor Charles Evans Hughes (Hellman 1972). The Wisconsin regulatory initiative was developed under Governor Robert M. La Follette, assisted by economist John R. Commons. Governor Woodrow Wilson of New Jersey also engaged the issue vigorously. These reform governors may have had as much concern with corruption of municipal franchising as with utility abuses. However, it is also significant that it was in the battles over state regulation that these governors—all future presidential candidates—developed some of their basic ideas on the need for regulatory standards for corporate enterprise (Anderson 1980).

An important early episode in the development of utility regulation came in 1905–1907, when the National Civic Federation (NCF) sponsored a major study of the subject (National Civic Federation 1907). Although divided in its con-clusions, the NCF panel ultimately proposed that ownership of utilities should devolve to municipalities or to the states if regulation of the private sector failed to halt price abuses and corruption and when cities were able to develop a competent corps of engineers and business managers (Hellman 1972).

The move toward state regulation, however, was ultimately not a move against corporate influence or consolidation. In fact, the electric utility industry's trade association, the National Electric Light Association (NELA), soon embraced a notion of state-level regulation that included licensing, rate-limiting rules, and prescribed standards of service. These standards, NELA proposed, would rationalize the hitherto destructive competition and corruption accompanying franchising and would establish public utilities formally as natural monopolies (Hellman 1972). Falling kilowatt-hour prices of electricity, along with the fact that the utilities were then constructing distribution lines along public rights of way, bolstered industry claims to natural monopoly status. ("The critical . . . characteristic of natural monopoly is an inherent tendency to decreasing costs over the entire extent of the market . . . as more output is concentrated in a single supplier"; Kahn 1988, *119*, quoted in U.S. Department of Energy 1996, *1*.) Between 1907 and 1927, the real price of electricity fell by 55% (Hirsh 2002.)

NELA's position was articulated by its Committee on Public Policy, made up of industry notables such as Samuel Insull of Commonwealth Edison; Charles L. Edgar, president of Boston Edison; and Midwestern gas and electricity entrepreneur Henry L. Doherty.

Also among NELA's goals was limiting competition from the independent industrial and commercial generation plants, which Doherty called a "terror . . . [that the private utilities must] greatly curtail" (McGuire and Granovetter 1998, *5*). Insull, acting as NELA's representative in the NCF forum, also saw state-level regulation as a barrier against public ownership, predicting that such regulation of "our public service corporations" would hasten the unhappy happy day "when few examples of municipal ownership and operation of a public utility for the supply of gas or electricity will be found in existence" (National Civic Federation 1907, cited in Hellman 1972, *17*).

By 1914, dozens of states had regulatory commissions, with the tendency to adopt regulations and rate limitations correlating strongly with the presence of large-scale industrial consumers of electricity as well as large electricity utilities and coal-mining interests (Knittel 2006). The utilities thus benefited from more stable relations with their major consumers, refereed by state commissions and courts.

Doherty and Insull appeared correct in their anticipation that state regulation would set ground rules facilitating the private utilities' dominance of the industry. Significantly, as well, although the utilities were technically susceptible to federal regulation under the Sherman Antitrust Act of 1890, they remained largely unhindered by federal law before the 1930s. The Republican administrations of Theodore Roosevelt and William Howard Taft left the electrical industry virtually untouched. Although Theodore Roosevelt in particular has enjoyed a reputation as a "trust buster," his role was rather that of an executive overseer of corporations. He created a Bureau of Corporations as an executive branch intermediary with corporations that sought to supersede the legislative and judicial roles in regulation. His basic vision, like that of his successors in office, was of the emerging corporations as the primary units of productive activity, subject to a degree of state and federal regulation; he never supported efforts to dismantle or hobble corporations in either domestic or foreign trade (Sklar 1988, *172*; Wolman 1992, Chapters 4 and 5).

Under the Woodrow Wilson administration, the Federal Trade Commission and Clayton Acts of 1914 allowed enforceable legal regulation of monopolistic competition (Sklar 1988, *169–173*). The Wilson period also saw the establishment of the Federal Power Commission (FPC) in 1920. However, the Republican administrations of the 1920s essentially skirted questions of federal regulation in the industry, and the FPC had no staff until 1928. The FPC did not begin to regulate interstate utility activity effectively until the passage of the Federal Power Act in 1935. In 1977, Congress reorganized the FPC as the Federal Energy Regulatory Commission.

In all, then, an array of technological, geographic, social, political, and economic factors favored consolidated corporate utilities. Nationwide, the continuation of mergers and establishment of holding companies swallowed up independent electric utilities. The first of these holding companies was United Electric Securities (later Electric Securities Company), formed in 1890 by Thomson-Houston, whose merger in 1892 with Edison Electric created General Electric. Other major holding companies followed, including American Gas and Electric (1906), American Power and Light (1907), and National Power and Light (1921) (Friedlander 1996, *78–79*).

By 1924, the top 16 holding companies controlled about 75% of U.S. electrical generation, and the top 7 controlled 40% of generation capacity (Hellman 1972; Rudolph and Ridley 1986, *38–41*; Friedlander 1996, *80*). By 1930, Insull's Commonwealth Edison had become a base for a $3 billion system of 60 local utilities, transit companies, and interstate holding companies generating 12% of U.S. electric power (Friedlander 1996, *80*).

As one historian of technology put it, the bulk of U.S. generation before 1930 flowed into "transmission and distribution to industry, traction, and the economic classes able to pay for luxury lighting" (Hughes 1983, *464–465*). As Hughes (1983) put it more gnomically, "Electric power systems before 1930 were mostly artifacts that manifested culture" (*465*).

The expansion, consolidation, and coordination of utilities between 1900 and 1930 had striking results in increasing urban electrification and in closing the gap between industrial and household electrification in the cities and suburbs. By 1930, U.S. urban and nonfarm households were almost 85% electrified (U.S. Census Bureau 1976, 827, series 111). But the picture of an efficient, rationalized, natural monopoly industry would change with the Great Depression, which witnessed the catastrophic failure of a number of major utility holding companies, including Samuel Insull's. The Depression would also help bring issues of social development and equity into the mainstream of policy debate, including the significant and largely unaddressed lag in U.S. rural electrification.

The Problem of Rural Electrification in the 1930s

In the period between the two World Wars, the rural areas of the United States still contained a substantial portion—about 44%—of the total U.S. population of 123 million (U.S. Census Bureau 1934). Some 50 million rural residents lived in unincorporated areas or in incorporated towns and villages with populations of fewer than 1,500. (This figure, derived from the Census of 1930, would become the statutory criterion of eligibility for the federal rural electrification program in 1935 [Muller 1944, 6].) More than 30 million of these residents were "farm dwellers." The rest were "village dwellers," who might be working in small industry or mining or other nonfarm activities (U.S. Census Bureau 1934). In 1935, only 12.6% of such households had electricity (Muller 1944; U.S. Census Bureau 1976).

U.S. rural areas were indeed more difficult to electrify than urban areas. One of the central reasons the rural sector lagged in electrification was its lack of major industrial demand centers, except for some extractive industries. Moreover, agriculture still depended more on petroleum fuels and even animal and human power than it did on electricity. Economic depression, rural poverty, and the unfamiliarity of rural populations with new electric appliances further inhibited demand (Muller 1944, 10–11).

Perhaps most important, rural enterprises and populations clearly lacked the demand and density of industries and cities. Much of the inhabited "rural territory" of the United States of the 1930s had population densities as low as a dozen or two per square mile, whereas a U.S. city might have a density of tens of thousands of electricity consumers per square mile and a concentrated array of electricity-using industrial and commercial ventures, as well as households. For example, in 1930, the city of St. Louis, Missouri, had a population of about 822,000 in a 61-square-mile area, and therefore a density per square mile of 13,475. In contrast, Benton County, a predominantly agricultural jurisdiction in central Missouri, had a population of less than 12,000 over a 745-square-mile area in 1930 and therefore a density per square mile of 16 (OSEDA 2000).

The 50 million rural residents of the United States in 1930 were also more distant from official and quasiofficial authority, compared with urban dwellers organized within neighborhoods, wards, and other urban and suburban jurisdictions. Institutions such as the National Grange, Farmers' Union, and Farm Bureau exerted some influence in rural areas, along with churches, existing farm cooperatives, and the Country Life Movement. Retail marketers were reaching rural citizens through catalog sales, and consolidated implement manufacturers were able to provide credit on good terms for better-off farmers who were purchasers of farm machinery and supplies. However, hostility toward government, xenophobia, racism, and weak connections with the cash nexus were factors that made many farm communities appear less than ideal as targets for long-term investment in infrastructure (Taylor 1942; Danbom 1979; Wolman 1992).

Problems distinctive to U.S. culture and the Southern economy of the early twentieth century also complicated the challenge of rural electrification. U.S. agriculture suffered from severe droughts during the 1920s and 1930s and from high levels of farm mortgage debt and foreclosures, which in turn fed into high rates of farm tenancy (42.1% in 1935) (Taylor 1942). The major farm organizations lost membership early during the Depression (Taylor 1942; Danbom 1979). Lagging education, health, and nutrition were endemic as well, leading some commentators, perhaps hyperbolically, to compare the living conditions of the

U.S. rural population unfavorably to those of the European peasantry (Box 10-1) (Keun 1937).

Much of the rural population unserved by electrification was in the U.S. South. Of the U.S. regions, the South Central—an area that historically lagged in both industrial and agricultural development in comparison with the rest of the country—was only 3% electrified in 1930, unchanged from 1920 (Muller 1944; Brown 1980).

The Early Role of Private Utilities in the Rural Sector

Even the earliest electrification pioneers, such as Thomas Edison, gave some thought to rural applications (Friedlander 1996, 37). But the self-contained DC generators Edison deployed in rural areas were intended to energize telephone and telegraph networks—not farmhouses, pump houses, or henhouses. The electric power industry's trade association, NELA, officially—if somewhat offhandedly—noted the lack of penetration of the rural market in 1911. NELA also encouraged the U.S. Department of Agriculture to urge farmers to consider the potential benefits of electrification and prompted the Census Bureau to collect statistics on rural use of electricity in 1913 (Brown 1980). These were little more

Box 10-1. Odette Keun on Farm Life in the United States, 1937

There are districts in West Virginia, East Tennessee, Kentucky, where the mode of material existence is not different from that of the first settlers, over a century and a half ago. (Useless to fall back upon the facile plea: "That's the South!" The South is American, isn't it?) Even when I visited the better-off farms, I discovered that a very large percentage of them had kitchens with ovens burning wood—the poor cooking in pots and pans over a little fire on the hearth, as in the Middle Ages; that they were lighted by dim, smoking, smelly oil lamps, that the washing of clothes was done by hand in antiquated tubs; that the water was brought into the house by the women and children, from wells invariably situated at inconvenient and tiring distances, for it appears to be one of the milder manias of the American farmer, to sink his well as far away as possible instead of near the front door, under trees, as the European peasant does. Ordinarily there is no icebox, so many products that might be grown to vary the horribly monotonous diet are out of the question: they could not be stored. (Nothing could be stored in the warm regions if it weren't for the springhouses. And not everybody has them!) Of the fifty million horsepower required by farms, 61 percent is still furnished by animals and only 6 percent by electric stations. About 90 percent of the citizens on farms, say the statistics, do not have the lighting and the simple comforts that have become a commonplace in most middle-class dwellings in urban communities. It's nothing to brag about. (Keun 1937; 29–31)

Note: Odette Keun (1888–1978), the French-based daughter of a Dutch diplomat, was a world traveler (notably in central Asia), leftist activist, novelist, and sometime companion of the writer H.G. Wells.

than gestures, however, and the private sector took no concerted action of its own on RE before World War I.

After the war, the industry's efforts to promote rural electrification (RE) progressed only slightly. In 1923, in cooperation with the American Farm Bureau Federation and a number of state agricultural colleges, NELA created the Committee on the Relation of Electricity to Agriculture (CREA) in an effort to test household and commercial prospects in the sector (Brown 1980, 4–5). CREA sponsored experimentation and investigation through 31 state committees on potential uses of electricity in agriculture, from lighting to water pumping to milking cows.

The first publicized CREA-sponsored electrification project was a group of dairy farms in Red Wing, Minnesota, which received an extension of the nearby (6 miles away) power line of the Northern States Power Company, along with the lighting of homes and barns. Major manufacturers supplied the appliances. The experiment confirmed the efficacy of electric power in an actual agricultural setting. Yet the Red Wing dairy farms were something of a Potemkin village because the real cost of the electricity for these hand-picked consumers was still somewhat high, at almost 6.5¢/kWh, and the community had the advantage of a free connection to the nearby grid (Brown 1980).

CREA's other major experiment, in Alabama, was more realistic, encompassing some 379 "dirt farms," producing mainly cotton and grains, along with a few dairy and truck farm operations as well as nonfarm rural customers, including 85 churches, schools, and other public or semipublic buildings. The Alabama experiment yielded some significant information for the progress of rural electrification because it showed that single-crop cultivation—cotton, in this case—was less conducive to efficient electricity use than was diversified agricultural production. Moreover, it also became clear that a potential major demand source, electric-powered water pumping for irrigation, was not a major feature of the local cultivation regime. In such cases, the farm and household uses for electricity did not appear sufficient to the industry to warrant investment in installation, maintenance, and billing (Brown 1980).

Thus, only in a limited sense did anything practical leading to expansion of rural electricity service result from the efforts of the private sector up to 1930.

The Roots of Public Rural Electrification

Although private utility corporations came to dominate the U.S. electric industry in the early twentieth century, public sector involvement in rural electrification is also traceable to this period. Then, as the United States ascended rapidly as an urban industrial and world power, even those with the most sanguine expectations of the country's global destiny fretted over the potential withering of its rural and agricultural activity.

The effects of industrialization and global trade in manufactures on U.S. resources and agriculture were characteristic concerns of U.S. thinkers of the early twentieth century. Among their worries was the danger of becoming, like Britain or Japan, excessively dependent on difficult-to-control foreign sources of agricultural raw materials and foodstuffs. The geopolitical notion that even the mightiest empires, if they failed to nurture a balanced development of domestic production and commerce, would fall into "desuetude" was a shibboleth of thinkers from Edward Atkinson and David Wells to Worthington C. Ford and Theodore Roosevelt's close friend Brooks Adams (see, e.g., Atkinson 1887). In a broad sense, U.S. policy intellectuals and historians such as Frederick Jackson Turner regarded the U.S. West and South as problems of development generally continuous with, rather than distinct from, global development (Sklar 1992, *46n6*). Among policymakers, then, the integration of U.S. agriculture into its modern economy had importance not only as a domestic priority but also as evidence of the qualification of the United States for leadership in primarily agricultural "neutral markets"—what we now call the developing world.

Theodore Roosevelt, one of the first national politicians to take action on this front, supported the fortuitously named Newlands Reclamation Act (1902), which promoted irrigation in the arid West through government dam construction and hydroelectric power generation using income from sales of public lands (U.S. Congress, House 1902). (Representative [later Senator] Francis G. Newlands [D–Nevada] gave his name to the act.) Among other activities, Roosevelt also appointed a Commission on Country Life, chaired by botanist–horticulturalist and rural development advocate Liberty Hyde Bailey of Cornell University and including progressive conservationist Gifford Pinchot and agricultural journalist Henry C. Wallace of *Wallace's Farmer*. (Bailey had initiated publication of *Country Life in America* magazine in 1901; Pinchot was a forester and future governor of Pennsylvania. Wallace was the father of Henry Agard Wallace, who would become Franklin Roosevelt's agriculture secretary and vice president and later, Progressive presidential candidate.)

The commission characterized the central issue as "the unequal development of our contemporary civilization" caused by rapid industrialization and urbanization. In particular, the commission called for expanding economic and technical knowledge in rural areas through extension work to counteract rural society's disadvantages with regard to "taxation, transportation, rates, cooperative organizations and credit" (Commission on Country Life 1909, *14*).

In a message accompanying the Country Life report, Roosevelt advocated sustaining agriculture and the rural economy because they helped guarantee U.S. food and military security. It is significant, however, that these concerns surfaced only after the peak of the Farmers Alliance cooperatives and the Populist insurgency of the 1890s. The Farmers Alliances had formed agricultural cooperative stores and mills, proposed monetary expansion and an extensive government commodity loan program (the subtreasuries) aimed at supporting proprietary farms' prices, and entered politics on a national scale (Goodwyn 1976; Kazin 1995). Roosevelt, in contrast, distrusted cooperatives and "socialism" and saw advocates of old-style proprietary enterprise as "rural Tories," much as Woodrow Wilson, even earlier, had castigated the Democrat-Populist candidate William Jennings Bryan and his followers as "retro-reformers" (Sklar 1992, *110n23*).

By the early twentieth century, however, with corporations, rather than farms and proprietary businesses, more firmly established as the leading organizational forces of the economy, Theodore Roosevelt felt it appropriate to cultivate an "effective cooperation among farmers to put them on a level with the organized interests with which they do business" (Commission on Country Life 1909, 9). Part of Roosevelt's explicit concern—as millions of urban European proletarians of questionable ideology and battlefield fitness poured into U.S. cities—was in maintaining a labor pool of sturdy rural individuals (whom he termed *buckwheats*, a contemporary variant of "hayseeds") for military service. More fundamentally, Roosevelt focused on stabilizing agriculture and promoting a balance between urban and rural people and between recent and older immigrants (Commission on Country Life 1909).

The earliest rural electricity cooperatives in the United States were formed in 1914, in Minnesota and Washington State, without public sponsorship. Along with several other cooperatives that originated in the upper Midwest during and after World War I, these early distribution cooperatives were small and often served a mixture of farm and vacation cottage customers (Brown 1980, *13–16*). In addition, these cooperatives often originated in efforts to extend distribution lines that private-sector utilities would not undertake because of the cost, and a number of them quickly turned over their lines and assets to the private utilities, either voluntarily or under pressure of denial of supply of electricity by the private generators serving their areas (Brown 1980, *15*).

Americans were aware that the United States lagged behind the comparably developed countries of Europe in RE for reasons that were not only geographical but also institutional and political. Sweden, for example, developed an extensive system of state-regulated and -subsidized consumer electricity cooperatives (Brown 1980, *17*). Canada, as well, had an efficient public electricity system, in which consumers owned the local distribution lines: the Hydroelectric Commission of Ontario. Founded in 1906, the commission managed the system and developed a gradually increasing rate structure promoting use by even the poorest households. These residents paid from 2 to 6¢/kWh, depending on their total consumption (this amounted to what would now be known as a lifeline rate structure). The Ontario system of broad public access to electricity impressed U.S. Senator George W. Norris, a leading Republican Progressive from Nebraska (see Cooke 1948). Experienced in the hardships of rural life, Norris recalled farm kitchens in summer, "where humidity and the blazing sun combined with the wood-burning stove to create unbearable temperatures" (Norris 1945, *318*, quoted by Brown 1980, *20*).

Along with Judson King, the long-tenured (1913–1958) director of the National Popular Government League, Norris battled during the 1920s to ensure public ownership of the Muscle Shoals hydroelectric power project in Alabama. Norris was also a major supporter of the TVA as a source of publicly generated power for the U.S. South. The effort was particularly notable because it came against the concerted opposition of the U.S. electric power industry and NELA (Brown 1980, *19–21*).

Giant Power versus SuperPower

Perhaps the greatest impetus to the development of public-sponsored electrification came from Morris Llewelyn Cook, a mechanical engineer and follower of Frederick W. Taylor's system of scientific management. Cooke had whetted his interest in alternatives to private-sector monopolies during his battle on behalf of the Department of Public Works in Philadelphia to regulate the activities and rate structures of the Philadelphia Electric Company. When the like-minded Gifford Pinchot became governor of Pennsylvania in 1923, the two men proposed a Giant Power Board as a state agency under Cooke (Cooke 1948).

Giant Power took the inspiration for its name from an industry-sponsored effort called SuperPower. Although the two proposals had some similarities—for example, both were scientifically precocious, even anticipating harnessing wind and tidal power—their methods and objectives diverged radically. SuperPower was a plan of the electrical industry, modeled on the cooperative efforts between industries that had prevailed during World War I, when the federal government had relaxed antitrust restrictions on industrial combinations in production and trade. Essentially, SuperPower sought to extend and increase the efficiency of electricity distribution by coordinating the efforts of diverse generators, including coal, hydroelectric, and oil-fired plants; harmonizing technologies and taking advantage of innovations; and pooling output. Supported by Commerce Secretary Herbert Hoover in Washington, D.C., as well as by banking, mining, and power industry interests, SuperPower envisioned a regime of minimal federal regulation, the continued development of holding companies, and a primary focus on ample returns on stockholder investment by providing power to industry and the cities. SuperPower did not aim at rural electrification or at redressing inequities in provision of service but rather at rationalizing corporate operation of the electricity sector. Rural and agricultural applications would come only after "getting our primary system onto right lines," as Hoover put it (Brown 1980, *22, 27–29, 123*).

Cooke and Pinchot shared the industry's goal of expanding the electrical system and coordinating it technologically. However, their Giant Power plan focused heavily on public service, on providing more equitable distribution of electric power, particularly to rural consumers. Like his friend and fellow engineer Hoover, Cooke focused initially on rationalizing power generation and distribution through coordinating the activities of private utilities. However, Cooke's Giant Power system proposed regulating private generation, transmission, and distribution of electricity closely, and by promoting appropriate siting of plants, serving both urban and rural consumers (Brown 1980). But the essence of the difference in Giant Power's proposal was that it held that public interest required "prudent investment" in distribution lines but did not need to guarantee that all distribution would meet the private market's standards of return. Rather, Giant Power proposed to accommodate public and private objectives by pooling the risk of serving needy but less profitable areas among the companies.

The Giant Power proposal had the support of labor and of some politicians (including Democrat Alfred E. Smith, then governor of New York). Economists working for Cooke investigated cooperative efforts in Wisconsin and Ontario and issued a report indicating their general cost efficiency. They also planned to issue securities to finance a cooperative effort. At the same time, Judson King revealed that several academics had accepted payments from the electric utility industry to attack Giant Power and that at least one had done so in terms flatly contradicting his own published work (Brown 1980).

Cooke and Pinchot saw their ideas as blueprints of a nationwide public–private partnership for rural electrification. In the expansive economy and generally anti-regulatory political atmosphere of the 1920s, however, the Giant Power plan failed in the Pennsylvania legislature (Brown 1980, 23).

Nonetheless, the Giant Power studies had a significant effect. They showed that rural electrification could be accomplished in an economically and technologically sound fashion, brought public regulation and the cooperative idea into greater view, and allowed Cooke to move on, in 1928, into an alliance with Franklin D. Roosevelt, the newly elected Democratic governor of New York. Cooke became a trustee of the New York Power Authority in 1931 and helped develop the St. Lawrence hydroelectric plant as a source of energy for rural distribution. Cooke's contemporary studies of rural electricity lines in New York indicated that distribution lines cost much less per mile to construct than the power industry claimed. By this time, the Great Depression was under way, and although Cooke entreated now-President Hoover in 1931 to consider the social as well as economic and employment benefits of using federal funds to promote rural electrification, the president rejected Cooke's proposal for an emergency nonprofit agency. Cooke then joined with progressive senators such as Robert M. La Follette, Jr., Republican of Wisconsin, and Robert F. Wagner, Democrat of New York, as well as with George Norris, in supporting the federal project at Muscle Shoals (Brown 1980, 34, 125). Cooke also broke formally with Hoover by backing Roosevelt's 1932 presidential bid. The similarities in objectives between the Republican reformer Cooke, Progressive Republicans, and reformist New York Democrats such as Franklin Roosevelt and Robert Wagner presaged the coalescence of support for a greater federal role in promoting, developing, and regulating national economic growth, including that of rural areas.

In addition to the positive examples of existing cooperatives and the impetus of Giant Power and other studies, the increasingly negative relationship between rural consumers and private power companies advanced public rural electrification. Most power companies considered rural areas as sources of too little demand to be profitable. The power companies continued to expect farmers to finance the extension of distribution lines at high cost, and they were unbending on connection costs. Hence, although electrification rose from 1.6% of all farms in 1920 to 9.2% in 1929, the bulk of this growth occurred in the higher-profit areas, close to towns and cities (Muller 1944, 16; U.S. Census Bureau 1976).

A substantial number of farmers saw their exclusion from service as arbitrary and unfair: "Utility wants too much—old story" summed up the myriad complaints farmers had accumulated in the years up to 1935 about installation and kilowatt-hour charges (Muller 1944, 17n24). In fact, the costs per mile of construction that the companies expected farmers to pay—in the range of $1,500 to $2,000 per mile—were much greater than what independent studies indicated (and greater than the per-mile costs well under $1,000 that cooperatives would realize under the REA). In 1939, the single-phase line construction cost of $825 included the cooperatives' overheads; actual costs that year were only $538 per mile (Pence 1984).

In turn, under the private utilities, rural electricity rates had averaged some 10 to 15¢/kWh—far more than the 4.6¢/kWh average urban and town rates at 100 kWh per month consumption in 1935 (and much more than the 5 to 6¢/kWh that REA cooperatives would soon realize at those consumption levels) (Muller 1944, *16–17, 144–145*; U.S. Census Bureau 1976, 827, series 113).

If the Republican administrations and utilities soft-pedaled RE during the 1920s, the incoming Democratic administration, with a power base that included the rural South, struck a distinctly more assertive chord in the 1930s. The continuing Depression also helped to facilitate RE by bringing it into the sphere of President Franklin Roosevelt's emergency relief and public works programs. In addition to their humanitarian aspects, the relief programs' targeting of poor customers now became a path of lower political resistance. That is, because the utilities had never sought out low-income rural consumers, the administration could sponsor public and cooperative enterprises as relief without directly threatening the private utilities' existing investments (Muller 1944, *38, 40*; Brown 1980, *41*).

Founding of the Rural Electrification Administration in 1935

Serious movement toward rural electricity distribution did not begin immediately after Franklin D. Roosevelt took office. Having just supported the Tennessee Valley Authority and the Agricultural Adjustment Act, Roosevelt hesitated to put more emphasis on rural development programs so early in his tenure. But Roosevelt did employ Morris Cooke under the Public Works Administration (PWA). From that position, Cooke pressed Roosevelt to make electric distribution an integral component of the national recovery program. Cooke's original proposals to Roosevelt echoed his Giant Power plan in seeking to create a joint government enterprise with private power generators (Muller 1944; Brown 1980). By 1934, however, circumstances had changed.

The Alcorn County Electric Cooperative began distributing TVA power operation in 1934 in northeastern Mississippi. Its success soon constituted an important instrumentality for expanding the distribution network, at least in the U.S. South, as more TVA generation capacity came on line. The Alcorn Cooperative originated in the back of a furniture store in Corinth, Mississippi, at a small meeting between local residents and TVA officials. In fact, the TVA leadership had planned the cooperative to originate within the district of John Elliott Rankin (D–Mississippi), a vigorous proponent of public power and the House cosponsor of the TVA Act of 1933 (George Norris was the Senate sponsor) (Brown 1980; TVA 2002).

The Alcorn County Cooperative had more than local or even regional support. In its connection with TVA Director David Lilienthal, it had the advantage of a national figure with creative ideas for expanding rural service. An energetic Harvard Law School graduate who had represented the city of Chicago in several important public service cases, Lilienthal accepted Franklin D. Roosevelt's appointment in late 1933 as one of the TVA's three original directors and served as chairman of its board of directors from 1941 to 1946. Early on, Lilienthal was responsible for the TVA's expansion. He saw cooperative distribution systems such as Alcorn as natural consumers of TVA power (Brown 1980; Electric Power Associations of Mississippi 1998).

Apart from encouraging the organization of cooperatives, perhaps Lilienthal's most crucial early act was to persuade Roosevelt to create the Electrical Home and

Farm Authority (EHFA) as an adjunct to the TVA to subsidize consumers' acquisition of home appliances. The EHFA did this by stretching out their payments from the then-standard consumer credit term of 24 months to between 36 and 60 months and by collecting the payments through a monthly surcharge on electric bills (Brown 1980, *38–39*). In the Alcorn Cooperative, then, many of the basic elements of the U.S. rural electrification program first came into place. The affordability of appliances was central to the Alcorn Cooperative's effort to sustain demand and revenue, and in 1935, the administration extended the EHFA's lending power to 1943.

Of course, Alcorn had several advantages over the self-constituted cooperatives of the previous two decades, in that the TVA furnished electricity at half the rate that the local utility (Commonwealth & Southern) would have charged according to its own calculations of cost and profit. Moreover, the TVA levied no distribution surcharges for rural consumers, and the territory also covered the town of Corinth, where demand was more concentrated. Like Red Wing, Minnesota, Alcorn County, Mississippi, was thus something less than a test under pure market conditions.

However, as Cooke would later recount, the Alcorn cooperative would show that farmers, once equipped with home appliances and using electricity in various agricultural applications, purchased more electricity than did townspeople—enough to amortize the true costs of constructing rural lines. Within six months, in fact, the cooperative proved itself as having the potential to become self-supporting within 5 years, as opposed to the 12–14 years originally projected. Its success served as a precursor to the rapid expansion of REA-financed cooperatives (Brown 1980).

By 1934, the Roosevelt administration was ready to take more decisive action on rural electrification (Muller 1944, *39*; Brown 1980, *40*). Although Morris Cooke believed in and proposed cooperation with the private utilities in developing a rural distribution grid, Roosevelt's interior secretary, Harold Ickes, like Rankin and many Democrats in Congress, had little confidence that the private utility industry would cooperate. Cooke, as chairman of the Mississippi Valley Committee, a special panel of the PWA that Roosevelt had created in 1933 (later superseded by the National Resources Board), undertook studies of the potential for coordinating the development of water power, flood control, and electrification in the area. The Civil Works Administration's Farm Housing Survey substantiated the widespread lack of rural service. Interior Secretary Ickes, meanwhile, began to use PWA funds to extend some rural lines and to build small generation and street lighting schemes in some towns. In addition, the federal Emergency Relief Administration surveyed rural areas to determine the feasibility of building lines. The vigor—and the redundancy—of these activities, together with the growth of TVA power output, underlined the administration's interest in pursuing sustained development of RE and testified to the need for better overall coordination.

Thus, on May 11, 1935, Roosevelt created the Rural Electrification Administration (REA), with Cooke as its head, as an emergency relief organization. Cooke soon found the relief requirements onerous because they required 25% of the organization's funding to go to employment, and the unskilled labor of the majority of the unemployed was not effective in accomplishing the technical work of wiring and construction that a distribution network required. Roosevelt's executive order soon relieved Cooke of the restrictive labor requirements, and he began to recruit trained engineers and specialists—notably staff who were both experienced in the industry and sympathetic with the REA's objectives (Muller 1944).

As Cooke had wanted, he was empowered to make loans to power companies as well as to local and public organizations. However, in what was to prove a turning point, the private utility companies vigorously opposed such arrangements. Their concerns were little different from those that had animated Insull and Doherty in 1907: acceding to the federal "yardsticks" for electrification that came with government loans (rather than to the friendly state regulation of private corporations as natural monopolies) would constrain the utilities' freedom of operation. The utilities rejected expansion of RE for social or developmental purposes, arguing that lending on current commercial terms and sales at current prices had already electrified all the farms that could amortize the investment in distribution and home wiring. Mostly these were farms that used electricity on a large, "industrial" scale, such as for poultry raising, dairying, or pumping for irrigation.

At the same time, the power industry was fighting the passage of the Public Utility Holding Company bill (later the Wheeler–Rayburn Act), which proposed to eliminate large-scale holding companies' control of multiple utilities when it led to anticompetitive practices. The Wheeler–Rayburn Act (after Senator Burton K. Wheeler of Montana and Representative Sam Rayburn of Texas) emerged from a Federal Trade Commission investigation that had begun in 1928. Although not outlawing holding companies altogether, the act specifically aimed at undercutting utilities' ability to delay development of the rural market. The holding and power companies resisted strongly (Funigiello 1973, *98, 113*).

Cooke's entreaties to the utilities to take advantage of federal lending for RE were ineffective against their hardening opposition to the Democratic administration and Congress, which they saw as perpetrating an "unholy trinity" of intrusive regulation and public social engineering made up of the TVA, the Securities and Exchange Commission, and the REA. Cooke was thus left almost wholly dependent on cooperatives for implementation of public–private collaboration in rural electrification (Brown 1980).

The utilities' focus, according to Brown (1980), was primarily on commercial agriculture rather than on use in households or for small cultivation in the South, where it had to compete with mechanized, animal, or human labor. Cooke disagreed, and in Brown's view, the electric companies probably underestimated the increase in consumption that home appliances would yield even in the smaller farms and rural homes.

Objectives and Activities of the REA

As an emergency agency, the REA had a relatively meager allotment of $10 million and spent most of 1935 getting established and waiting for the agency to gain a permanent mandate. The Rural Electrification Act, under the sponsorship of Republican Senator George W. Norris of Nebraska and Democratic Representative Sam Rayburn of Texas, passed on May 20, 1936 (Brown 1980, 58–65). Although, as noted, the private utilities were permitted and even encouraged to participate, hardly any did so. The passage of the act thus focused the REA on financing cooperatives and on providing them with technical assistance in the construction, operation, and maintenance of rural distribution lines.

Scope and Objectives

The Rural Electrification Act of 1936 codified the terms by which the REA could finance rural electrification. Funds would go to the states in proportion to their percentage of unelectrified farms (only Connecticut and Rhode Island received practically no funding, largely because of their extremely high rates of electrification). The act empowered the Reconstruction Finance Corporation (RFC) to lend the REA $50 million in the current year (and $40 million for five years thereafter) at 3% annual interest (or the average rate of long-term U.S. obligations of the preceding year). Thus, the loans did not incorporate a significant subsidy. The REA would in turn lend to cooperative borrowers in unserved rural areas. These areas were defined by reference to the criteria of the 1930 census as including unincorporated areas and those incorporated rural areas of 1,500 inhabitants or fewer. The population included more than 30 million individuals actually living on farms in 1935 and perhaps another 20 million consumers in towns or villages of populations less than 1,500. Despite the large numbers, the density of the REA-served areas would remain quite low, averaging fewer than three consumers per mile of line in 1941 (U.S. Census Bureau 1934; Muller 1944, 6).

The REA was technically empowered to lend for construction of generation and transmission capacity as well as for distribution lines and home wiring, but early on it disbursed funds largely for the latter two. The lending program could give preference to cooperatives and nonprofit public associations, but all lending had to come from funds the RFC had lent to the REA—not from any revenues or operating income of the cooperatives, which had to go strictly toward amortizing cooperative debt and maintaining cooperative facilities. The terms of the distribution loans were 25 years, but the REA administrator could extend them by 5 years. The terms for interior household electrification loans were for a maximum of 66% of the lifespan of the appliances financed but with a cap of 5 years in all. The TVA provided power to many of the first cooperatives (Muller 1944).

The focus of the REA was not principally on the poorest farms, however. The act made "high line" service (i.e., connection to the distribution grid) available only through self-amortizing loans. Therefore, the program focused primarily on electrifying cooperatives consisting of large numbers of middle-income rural farms, with poorer farms benefiting from their inclusion in the overall pattern of the distribution network. Understanding that if cooperatives failed, the private utility companies could easily take over the distribution lines and raise rates, the early cooperatives scrupulously collected payments and kept their finances as well as their maintenance and operations activities in order. Moreover, some cooperatives sought to ensure their longevity through activism on a broader scale on behalf of rural development. For example, they exerted pressure within their respective states not only for their own operations but also for regulation of bulk power purchases, farm legislation, and tax rates (Muller 1944, *35*).

Part of what goaded the cooperatives into expanding their own activities was the concerted opposition of the private utilities, which included active subversion of the cooperatives (Box 10-2). For their part, the cooperatives sought to "fight the companies with their own techniques" by improving on existing technologies and innovating lower cost solutions, such as simple pole and transformer assemblies, that suited local conditions and finances (Muller 1944, *38*). In all, the hostility of the private utility generators probably contributed to the sense of commitment among the participants in cooperatives. The sense of a mutual effort against the dual adversaries of the Depression and the private utilities motivated cooperative members to value their system, vote for pro-REA politicians, pay their bills on time, and refrain from vandalism or theft of services and equipment.

To some extent, the conflict between the utilities and the cooperatives was not really over the effects of the TVA and REA on overall utility profits. Even in 1941, private utilities continued to generate about half (49.7%) the power that the cooperatives bought and to electrify the higher population rural areas. Some utilities even assisted cooperatives with "promotion, load-building, and engineering" (Muller 1944, *38*). The opposition was perhaps more about the countervailing or alternative examples of public rulemaking; public, nonprofit power generation;

Box 10-2. The Utilities' Attitude toward the Cooperatives: From Subtle Subversion to Spite Lines

According to early REA documents, the private utilities pursued several subversive tactics:

- Building so-called spite lines through the more profitable portions of cooperative territories, leaving only the less economic territory open to prevent cooperatives from rationalizing distribution and becoming viable. Some state power commissions assisted the utilities in this practice by holding that cooperatives could not construct lines within a mile of existing utility company lines. In Louisiana, a private utility built a pattern of lines radiating from a center that created "pie slices" of territory that could only be served by the utility. The REA responded by pursuing such cases in the courts, but at least initially this targeted obstruction had some effect in inhibiting the formation of new cooperatives (Brown 1980, 71–72).

- Attempting to offer potential leaders of cooperatives reduced-rate utility service if they abandoned the cooperatives.

- Mobilizing agricultural extension agents, the professoriate, and others who had connections with the utility industry to lobby farmers against the cooperatives. Some county agents even gave the utilities maps showing projected cooperative power routes (Brown 1980, 69).

- Raising concerns through advertisements about the engineering adequacy and safety of REA-sponsored cooperatives' distribution facilities and lines (Muller 1944, 37–38).

and the federal–cooperative, public–private relationships that the TVA and REA established and sustained. As Muller commented, "The cooperatives have, in fact, created a new indirect market for private suppliers. Many companies recognize this . . ." (1944, *38*).

In addition, the differences between the utilities and the REA and cooperatives on the other hand extended to the developmental role of rural electrification. The utilities viewed broad coverage of rural areas as "a relatively unimportant and very difficult business undertaking" if not an outright "utopian ambition to remake rural life." As one utility executive opined, it would be unfortunate "to raise in the breast of the farmer any false hopes with regard to electrical supply" (Muller 1944, *23n44, 16n24*).

In contrast, if Cooke himself had been initially unsure of the viability of cooperatives as a vehicle for RE, the REA, once under way, vigorously embraced a broad social and developmental conception of RE. An official of the U.S. Department of Agriculture expressed this view in fervent terms after the REA came under that department's purview in 1939. The REA, he said, was for "agriculturally minded people. . . . Lines running from one small town to another are not rural lines. Lines running to gins and other processing plants, no matter where located, are not rural lines. Lines running into thinly populated farm localities are not actually rural lines; they represent simply a commercial projection into an attractive market, and do not reflect plans of a real service to a substantial and characteristic agricultural population" (Muller 1944, *65n43*).

In its first five years, the REA in fact focused, as its mandate to serve the less populated and less electrified rural areas demanded, on groups of low-electrification states largely in the Midwest and South. On average, the REA did its heaviest lending, construction of distribution lines, billing, and other activities in Texas, Minnesota, Iowa, Indiana, and Georgia (derived from Muller 1944, Table 6, *32–33*).[1] Another way of looking at REA activity would be to say that in mileage of lines built and number of consumers "energized," about half of the REA's activities focused on nine midwestern and southern states as "nuclei" of four distinctive regions: "(1) Ohio, Indiana, Kentucky, and southern and central Illinois (running over into southern Michigan, Tennessee, eastern Missouri, and northeastern Arkansas)." This region had 19.3% of the distribution mileage and 20.7% of the consumers. "(2) Wisconsin, Minnesota, and Iowa (running over into eastern Nebraska)." This region had 17% of the mileage and 36.3% of the consumers. "(3) Georgia (considered as the center of an area running from North Carolina into Alabama)." This region had 5.8% of the mileage and 6.5% of the consumers. "(4) Texas (and over into Oklahoma)." This region had 9.8% of the mileage and 8.3% of the consumers (Muller 1944, *32–33*).

What the REA put into action, then, was a fulfillment of its egalitarian mandate of "furnishing electricity to persons in rural areas"—that is, a proactive policy of *full area coverage,* as opposed to cream-skimming or cherry-picking of high-profit consumers. In this, the REA aimed not only at giving its staff the satisfaction of "helping farmers group together for stringing distribution lines into their homes" but also at "the broader role of cooperative economic development in rural areas" (Muller 1944, *95; 35n70*).

Leadership, Structure, and Staff

The REA's own leadership showed its desire to move federal funds effectively into the field for rural electrification. Morris Cooke, however, was more theorist and organizer than manager. Soon after the Rural Electrification Act passed in 1936, Cooke appointed John M. Carmody, a former mining engineer and chief engineer of the Civil Works Administration, as deputy administrator and de facto chief of operations. Cooke formally turned over the leadership in August 1937.

The REA's structure reflected its principal role as a lending agency for disbursing funds and recouping investments. Among the REA's units, the Applications and Loans Division evaluated cooperative proposals for creditworthiness and issued loans, and the Finance Division monitored and audited compliance with the terms of the loans as a way of securing them. A Design and Construction Division provided cooperatives with technical advice on the initial planning and setup of the distribution infrastructures. Then, the Cooperatives Operating Division assisted with the physical operation and maintenance of the distribution systems. It soon absorbed the administration's Utilization Division, which included field specialists working on load building and promotion of appliances. This was not only a technical task but also one that required knowledge of and ability to communicate with local residents (Box 10-3).

An Engineering and Operations Division (created in late 1939 by merging the Technical Standards and Operations Supervision divisions) assisted cooperatives in effectively managing their own budgetary, rate setting, and auditing functions and helped with research engineering and testing of new equipment. A Management

Box 10-3. The REA's Traveling "Electric Circus"

To help educate rural people in the 1930s about how they could use electricity in their homes and on their farms, REA sponsored a traveling road show, which became known as the "electric circus." Louisan Mamer was one of REA's first employees, hired in 1935 to help stage those road shows.... During the [REA's 65th] anniversary ceremony ... Mamer recalled being intrigued by an REA advertisement seeking people with "a pioneering spirit." Born in 1910 and raised on a farm in southern Illinois, where her father cleared 1,000 acres of Illinois River bottom land, Mamer said she knew well the hard labor of rural life. So when the chance came to leave home to attend the University of Illinois at Urbana, she took it. The REA road show used two big circus tents, one for a general meeting and the other to demonstrate electrical appliances and farm equipment, Mamer recalled. One of her main duties was to speak to farm wives to help them "convince their husbands to pay to join a cooperative." Small radios and electric irons were among the first appliances sold. In the North, washing machines were in big demand, while refrigeration was more of a priority in the South. Mamer also trained other instructors so that they could conduct workshops, and she developed training materials, remaining with REA until her retirement in 1981.

Source: Campbell 2000.

Division was responsible for the REA's own accounts, procurement, policy, and relations with the TVA as a purchaser of its generated power.

Finally, the REA's Information Division published *Rural Electrification News* and promoted information sharing—publicizing the experience and practices of other cooperatives (Muller 1944, 51).

The profusion of departments within the REA marked it as one of the "alphabet soup" of bureaucracies in the Depression-era federal government. Still, the organization actually employed a relatively lean staff of fewer than 1,000 after five years of operation. About half of these employees were professionals, including financial specialists, lawyers, and agricultural experts, and half of the professional employees were engineers—construction, electrical, or mechanical—the majority with private-sector experience (Muller 1944).

Starting in 1936, the REA began year-long training internships for young administration and engineering graduates. The REA was able to expand an initial staff of about a hundred tenfold into a substantial cadre of expert and loyal specialists within five years. The strategies the REA pursued in this regard included an initial vetting to select individuals sympathetic to the REA's mission, followed by training in the goals and procedures of operating cooperatives. The REA emphasized promoting from within and began incorporating staff into regular civil service positions in 1941. It also allowed affiliation of the staff as a local of the Congress of Industrial Organizations' United Federal Workers of America (Muller 1944).

Functions and Structure of the Cooperatives

Many cooperatives had unique or distinctive elements that reflected their geographic locations, sources of generated power, and requirements for constructing distribution lines. Almost all operated as nonprofits, however, and followed the Rochdale Principles of Co-operation, the historic basis of the cooperative movement.

The Rochdale principles originated with a group of cotton weavers who created a food cooperative in the English town of Rochdale in 1844. Cooperatives have often reframed the principles, but almost all have come to endorse the same basic elements, the first four relating to internal governance and the last three to relations with external organizations—all of which contribute to placing cooperatives effectively within the framework of democratic market societies:

1. *Open membership* is voluntary and formalized by the purchase of a nominal holding of share capital, with limited or no return on equity.
2. *Democratic control* is exerted by elected or consensually appointed managers, operating on the principle of "one member, one vote."
3. *Equitable economic organization* involves fair treatment of employees according to the best prevailing and possible social norms; it also means sharing of surplus from operations among members, who may set aside profits for development of the cooperative, use them to provide services, or distribute them to members in proportion to their transactions with the organization.

4. *Education* means inculcating in members the social and philosophical principles of cooperatives and their techniques of governance and operation.
5. *Cooperation among cooperatives* involves sharing of knowledge and undertaking of joint initiatives.
6. *Autonomy* establishes for cooperatives no less independence from governmental, social, and economic institutions than what prevails for other business enterprises.
7. *Community* emphasizes the responsibility of cooperatives to the surrounding social setting (Schuller 1985; McPherson 2002).

In addition, most cooperatives were effectively independent of the jurisdiction of local and state public service commissions, which were mostly sympathetic to the large utilities. The relative independence of the cooperatives from the state commissions probably derived from the leverage of the federal government and the federal REA. Nonetheless, the federally financed and supervised cooperatives remained private entities and operated on a small-scale commercial basis (Box 10-4).

By 1950, with 77.7% of farm households served overall, the average monthly consumption of REA members was 180 kWh, and that of residences was 147 kWh, excluding irrigation, with rates falling commensurate with consumption. By 1955, with 94.4% of all farms electrified, the average monthly consumption of REA members had grown to 312 kWh and that of residences to 242 kWh, excluding irrigation (U.S. Census Bureau 1976, 827, series 110 and 830, series 156, 157).

Box 10-4. Typical Features of Rural Electric Cooperatives in the First Decade of Operation

Typical cooperatives of the mid-1940s had the following features:

- Capital investment of about $100,000.
- A distribution system covering approximately 425 miles of line.
- Consumer membership of about 1,000 (about 80% of them farmers).
- Average monthly bills for 100 kWh of between $5 and $6 (i.e., 5 to 6¢/kWh).
- Minimum monthly consumption requirements of about 60 kWh. These were higher than those of the low-income consumers in cities, who might purchase 25 kWh or less, because of the need in rural settings to amortize new equipment and appliances and because of the continued (though lower and falling) differential in the distribution costs.
- Availability of plans for the lowest income households. In addition to steering consumers to finance appliance purchases through the EFHA, many cooperatives replicated a system that an Arkansas cooperative first designed. Known as the Arkansas Plan, it involved installing an electrification system designed for small houses and available on a down payment of $1.00 (the full cost of the home wiring was $10). Recipients paid a $5 cooperative membership fee and amortized the wiring through affordable payments of 10 cents a month (Muller 1944, 73, 144; Brown 1980).

Scope, Leadership, and Staff

The Rural Electrification Act restricted cooperatives to serving unincorporated areas and incorporated rural areas of 1,500 inhabitants or fewer, thus ensuring the cooperatives' predominantly rural character and focus on the farm community. On average, 80% of cooperative members were farm consumers. Nonfarm members usually were small-scale local merchants, trades people, or professionals whose livelihood depended on that of the surrounding agricultural community and who therefore largely identified with farm viewpoints (Muller 1944, 73).

In general, the poorest farmers did not lead the cooperatives. Poor farmers had the far more pressing problem of survival during the Depression (Muller 1944, 72). Poor farmers typically received service by being part of the logical distribution routes for full area coverage. The poorest farmers contributed to the development of the cooperatives through "sweat equity"—by trading their labor in the construction of lines for their initial membership fee (usually about $5) and sometimes for electrical service. They generally paid minimal use charges of $1 per month for the first 11 kWh. With payments for wiring and appliances such as radios and irons, their bills typically came to about $2 per month (Brown 1980, 70).

The leadership of the cooperatives generally came from among the middle-income farmers, especially those who could adopt electricity for both household and agricultural use. These farmers generally had experience with extension work and community involvement of various kinds, such as crop and farmer associations. Middle-income, "progressive" farmers (i.e., those sufficiently able and disposed toward adopting machinery and "modern" agricultural methods, as opposed to relying on hand and animal labor) also contributed to the cooperatives by devoting unpaid time to serve on elected boards of directors that met once a month or so to review operations. These generalizations probably held true across agricultural or rural communities in various regions—even where the particular wealth and occupations of the community might differ radically: for example, the cotton tenancy in the Deep South versus cattle ranching in the Plains and Mountain West (Muller 1944, 73). Undoubtedly, as well, farm families developed a strong loyalty toward cooperatives for giving them access to a heady mixture of new household, community, and farm applications for electricity (Box 10-5).

The members did not handle the cooperatives' day-to-day operations. This was the work of a permanent staff, typically between 7 and 20. They might include a manager, bookkeeper, office clerks, head lineman, and line crew. In the period of building distribution lines, a construction boss and electrification specialist from the REA might join this group. Local staff was neither highly skilled nor highly paid. The exception might be the manager, who generally had some previous management experience, although frequently not in the electrical industry (Muller 1944).

The boards appointed the managers, but the REA had power of approval. In the view of one contemporary source, most cooperative members had a rough idea of the structure and function of the enterprise, particularly that it was assisted by the federal government and had a nominal 25-year amortization period. Consumers'

Box 10-5. Excerpt from the Diary of Oklahoma Electric Cooperative Members Elbert and Erma Cassel of Verden, 1939–1950

(After carbide light and wind charger)

July 27, 1939—	Signed for [cooperative membership].
June 1, 1940—	Uncle Lark wired house.
August 8, 1940—	Got electricity.
August 26, 1940—	Bought a washer.
October 29, 1940—	Bought a radio.
January 6, 1941—	Wired milk barn.
July 25, 1941—	Bought iron.
September 19, 1941—	Bought refrigerator.
January 21, 1942—	Wired Dutton Church.
December 3, 1942—	Bought mixer, radio.
December 31, 1948—	Bought deep freezer.
August 30, 1950—	Bought sewing machine.

Source: Oklahoma Electric Cooperative 2002.

reactions to cooperatives generally reflected their familiarity with them. Where cooperatives had operated for longer periods, the public viewed them more favorably. Ultimately, it was probably the price and reliability of the service that shaped the attitudes of most consumers (Muller 1944, *81*).

If most consumers who joined cooperatives voiced positive or generally benign attitudes toward their organizations, the privilege of voicing a member's opinion was not open to all in states where racism and segregationist policies excluded black farmers from governance of, if not membership in, cooperatives (Muller 1944, *86*). For its part, the REA encouraged broadening of membership across racial lines. In the 1930s, however, the federal government had not yet mandated against segregation policies or extensively opposed the denial of service to black farmers. In fact, John Rankin, one of the REA's most committed supporters in Congress, was an outspoken white supremacist (Brown 1980, *39*). Discriminatory practices became far less prevalent after the advent of the civil rights movement of the 1960s, but the possibility of such manipulations has a cautionary value for systems in which racial or ethnic conflict could come into play.

In the experience of one who worked for electric cooperatives in North Carolina in the 1960s, the cooperatives of that time had black members who, in accordance with the Rochdale principles of "one member, one vote," theoretically had a say in cooperative affairs. However, through various manipulations, black members were sometimes excluded from full participation. One way white-dominated boards did this was by holding annual meetings at remote sites to ensure absence of a quorum and hence continued rule by the incumbents (Richard A. Pence, personal communication, June 12, 2003).

Relationship with REA: Financing and Technical Assistance

The supportive relationship between the federal government and the rural cooperatives led opponents (and even some supporters) of the cooperatives to characterize the REA as a kind of federal government "holding company" for the rural cooperatives (Muller 1944, 89). If such charges had been accurate, the REA would have been violating the spirit if not the letter of the Wheeler–Rayburn (Utility Holding Company) Act of 1935. But the REA did not hold equity in the cooperatives, so it was not in fact a holding company. However, the REA did serve as a holder of cooperative indebtedness, a function a holding company might play, and in providing technical, managerial, and supervision services, it also acted, as did some holding companies, as (in today's terms) an energy service company. Of course, the REA came nowhere near the scale of the holding companies of the 1920s. Even so, the degree of interest and opposition it engendered suggests the substantial effect it had in rural areas in the extent of its loans, lines, and consumers served in operations from 1935 to 1950. Table 10-1 shows

Table 10-1. *REA and REA Borrowers in the United States, Summary of Operations, 1935–1950*

	REA				REA Borrowers			
	Loans Approved	Number of	Number of Consumers	Miles Energized	Electricity (million kWh)			Revenue
Year	($ millions)	Systems	(thousands)	(thousands)	Generated	Purchased	Sold	($ millions)
1935	7	2	0	0	—	—	—	—
1936	44	29	8	3	—	—	—	—
1937	82	126	44	17	—	—	—	—
1938	181	350	176	67	—	—	—	—
1939	268	548	436	181	—	—	—	—
1940	351	685	674	268	34	402	311	17
1941	434	773	902	348	83	854	724	35
1942	460	803	1,012	378	131	1,305	1,151	47
1943	474	811	1,088	390	199	1,721	1,572	54
1944	518	826	1,217	410	213	1,974	1,795	63
1945	667	848	1,409	450	258	2,159	1,951	71
1946	958	869	1,684	507	320	2,497	2,244	87
1947	1,191	911	2,046	603	443	3,379	3,056	111
1948	1,575	952	2,518	759	718	4,514	4,252	145
1949	1,999	995	3,040	943	903	5,879	5,564	188
1950	2,312	1,007	3,413	1,089	1,077	7,270	6,884	229

Note: Cooperatives represented virtually all borrowing entities.

Source: U.S. Census Bureau 1976, 830, series S 148–155.

some details of the REA and of the cooperatives' generation, consumption, and financial dimensions.

During its first years, the REA's precise structural role in the energy sector remained unclear because it largely cultivated hitherto neglected markets for which few real standards and statistics for service yet existed. After several years, however, the cooperatives developed their own records and benchmarks. As Table 10-2 shows, electrification progressed rapidly under the REA. By 1939, 27.4% of the nation's farms had grid-based electric service, up markedly from the 10.4% of

Table 10-2. *Household Electrification in the United States, Farm, Nonfarm, and Total, 1930–1956 (%)*

Year	Nonfarm (Urban and Rural)	Farms	All Households
1930	84.8	**10.4**	68.2
1931	—	**10.7**	67.4
1932	—	**11.2**	67.0
1933	—	**11.8**	66.7
1934	—	**12.1**	67.1
1935	83.9	**12.6**	68.0
1936	—	**14.5**	70.3
1937	—	**18.3**	73.1
1938	—	**23.9**	74.9
1939	—	**27.4**	77.3
1940	90.8	**32.6**	78.7
1941	—	**35.0**	80.0
1942	—	**37.8**	81.2
1943	—	**40.0**	81.3
1944	—	**42.2**	84.0
1945	93.0	**48.0**	85.0
1946	—	**53.3**	85.5
1947	—	**60.2**	86.2
1948	—	**66.8**	89.6
1949	—	**72.9**	93.0
1950	96.9	**77.7**	94.0
1951		82.2	95.2
1952		86.9	96.1
1953		91.4	97.2
1954		93.0	97.9
1955	98.8	94.4	98.4
1956	99.2	95.9	98.8

Note: Bold figures highlight the early period of REA operation.

Source: U.S. Census Bureau 1976, 827, series S 109–111.

1930 (U.S. Census Bureau 1976). In 1939, some 417 cooperatives serving 268,000 households were operating and received loans of more than $3.6 million for household wiring and plumbing (Brown 1980, 75). The 1940s promised similar progress, but it did not come without controversy.

Toward Completion of Rural Electrification in the Early 1950s

The REA operated as an independent agency until July 1939, when the Department of Agriculture formally assumed oversight. This step regularized its operations within the framework of government departments to some degree, but it also led to the resignation of the very able John M. Carmody, who was concerned about the REA's loss of independence. In September 1939, Harry Slattery took over Carmody's position. An attorney who had worked for the Inland Waterways Commission, the Forestry Service, and the Interior Department, Slattery had been associated with natural resource causes for more than 20 years. Early during World War II, the agency underwent another major change when it moved from its Washington, D.C., offices to St. Louis, Missouri.

Although a veteran administrator, Slattery was far less technically adept than Carmody, and his age and poor health may have compromised the REA's performance. His tenure, which lasted until 1944, saw a continuing conflict with the emerging National Rural Electric Cooperative Association (NRECA), under former Arkansas Representative Clyde T. Ellis. The NRECA organized itself as a trade association to promote and protect the cooperatives, particularly by mobilizing federal aid to help them build their own power generation facilities and by arranging for lower cost insurance for the hazardous construction and maintenance operations of the cooperative electrical systems. In this regard, the NRECA's Ellis followed a pattern akin to that of his former allies in Congress, George Norris and John Rankin, who saw public authority and activity in all aspects of the power industry as a way to protect rural residents. In contrast, Slattery, a member of the RE "old guard," like Cooke, did not see promotion of cooperatives as an end in itself and envisioned public and cooperative activity in the sector as best limited to distribution, particularly where cooperatives could buy power from the private sector at wholesale rates (Brown 1980, 87, 90–91).

President Franklin Roosevelt, however, found the expansion of the cooperatives impressive. He was also cognizant of the widespread support for expansion of public power generation in Arkansas (an important REA state) and in Speaker Sam Rayburn's Texas (the leading REA state in several categories, such as funds advanced, miles energized, and consumers connected) (Muller 1944, Table 6, 32–33). Irritated as well by Slattery's hostility to his administration, Roosevelt ultimately supported this expansion. Slattery finally resigned, shortly after Roosevelt's reelection in November 1944. Slattery's exit followed protracted public hearings involving charges of inefficiency, ineptitude, and favoritism (of which he was exonerated), and countercharges against Ellis and the NRECA of attempted profiteering in collusion with the insurance scheme (of which they were also exonerated) (Brown 1980).

The overall triumph of the NRECA point of view and of a "second genera-tion" of cooperative and public power leaders was apparent even before Slattery's resignation. This triumph became clear when Congress passed the Pace Act in 1944, which set the interest rate for REA loans to cooperatives at a low 2%, changed the loan amortization period from 25 to 35 years, required area coverage, and facilitated cooperatives' ability to expand into power generation. The Pace Act extended the rural electrification program indefinitely, beyond the 10-year scope originally planned in the Rural Electrification Act, and thus facilitated the near completion of the electrification of the rural United States in the 1950s (Brown 1980, *95–98*). We say "near completion" here because some U.S. rural communities, particularly some communities of Native Americans in the arid Southwest, remained without grid service even into the 1990s (Glaser 2002, Chapter 9, para. 13).

As Figure 10-1 shows, the generation capacity of the cooperatives grew signifi-cantly after passage of the Pace Act in 1944.

Lessons from the Program

The rapid and timely electrification of rural areas in the United States required more than a willingness to let the evolution of the market take its course. Private utilities, even where they chose not to invest, continued to subvert rural develop-ment as a way of opposing nonprofit enterprise and maintaining future preroga-tives to electrify territories they considered uneconomic at the time. As a utility

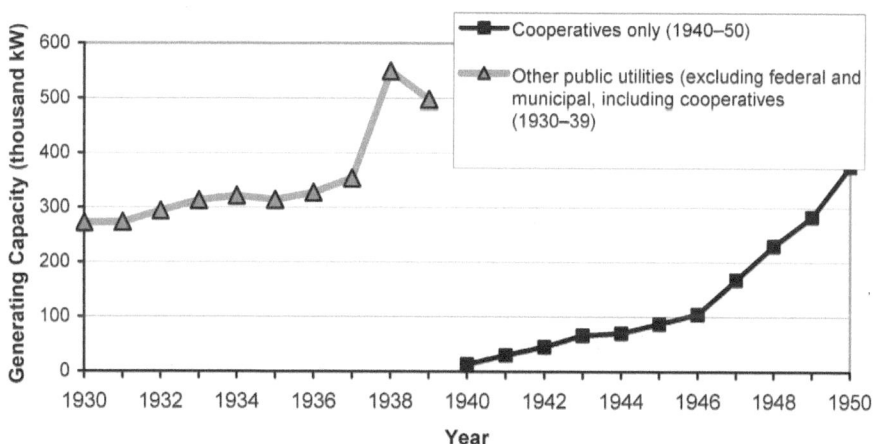

Figure 10-1. *Generating Capacity of Public and Cooperative Utilities in the United States, 1930–1950 (thousand kW)*

Note: U.S. government statistics combined data on cooperatives with data for all public utilities until 1940, when cooperatives began acquiring significant generating capacity in addition to their distribution activities.

official put this viewpoint, "Unless rural service is worth more than it costs it should not be supplied" (Muller 1944, *16n24*). Rather, indispensable to RE in the United States was a developmental and, to an extent, an egalitarian mindset among both national and local leaders. In the U.S. case, the confluence of several important factors in the 1930s helped translate a mindset into a practice. These factors included the precedents of federal activism in rural environments from the time of Theodore Roosevelt; a dedication to public service, efficiency, and conservation characteristic of Republicans such as Morris Cooke and Gifford Pinchot; a regulatory activism more characteristic of the Democrats Woodrow Wilson and Franklin Roosevelt; and the support of Progressive and populist movements that helped to mobilize rural communities in favor of imaginative, large-scale public development programs over the resistance of private utilities that appeared intent on exploiting farmers and rural residents.

RE in the United States required government financing and technical assistance, but it did not require massive subsidization. It also did not require the federal government to arrogate for itself a large or permanent part of the distribution sector to make power available. The guarantee of lending at low rates was necessary to overcome private finance's unwillingness to support investments it considered marginal from a profit viewpoint. In addition, accomplishing RE appears to have required the government to operate as an entrepreneur (i.e., through the TVA) at least in the mid-South to benchmark wholesale pricing and ensure the availability of power to cooperatives (Muller 1944, *2–3*). It also seems to have been prudent for the federal government to guarantee the availability of power to cooperatives by underwriting their expansion from distribution activities into generation and transmission through the Pace Act (1944).

Among the more important government roles was in taking the initiative, through the Electrical Home and Farm Authority, pioneered by David Lilienthal of the TVA, to extend generous repayment periods for appliances and to enable repayment through the household electric bill during the 1930s and early 1940s. This plan took advantage of the new popularity of electric home appliances in the cities to boost rural household consumption into the effective complement to commercial and farm consumption that it had not been during the 1920s. Although the pairing of nonresidential and residential consumption in rural areas would never create the volume of demand it had in the densely populated cities, the coordination of the two helped to boost demand in rural areas rapidly and substantially enough to make nonprofit electrification viable. After 1950 (with 77.7% of rural households electrified and household electricity prices in the 3–4¢/kWh range for 100–250 kWh per month residential consumption), the Census Bureau ceased distinguishing average rural and urban electricity consumption (U.S. Census Bureau 1976, *827n1*). That change alone might be considered something of a milestone.

The Challenge for the Cooperatives

The cooperatives constituted a minority, though rapidly growing, share of the early rural market, rocketing from virtually zero in 1935 to about 30.5% in 1941

(Muller 1944, *36*, Table 7). Cooperatives and public generation also represented a tiny part of the total generation sector. In 1950, for example, cooperatives accounted for only 0.26% of the total power generated and, with all public power, just 16% (calculated from U.S. Census Bureau 1976, 821, series 46, 47). Despite programmatic support from the federal government, then, the RE program faced substantial challenges as a niche operator in the system.

One of the cooperatives' chief obstacles was their vulnerability to opposition from powerful electric utilities. Power companies obstructed the cooperatives by constructing spite lines to cherry-pick the market of its economically better off customers, and to impede the cooperatives' building of cost-efficient distribution systems. Many utilities did so extensively, and the industry's trade associations and the state power commissions they often dominated conducted heavy opposition to cooperative rural electrification in its crucial early years.

Among the manifestations of utility opposition were the cooperatives' difficulties in securing services and support from hostile state and federal institutions. Cooperatives often found it difficult to get competent legal representation because many lawyers in rural areas were on retainer for utility companies. State public utility commissions also had been dominated by utility companies and at times refused to grant incorporation to cooperatives. For similar reasons, even state rural electric bureaus, agricultural extension services (part of the U.S. Department of Agriculture, and thus representing a form of internal opposition), and university economists were hostile to cooperatives and engaged in public opposition to them.

Lack of time and a pool of skilled workers and managers were additional challenges. Farmers and rural residents, although eager for electrification, were hard hit by the agricultural and general depression of the period, often migrated to cities for work, and were generally poorer than urban residents. They often lacked time to manage cooperatives effectively. The tendency of some cooperatives to exclude black Americans (or, for example, Hispanic and Native Americans in the Southwest) from participation was an unsavory aspect of the regionalism and parochialism that provided solidarity to the organizations. Hence, the melding of federal and local private participation was not smooth and equitable for all.

Programs to Meet the Challenges

Overall, the successes of the cooperatives as the engines of rural electrification in the United States could be attributed to several factors.

Simplicity, Popularity, and Government Assistance Made Cooperatives Attractive. The Rochdale principles were easy to understand as an operating and governance model, and the desire of independent farmers to create an organization largely under their own control contributed to the popularity of cooperatives. Many farmers saw the cooperatives as the only way they were likely to get power and as a door into a modernity and convenience now increasingly apparent in the cities. "We don't see how we lived without it" was a frequent refrain, along with "We would never have had it if it hadn't been for the cooperative" (Muller 1944,

81; cf. Pence 1984, *99–117).* Electrification made a real difference in people's lives in the United States then—particularly the lives of low-income people—just as it does in developing countries now, with incandescent lighting enabling the extension of family life, education, and social activity past dark. Members of cooperatives in Kentucky, Nebraska, Pennsylvania, Texas, and elsewhere carried out mock funerals for kerosene lanterns in celebration of their connection to the grid (Pence 1984). At the same time, the continuing hostility of the consolidated utilities probably reinforced cooperative members' determination to make their systems work (Muller 1944, *81).*

The Security of Federal Financing and Technical Assistance. The TVA and REA ensured that the cooperatives could obtain low-cost financing and necessary technical skills to run an electricity business. This assistance helped ease rural residents' suspicion of authority while providing protection and expertise they lacked. Cooperatives operated by a dual "top-down" and "bottom-up" principle. The top-down part, the central REA, supplied financing and expertise in system design, construction, and management. The REA's customized attention to cooperative construction, governance, accounting, operation, and maintenance was itself a form of education that private utilities by definition did not provide. The bottom-up part, local involvement, was complementary to the REA's role, as members or employees became indispensable in building the distribution lines (including creating rights of way, tree trimming, and brush clearing) and in managing and maintaining the system according to strict economies because of the low marginal profit rates and the REA requirement that cooperatives use operating income strictly for maintenance, operating expenses, and debt amortization (Muller 1944, *72–73).*

The REA Program's Economies of Scale. The REA encouraged consumers' self-reading of meters and reporting by postcard; replication and standardization of other practices across cooperatives; and some collective bargaining and common contract negotiations for line crews (usually in the upper Midwest). Its own adherence to the federal Wages and Hours Act of 1940 led the REA to suggest, although not require, that cooperatives comply. In 1941, the REA developed a joint Framework of Operations guiding negotiations between the Brotherhood of Electrical Workers and REA-financed cooperatives. It included minimum wages and rules for hours and overtime. At the same time, the organization of the cooperatives reflected the social system of the local areas—and the assistance of local chapters of farm bureaus, in particular in Illinois, Indiana, Minnesota, and Ohio. Most significantly, the cooperatives were largely able to escape the influences of the state public service commissions, which the consolidated utilities continued to dominate (Muller 1944, *82, 121).*

The Program Took an Egalitarian Area Coverage Approach. The program encouraged the use of sweat equity to enhance participation of low-income households. Cooperatives allowed poor farmers to trade labor in-kind for investment in cash. Innovations at the household level also assisted cash-poor consumers—for example, through the widely replicated and inexpensive Arkansas Plan for electrifying poor households.

The REA and TVA Had Access to Low-Cost Power. The TVA provided power to 27% of the existing cooperatives in 1941 at wholesale costs of 2–3¢/kWh. The TVA's presence in the market encouraged private lines to supply electricity at more competitive rates and terms for distribution, and the REA exerted some pressure on private utilities with regard to wholesale electricity sales to cooperatives (cooperatives not served by the TVA generally paid utilities wholesale prices of 3–5¢/kWh at the time). The REA could make loans to a cooperative to build generation facilities in the face of utility recalcitrance to sell at standard wholesale rates and did so in several cases (e.g., in Michigan) even before the 1944 Pace Act. After the Pace Act, the REA financed generation on a more extensive scale, which enhanced the economies of providing wholesale power to the cooperatives.

Conclusion

The REA–cooperative effort represented only a small part of the electricity sector as a whole, but its accomplishments between 1935 and 1950 went beyond its size in kilowatts generated, distributed, or billed, particularly by contributing to the revival of an effective public and regulatory presence in the sector; to developing an effective and creative mode of new entry; and to establishing viable, democratic public–private cooperation in electrification. That sort of developmentally minded, equitable approach to electrification had almost vanished by 1929, the last non-Depression year. (Publicly owned generation capacity, for example, which then included existing cooperatives, amounted to just 4% of the total in that year [calculated from U.S. Census Bureau 1976, 821, series 46, 47].)

Rural electrification in the United States has the well-deserved aura of a heroic effort of the "little person" fighting against the large corporations and the odds. A word of caution is also in order, however. True, the private utilities often resisted or obstructed the cooperatives' expansion of distribution to poor consumers and opposed the Roosevelt administration politically. However, the private utilities soon became significant participants in the market for generation and transmission of power to the cooperatives, supplying almost half of the cooperatives' power and even assisting some cooperatives with load promotion. Recognizing the cooperatives' financial viability, the private utilities, at least temporarily, tempered their overall opposition to a utility system including cooperatives. It is thus probably fair to say that in some respects U.S. rural electrification took place when it did despite the private utilities. However, it is also realistic to say that it took place with their considerable involvement and tacit support, at least on the generation and transmission side.

By 1956, the last year for which the Census Bureau reported separate statistics for farm electrification, almost 96% of farm households had electricity (versus 99.2% for urban and rural nonfarm residences). More than 1,000 REA cooperatives were serving some 4.4 million households and had energized some 1.4

million miles of distribution line. Accompanying this public–private cooperation were significantly improved conditions of life for rural residents. These were no small achievements amidst the Depression and then world war, and they stand as a reminder from the past of the options for shaping the social, political, and economic dimensions of national as well as rural life.

Acknowledgments

The author thanks Doug Barnes, Norma Adams, Paul Clark, Richard A. Pence, Eleanor Miller, and coparticipants in a panel at the NRECA International, Ltd., forum, Sustainable Rural Electrification in Developing Countries: Is It Possible? The coparticipants were Dan Waddle, Joy Dunkerly, and Tom Carter. The author remains solely responsible for any errors in this chapter.

Notes

1. Ordered from most to least as an average of nine types of activity: borrowers, allotments, fund advanced, weighted miles constructed, systems energized, miles energized, consumers connected, kilowatt hours (kWh) billed, and gross revenue.

CHAPTER 11

Electricity for Social Development in Ireland

Michael J. Shiel

THE FIRST POLE OF THE RURAL ELECTRIFICATION scheme in the Republic of Ireland was erected on November 5, 1946. It was a considerable act of faith on the part of such a small and impoverished country. But the ensuing results were dramatic, and Ireland's rural electrification program was officially completed in 1976 as a resounding success. In little more than 30 years, the number of connected rural consumers had grown from a few thousand to over a half million customers in 1976, extending service to even the most remote rural areas. The program amply justified the faith of its initial planners and promoters.

As a young, new nation in 1946, Ireland had been unable to make significant economic development strides since its independence in 1921. Instead, it had experienced almost a decade of civil war, followed by damaging trade disputes with the United Kingdom, its major trading partner, and the deleterious effects of the economic depression that gripped most of the Western economies in the 1930s. In 1946, the Irish economy was also suffering from the deprivations and disruption of trade that came with World War II. Agriculture, the country's major source of exports and the source of livelihood for 60% of the Irish population, had been particularly disrupted.

The Irish government was nevertheless committed to doing what it could to improve the living conditions of Ireland's rural population and stem the traditionally high levels of emigration from the countryside. It realized that improvements in the rural infrastructure were crucial to increasing farm productivity, and thereby improving the country's economy and overall export position. For these reasons, despite the daunting financial and other obstacles to such an effort, it committed itself to a program of rural electrification that, after 30 years, resulted in almost 100% electrification of the country's rural areas.

From the outset, government and institutional leaders at all levels recognized the program's social component. They viewed electricity not only as a way to improve rural productivity, but also as a way to free rural people, particularly women, from the age-old drudgery of farm life. Electrification would not only make women's work easier; it would change their role in the household. Both individually and organizationally, rural women proved that they were highly effective in demanding, and eventually gaining, the means to realize a better life for themselves and their families.

What Ireland's rural electrification program achieved cannot be measured solely in terms of changes in the physical environment of the household or farm labor-saving devices. Rural electrification, particularly in the early years, was of immense social and psychological significance. It gave rural people a belief in themselves and their potential, which had hitherto been lacking. As one farmer in Wicklow County said, "Before the electricity came, we knew about it and we saw it in the towns, but we never dreamt that it could come to us in the country; not in our lifetimes."

The parallels between Ireland in 1946 and many developing countries today should not be overstated, but the fact remains that there were high levels of poverty in rural Ireland when the program began. In this regard, the electrification effort in Ireland had to contend with many of the same difficulties as in developing countries today. The program was successful in no small part because it was creatively and thoughtfully designed for the highly specific conditions and circumstances that confronted those who developed and implemented the program. This chapter first examines the historical context of the program and then turns to how the program achieved such success.

Electrification before 1946

The Dublin Electric Light Company, established in 1880, provided the first public supply of electricity. Slowly at first, then more rapidly, electricity spread to all major towns. By 1925, the country had 161 power generation and distribution undertakings. These operated on multiple standards and voltages, and nearly all were direct current systems. In 1927, the Electricity Supply Board (ESB) was created to operate the Shannon Scheme, a 90-megawatt (MW) hydroelectric station and the country's first large-scale generating plant. The ESB was also responsible for national distribution and sales of the station's output.

By 1929, when the Shannon Scheme came on line, the ESB was well along in establishing a basic transmission network. The existing urban and municipal utilities, including the Dublin Corporation's 28-MW coal-fired station (subsequently updated to 78 MW) and a 5-MW station in Cork were taken over. With a total capacity of 173 MW, these stations met the demands of the entire ESB system through the end of World War II.

During the 1930s, the ESB set up a sales organization, which vigorously promoted electricity use. Outlets for selling tested electrical appliances were opened in all major cities and towns. Electricity was sold to customers on a nonprofit basis, after providing for interest, a sinking fund, depreciation, and other charges. Before establishing the ESB, the retail price of electricity was about 1 shilling. In modern terms, this was amazingly expensive—about US$2.50 (1985) per kilo-watt hour (kWh). However, by 1930, the price had fallen to about one-fifth of this amount.

By the time war broke out in 1939, 170,000 consumers—almost all located in cities and large towns—had been connected. Even during the war, system growth continued, and by 1946, the total number of consumers had reached 240,000. However, consumption was severely restricted by wartime fuel shortages and elec-tricity rationing.

In 1946, a high-voltage grid linked the main cities of Dublin, Cork, Limerick, and Waterford, with extensions to Dundalk in the north and Carrick-on-Shannon in the west. The three main generating stations supplied electricity to this grid. The transmission grid fed a subtransmission system through seven substations. These, in turn, supplied the distribution system through 100 smaller substations. During the postwar decades, the supply system quickly expanded as the rural elec-trification program got under way, and demand rose rapidly in already electrified urban areas as the nation recovered from wartime austerities.

The Irish Rural Electrification Program

Persistent poverty and decades of overlooking rural development were ongoing concerns of the Irish government. It fully recognized that, if the quality of farm life and the low productivity of most farms failed to improve, the country's main rural export would continue to be its young people rather than agricul-tural goods.

Initial Planning Stage during 1939–1945

As early as 1939, the government asked the ESB to prepare plans to supply rural areas with electricity and produce proposals for how the program should be financed. Work on drawing up these plans was interrupted by the outbreak of war, and the ESB became preoccupied with keeping the supply system functioning under difficult circumstances. However, the government was serious about rural electrification and instructed the ESB to recommence its work on the program during the war. The ESB studied the available accounts of rural electrification programs elsewhere, especially in the United States, Canada, and Scandinavia, and considered a variety of technical and institutional options.

The ESB submitted its proposals to the government early in 1943, and they were quickly approved. A year later, in 1944, a detailed white paper was pub-lished, setting out plans for a program to be implemented as soon as the materials became available. From the beginning, the social element in the program was recognized, and electricity was seen not just as helping improve rural productiv-ity but also the lives of women.

Although there was virtually unanimous political support for rural electrifica-tion, there were major problems to be surmounted. An obvious issue was how to finance the program in the face of a conservative rural population. The considerable inertia of the rural people would also have to be overcome if the scheme were to achieve its social and financial objectives. Rural society in Ireland was deeply conservative, and incomes and expectations were low.

Many farmers with small farms and uncertain incomes needed to be convinced of the benefits of electrification. Generations of insecurity had resulted in a near obsession with the need to have secure possession of land and to expand farm holdings, but there was considerable reluctance to invest in land improvements. The crucial question was whether there would be markets for any increased production they might manage. To many, electricity appeared merely as an expensive alternative to the traditional oil lamp or candle. Their forebears had survived without it, and they could see no great advantage in hurrying to involve themselves in this new expense. The question being asked in Ireland at that time is the same one that is being asked in many developing countries today: Would the rural populations be willing to pay for electricity service?

Exact farm income figures at the time are impossible to establish. It is notoriously difficult to establish the level of farm incomes in any traditional rural society, and Ireland is no exception; nor is it a simple matter to make meaningful comparisons over a time span of 40 years. One of the few available sources of data is a survey of 20 farms ranging in size from four to 40 hectares, which was carried out in 1945–1946 in County Roscommon, one of the more prosperous western counties. This survey found that the average value of the total annual output of the farms, excluding the produce consumed at home, was about US$5,650 (IR£320). When the costs of seeds, hired labor, and other inputs were deducted, the average net family income was US$3,175 (IR£180). Breaking this overall average down further, the survey found that the average annual cash incomes in farms under 20 hectares was under half this figure, or US$1,500 (IR£85) per year. With an average family size of just under eight, this is about US$190 per person. Although this figure does not take inflation into account because for such a long period this would be quite complicated, it is typical of the income levels per person found in small-scale cash-crop farming areas in many parts of the developing world today. Also, many farm families had considerably less than this amount, particularly in the poorer western counties of the country.

Support through Capital Subsidies

Financing the program was a major issue from the beginning. Some help was available from the Marshall Plan, under which the U.S. government provided aid for postwar reconstruction in Europe, but the vast bulk of the expenditure had to come from the meager resources of the Irish state.

The existing tariff structure was designed to cover operating and debt-servicing costs incurred by the ESB in supplying its predominantly urban consumer base. Rural electrification, with its high investments and small loads, was more difficult. A tariff high enough to service the full capital investment cost of rural electrification was beyond the means of most rural householders.

To reduce the initial investment costs, the government decided to exclude from the program the 30% of rural households living in the most remote rural areas, where the investment costs were highest. Even so, the estimated capital cost for the remaining 280,000 households was IR£21 million at 1946 prices, about US$370 million at that time, and was expected to escalate considerably in the course of program implementation.

In some other countries that had carried out rural electrification programs, the costs had been covered by cross-subsidies from urban to rural consumers. However, in Ireland at the beginning of the program, there were only 240,000 urban consumers; they could not be expected to subsidize 280,000 rural consumers to any significant extent. The government therefore agreed to contribute 50% of the capital cost of the scheme.

During program implementation, which ultimately was extended to cover almost 100% of the rural population, the original 50% capital subsidy varied with the ebb and flow of national finances. In 1955, the subsidy was withdrawn completely for a period of three years. By 1980, when 468,000 rural dwellings had been connected for a total outlay, uncorrected for inflation, of IR£109 million, government subsidies had provided IR£28 million, or 26% of the total. The availability of the capital subsidy in the early and difficult years of the program had afforded a lifeline to the ESB when it was most needed.

It is almost impossible to establish the operating costs of rural electrification as the distribution system was being expanded under the program. The ESB was reluctant to produce separate accounts for the rural areas because of the difficulties in deciding on the proper allocation of overhead costs, interest charges, and the costs of peak and base load consumption.

Nevertheless, under government pressure, the ESB produced rural revenue accounts for the years from 1950–1951 to 1970–1971. These accounts showed a small operating surplus in the beginning, but as the program expanded into the less economically viable areas, this turned into a deficit which reached IR£2.3 million in 1970–1971. This year was the last for separate rural and nonrural accounts. The total accumulated operating loss on the program was IR£19 million at that time. This was, in effect, the total cross subsidy by nonrural consumers.

Organization and Management of Rural Electrification

A special organization named the Rural Electrification Office (REO) was set up by the ESB solely devoted to rural electrification. The ESB appointed William F. Roe, one of its senior engineers, to take responsibility for the whole scheme. The REO had a small specialist staff at the ESB head office in Dublin concerned with overall supervision, strategic planning, monitoring progress, materials procurement and distribution, construction methods and standards, terms for supply and rates of charge, and the investigation, development, and promotion of electricity uses for rural farms and households.

Roe believed that the rural electrification task could only be carried out by mobilizing the enthusiasm of the staff and by involving the rural communities in the planning and execution of the work to the greatest extent possible. These beliefs eventually proved to be important aspects of the rural electrification program's success.

The construction work and face-to-face contacts with householders were decentralized to the 10 administrative districts that had been set up in 1929 to provide electricity service to the towns and larger villages throughout the country. Staff in these districts had already acquired expertise in line construction, project management, cost control, and direct consumer contact. The construction crews required for the rural electrification program were thus easily created around the nucleus of skilled and experienced personnel in each existing district organization.

The Process of Planning

To ensure a fair spread of the benefits of the scheme as it was being implemented, the work was carried out on an "area" basis, rather than working in a "ribbon" from existing centers. The division of work between the districts helped this, but the adoption of the ecclesiastical parish as the basic unit for development was also an important factor. There were approximately 800 such parishes in all, and the order of selection for development depended to a large extent on the proportion of householders agreeing to sign up for connections. This system developed a healthy rivalry between parishes, with each striving to attain as high a place as possible on the necessarily lengthy construction program, by encouraging a high initial sign-up rate for electricity service.

The ESB gave William Roe a considerable degree of autonomy. He in turn delegated responsibility downward to a degree hitherto unprecedented in the organization. Even the young engineers, most of them straight from engineering school, who were in charge of area construction crews, were given a high level of authority and responsibility, with corresponding accountability for results. Each was trained in his job and then left to get on with it, subject to regular monitoring of quality, quantity, and the costs involved. This decentralized system was responsible in no small way for the speed with which the scheme swung into action and achieved its objectives. It also provided an excellent training for these young men in job control and project management.

Before the ESB would consider providing electricity to an area, it required a written statement signed by all householders who would be prepared to accept an electricity supply if it were offered. The local people themselves elected a committee to solicit whether their neighbors would be adopting electricity service being offered to them. Thus, from the beginning, local people were involved in visiting every house, explaining the benefits of electricity, and persuading those who were reluctant to sign up to do so. Those areas with the highest preliminary sign-up were then selected for a more through investigation by engineers and a more formal survey of potential customer intentions. Those areas showing the best return on capital and the highest percentage of acceptors were placed highest on the list in the construction program.

Emphasis on Communication and Local Support

In tackling what was one of the largest development projects ever undertaken in Ireland at the time, William Roe was keenly aware of the importance of good communications with the staff who were to carry out the work. If a high standard of performance was to be achieved, they needed to be well briefed and motivated at the start, but also to be refreshed with information on the progress of the program, advised of developments in all aspects of their work, sustained when difficulties arose, and motivated to give their best at all times.

At the peak of activity of the scheme in the early 1950s, there were 40 separate units, each of up to 100 people of various disciplines and skills, working in different parts of the country, some in remote locations. Day-to-day personal contact was out of the question. The solution was the *REO News,* a monthly stenciled magazine, which was sent to all areas and avidly read by all levels of staff. In the course of its 14-year life, from December 1947 to November 1961, the *REO News* played a highly important management role in informing, educating, and motivating the widely dispersed staff, countering their sense of isolation, and building a team spirit. It provided a vehicle for an often-spirited exchange of views and for criticism of performance of management and field workers alike. It acted both as a suggestion box and a safety valve, as its columns were open to all rural staff to make suggestions and criticisms, or simply get rid of steam.

A particularly popular regular item was the Top Ten, in which the cumulative performance of each crew from the beginning of each year was logged. This was one of the first features to which engineers and supervisors turned on receipt of their copy of *REO News*. Although this may appear naive in these days of sophisticated management techniques, and indeed, more sophisticated worker response, it was effective in keeping the importance of high productivity constantly in the forefront. The ESB board of directors was extremely conscious of the need to observe the stringent budgetary limits under which the scheme was being carried out, and the *REO News* played an important role in keeping cost-consciousness and the need for efficiency uppermost in the minds of the work crews.

One of the most important members of each crew was the area organizer. The area organizer was responsible for promoting the use of electricity and developing a good relationship between the ESB and the local community. They carried out interviews with potential consumers and in the process were involved in the persuasion of reluctant householders of the benefits of electricity. They also served the legal orders giving the ESB access to land for poles and lines and often acted as intermediaries in disputes. They organized demonstrations of electrical equipment, advised on the selection and installation of electrical appliances, and kept in close contact with local community organizations

Because most of these area organizers were themselves from farming stock, they were easily accepted by local people, and through their frequent visits to every house, quickly became the best-known members of the ESB area crews. They played a vital part in the success of the scheme by presenting the ESB as a human and responsive organization, rather than a faceless bureaucracy.

There was little mechanization of line construction in the early stages of the program, and large numbers of unskilled laborers were recruited locally as the construction crews moved from area to area. This local recruitment also did much to achieve acceptance of the scheme because rural families could see it being carried out by their own fathers, brothers, and sons, and not by strangers sent in by a remote government organization. For the duration of the work, regular wages arrived in homes where they had rarely, if ever, been seen before, which also worked to generate good will toward the program.

Connection Charges and Tariffs Rolled into Electricity Bill

A basic principle of the program from the outset was that to the extent possible no large initial charge would be made for a connection. Although this system had to be modified in the later stages, it applied to the vast majority of early connections. It was feared that a connection charge would discourage consumers from signing up for electricity service. Such low consumer response would mean that the program might have trouble taking off in its initial stages. Instead, on the consumer's electricity bill there was a two-part tariff. The first part was a modest fixed charge to cover maintenance, depreciation, and all capital and standing charges, and the other was the per-kilowatt-hour charge for the electricity actually consumed. The allocation of the fixed charge was on the basis of the size of the dwelling and outbuildings; outbuildings were charged at a reduced rate. Bills were issued every two months.

A typical fixed charge in 1946 was IR£1.0–1.5 (US$18–26 in 1985 prices) for a two-month period. The electricity charge was 2.5 pence per kWh (18 U.S. cents in 1985 prices) for the first 80 kWh, with a follow-on charge of 1 penny (7.2 U.S. cents) per unit for the next 280 kWh. Consumption over 360 kWh was charged at 0.75 pence (5.4 U.S. cents) per kWh.

The ESB provided a fuse board and meter inside each dwelling. Private electricians were hired to do the house wiring. In the vast majority of cases, the technology required was minimal; wires were simply attached with clips to walls or strung from ceilings. The only restriction was that approved wiring materials had to be used. In more complicated cases and where local skills were not available, the ESB carried out the wiring on a contract basis.

Lowering System Costs through Manufacturing and Single-Phase Distribution

The decision to use single phase for the supply lines was based on the lower cost. At that time, most of the farms being connected were small, and the general levels of farm activity and standards of living were low. An average annual consumption of 800–1,000 units was assumed for design purposes. The problem was to devise a system that would adequately meet the initial demand and require the minimum investment, but that would be capable of expansion at reasonable cost to supply a growing load.

The foundation of the rural supply system was a mesh of three-phase 10-kV lines connecting the 38-kV substations. Ideally, two such lines connected neighboring substations 40–50 kilometers apart and were routed to run through the centers of four to six rural areas, forming the distribution backbone for each area.

These distribution areas were 60–90 square kilometers and were usually based on the ecclesiastical parishes. Halfway along each backbone line, a "normally open" point was positioned, generally by means of a three-phase air-break switch locked in the open position. This system ensured that the areas fed from each half of the line were under the control of the nearest substation, while allowing alternative feeding in an emergency.

From the backbone lines, single-phase, 10-kV spur lines of varying length fed groups of consumers through pole-mounted, single-phase, 10-kV/220-V transformers. Large users, such as creameries and factories, were provided with a three-phase supply under special contracts, but more than 95% of consumers were on single phase.

The solution of having a three-phase backbone line with single-phase spurs met these low-cost requirements. In most instances, the single-phase lines could be converted by adding a single conductor. In the case of large groups of houses, the low-tension lines were designed so that they could easily be converted to 10 kV. This adaptability would enable a group of houses initially fed by one transformer to be split later into one or more transformer groups. A three-phase supply is not, of course, required for normal domestic purposes. Its main advantage from the farmer's point of view is that three-phase motors are simpler, cheaper, and easier to maintain. But the single-phase motors of up to 2.5 kW were easily available to farmers. Although they were somewhat more expensive than three-phase models, the price difference was only a small fraction of the supply network savings, and they were adequate for the vast majority of Irish farm uses, as indeed they still are.

Poles were generally 10 meters long with a top diameter of 15 centimeters. Taller and heavier poles were used at angles and end positions and were provided with suitable staying arrangements. Vacuum creosoting was found to be the most effective preservative treatment. Only a small number of poles were available from the country's own limited wood resources; the majority were imported, almost all from Finland.

However, for most of the remaining parts of the system, manufacturing facilities were encouraged to lower costs. The conductor was initially hard-drawn copper, but when the price of copper escalated in 1947–1948, a switch was made to steel-cored aluminum. This aluminum was originally imported, but an Irish factory was quickly set up by private enterprise to supply all the conductors required. Transformers were all imported in the early stages but were manufactured in Ireland from 1949 onward. By 1957, about 70% of the expenditure on materials went to Irish manufacturers, and by 1965, this had increased to 84%. In addition to reducing the burden on the country's balance of payments, local manufacture considerably simplified the tasks of quality control and scheduling of deliveries.

The Vital Role of Grassroots Organizations

William Roe himself was the first to recognize that a finely honed construction organization was not in itself sufficient to guarantee the success of the rural electrification program. Realization of its social and economic objectives, as well as its financial viability, depended on an enthusiastic response from the rural community.

This response was by no means guaranteed. The hundreds of thousands of small farmers who made up the bulk of the rural population were still living in a near-subsistence economy with strong folk memories of hardship and privation. Many were deeply reluctant to commit themselves to the regular payments that electrification would involve. There was also an emotional resistance to the fixed charge, partly because of its historical associations with the hated ground rent, but also because it had to be paid irrespective of the amount of electricity consumed and was effectively outside their control.

On the other hand, the fact that farmers were now the absolute owners of their holdings provided them with a solid incentive for investing in improvements in their land and buildings. A central program task was therefore to educate and persuade farmers to overcome their conservatism, inertia, and inherited feeling that the better things in life were destined to pass them by. At that time, some of the most powerful forces acting to break the old molds of thought in the rural community were voluntary grassroots organizations. These organizations had sprung up initially among the more progressive elements of the rural population. They now began to find that the advent of rural electrification provided them with an opportunity to develop and become more effective in their task of raising the general standards of rural living. Some examples of these organizations and their role in rural electrification are provided in this section.

Muinntir na Tire

Muinntir na Tire (People of the Land) was founded in 1931 by Reverend John Hayes, a rural parish priest. It was an organization committed to raising the standard of living in rural areas. Organized on a parish basis, it had become a powerful influence in rural Ireland by the mid-1940s. William Roe had himself been a prominent member of Muinntir na Tire since 1938 and was a firm friend of its founder. Both men saw rural electrification as one of the most effective ways of providing the stimulus to overcome rural stagnation.

Muinntir na Tire played a major part in the promotion of the rural electrification program especially in the difficult initial years of the program. In many areas, it helped set up local rural electrification committees to stimulate interest and approach the ESB. The initiative for a preliminary canvas of the potential support for rural electrification in an area often came from Muinntir na Tire and was carried out by its members.

When construction eventually started, the local committee helped recruit unskilled labor, procured lodgings for the mobile skilled crew of 10–15 men, found offices and stores, and generally smoothed over any difficulties arising from siting poles or damage to crops or property. The fact that more than 1 million poles were erected during the scheme, most on private property, with virtually no serious siting problems indicates the high degree of integration of the construction crews with the local community.

Irish Countrywomen's Association

The Irish Countrywomen's Association (ICA) was another powerful voluntary organization whose help was vital in the rural electrification program. It was founded in

1910 to help alleviate the poverty and destitution then so common in rural Ireland. Its members were mainly farmer's wives and daughters and housewives from the rural towns and villages. The provision of district nurses and adequate clothing and food—especially milk for rural children—were its early priorities.

By 1946, it was well established as an organization devoted to raising living standards in rural homes and easing the drudgery of women's work. Its emphasis had by then shifted to the provision of piped water into rural homes, the introduction of labor-saving devices, the provision of home advisers, and adult education and training. One of its campaigns, conducted in cooperation with the ESB, featured a modern, labor-saving kitchen, designed to fit within the traditional farmhouse, mounted on a trailer and displayed at agricultural shows and other rural events.

The ICA had an enlightened and energetic leadership and cooperated to a high degree with the ESB in promoting the rural electrification program. The influence of the ICA was such that rural housewives were usually far ahead of their husbands in appreciating the benefits of electrification.

Macra na Feirme

Macra na Feirme (Farming Youth) was a national organization of young farmers' clubs. It was founded in the early 1940s in response to the low level of formal education typical of Irish farmers. At the time, postprimary and higher education were reserved for children destined to leave the rural areas, whereas the son inheriting the farm usually received no education beyond primary school. This lack of formal education among most farmers was a prime factor in hampering progress in agriculture.

From the start, Macra na Feirme gave strong backing to rural electrification. It cooperated with Muinntir na Tire in organizing local canvasses and supporting the ESB in the initial canvas of areas where rural electrification was being considered. It showed young farmers how to apply the benefits of the latest agricultural research and how electricity could be used to improve the quantity and quality of their produce. By focusing on the young progressive farmers and farmers' sons, Macra na Feirme introduced scientific farming methods to replace the archaic and outdated approach of an older generation.

Macra na Tuaithe

In 1951, Macra na Tuaithe (Youth of the Countryside) was formed to cater to the 12–18 year age group. The emphasis of Macra na Tuaithe was on learning by doing, and it received strong support from the other rural organizations. The members were bright, forward-looking, and eager to experiment with new methods, which included the application of electricity to farming.

The ESB was one of more than 200 national firms that formed the National Youth Foundation to ensure adequate funding for Macra na Tuaithe. With this backing, it flourished. In 1959, there were 125 clubs. In 1963, the number was 231, and by 1966, there were 253 active clubs dispersed throughout the country.

Overcoming Problems of Rural Electrification

The Irish rural electrification scheme encountered the usual problems associated with electrifying impoverished rural areas with poor roads and infrastructure. In addition, there were problems specific to the conditions and circumstances of the rural areas of Ireland at the time.

With its tight construction schedules and even tighter budgets, the ESB had strong incentives to solve these problems as cost-effectively as possible. Moreover, having been given such a high degree of autonomy, the REO and its construction crews were in a position to take creative, imaginative approaches when necessary.

Promoting Increased Electricity Demand

The need to increase the demand for electricity in the rural areas and improve the economics of the program was a constant preoccupation. The ESB took advantage of rural shows and exhibitions to demonstrate the various uses of electricity on the farm. It also promoted a national breadmaking competition to counter the belief that traditional bread could not be baked in an electric oven.

One of the most serious problems of the early years was backsliding, when householders who had signed up for a connection—often under pressure from overenthusiastic local canvassers—changed their minds at a time when considerable expense had already been incurred by the ESB. Visits by the area officer, backed by demonstrations of electrical equipment and the example of neighbors enjoying the benefits of electricity, slowly eliminated the problem. In the first few years, however, the problem of potential consumers having second thoughts seriously threatened the viability of the program.

Controversy over the Fixed Charge

The two-part tariff, with its fixed charge and additional per-unit charge, was a major bone of contention. In the folk memories of the Irish farming community, the fixed charge, based on the size of the dwelling and its outbuildings, was akin to the ground rent exacted by the notorious absentee landlords who owned much of Ireland's farmland in the nineteenth century and was usually referred to as such.

Any increase in the fixed charge met ferocious resistance. An attempt to increase it by 10% in 1961, the second 10% increase since 1946, brought unparalleled vituperation on the ESB. One mass meeting of farmers in the western county of Mayo passed a resolution that was sent to the chairman of the board of the ESB and the prime minister. It included the following comments:

> The increase in the ground rent has brought the small farmer from Donegal to Kerry back to the days of the Clanricardes and the Boycotts [notoriously oppressive nineteenth century landlords]. It is unchristian that the lowest wage earners in Europe should have to pay more for the most expensive electricity west of the Iron Curtain. . . . To rob the poor to pay the rich seems to be

the new policy of your Board. What the crowbar brigade [who carried out forced evictions] failed to do, the ESB is helping to accomplish.

Even the efforts of the ESB to emphasize the term "fixed charge" instead of "ground rent" backfired. The farming community claimed that, if it was a fixed charge, by definition, it could never be altered. The secretary of one local guild of Muinntir na Tire wrote to the ESB saying:

> I am requested to write to you as regards the increase in the ground rent. In this area we were led to believe that the units may be raised, but the ground rent would never be increased. Therefore, if you insist on the increase, we request that the whole area be switched off.

The question of the fixed charge had been one over which the ESB had deliberated greatly in the beginning. Because of the anticipated low initial levels of consumption, an economically viable unit charge would have been a major obstacle to load growth and hence, would have undermined the long-term finances of the scheme. The fixed charge was decided on as a means of guaranteeing a minimum return on the heavy initial capital investments in the program.

The experience highlights the need for extremely careful consideration of the rural electrification tariff system and the need to tailor it to rural sensitivities. As growth in the rural load took place, the issue gradually lost its significance and eventually ceased to be a major bone of contention.

Pole Transport and Other Problems

Transport of poles was a serious problem, particularly in the remote western areas. The roads were extremely narrow and twisting and impossible for travel by the trucks and low loaders on which poles were normally carried. The solution was to use small coastal vessels, which sailed into tiny fishing harbors, many of which were silted up and unused. The risks that the shipping companies were prepared to take in this regard were remarkable. In response, the construction crews of local young men seemed to be almost obsessed with demonstrating how quickly they could unload a vessel and would have taken it as a defeat if a vessel missed a tide because of a delay on their part.

There were numerous other practical problems that were overcome by skill, enthusiasm, and commitment. Most were resolved at a local level. Even the ticklish issue of rights to run lines and plant poles on farmers' lands produced remarkably little contention. Most landowners accepted that the scheme was for their benefit, and a sympathetic attitude by local ESB staff helped greatly to overcome any opposition.

Everywhere, whatever the problems, they were all forgotten in the satisfaction of the "switch-on." This was a moment of enormous excitement for the community. The local paper carried this account of the switching on of the village of Ballyduff in 1948:

> Ballyduff was "lit-up" in every sense on the night of 20th August when the great "switch-on" ceremony of rural electrification was performed. The

ceremony was timed to start at 9:30 p.m. and, for some hours before, it was obvious that the people of Ballyduff and surrounding districts intended to make it a great night and one to be remembered. . . the culmination of many months of organization by the Ballyduff Guild of Muinntir na Tire and hard work by the staff of the ESB. One old resident said that he never saw a night of more spontaneous enjoyment. Everyone was happy and realized what a great night it was for the district when light and power, which is going to revolutionize country life, was made available.

Overall Results of Meeting the Challenges

As indicated, work commenced on the program in November 1946 in a rural parish not far from Dublin. The proximity of this initial work enabled it to be monitored closely by headquarters staff so that teething troubles could be sorted out at the beginning. Work in the next area began in February 1947 in Limerick on the west coast. By September 1947, the first area had been completed, and work was under way in nine others.

These initial projects were also used as training grounds for new crews to permit the work program to expand. At the peak of the program in 1955–1956, there were 40 crews working. In that year, they completed 99 areas and connected 34,257 new consumers. The rapid build-up of the program is demonstrated in Table 11-1.

The final total is considerably in excess of the number envisaged in the original rural electrification plans of 1946. One of the reasons for this is the large number of houses built in the rural areas around the cities and larger towns in the 1960s and 1970s. By 1976, the program was completed, with virtually 100% connections throughout the whole country.

Table 11-1. *Cumulative Numbers of Consumers Connected in the Irish Rural Electrification Program, 1948–1976*

Year	Total Cumulative Number of Consumers Connected
1948	2,203
1950	25,153
1955	128,634
1960	243,698
1965	296,205
1970	336,743
1976	505,890

Source: Shiel 1988.

Complement to Rural Development Programs

As the scheme progressed, there were marked changes in rural life. The most immediate was the increased comfort and convenience that came from having electric light available. During Ireland's dark winter evenings, this was a tremendous boon. Children in particular benefited from having light for reading and homework. The ability to carry out tasks around the farmyard after dark also freed the farmer for other work during daylight hours.

Promoting Rural Water Supply

The provision of water on tap in the home, in which electrification played a large part, got rid of one of the most onerous domestic tasks of rural Irish people: carrying water from the well or pump. Hot water in the house also eliminated the chore of boiling water in a large cauldron over the open fire for washing and cleaning.

The dairy farmer benefited particularly. A continuous supply of drinking water for his cows brought higher milk yields and better weight gain for beef cattle. It saved the labor of carting endless barrels of water for livestock; readily available hot water brought greatly improved standards of hygiene in milk production. Hygiene was a matter of growing importance because increasing public awareness of the need for higher standards of cleanliness in food production was reflected in legislation.

The ESB established a Pump Advisory Service, which developed into a comprehensive Water Advisory Service, providing free advice on all aspects of running water installations, from the location of water sources to the selection, installation, and maintenance of pumps and water systems. The service was used not only by private individuals, but also by local authorities, government departments, and nongovernmental organizations (NGOs).

In the early stages, three methods of providing a domestic water supply were demonstrated. The simplest relied on a pump driven by a small motor (0.25 kW), which delivered water through a 12.5-millimeter pipe to the household sink. It was switched on and off by a simple cord switch over the sink. A more elaborate system used a roof tank with a float switch, which controlled the pump; this system permitted installation of a hot-water cylinder with an electric immersion heater. The method that proved most popular over the years, however, used a pressure storage tank. This required no roof tank and was extremely simple to install.

The Department of Agriculture provided a grant of 50% for approved farm and domestic water supply installations. Yet, despite intense promotion by the ESB, the rate at which domestic water supplies were installed was slow. A newspaper report in 1959 said: "It is still possible to see a man balancing two buckets of water from the pump as he hurries home to be in time for some favorite television act. His house may have a vacuum cleaner, but not a tap."

By 1960, when almost 250,000 rural houses had been connected to the electricity supply, only 50,000 had piped water. The Irish Countrywomen's Association was extremely dissatisfied with progress and set up a Campaign for Rural

Water Supplies, which promoted and coordinated efforts to bring water to farm-ing families.

At one meeting, the audience of women was urged, "Many men in rural Ire-land are not yet fully aware of the advantages of piped water supply. Very often they will get a water supply with the advantage of the stock in mind rather than the housewife, and very often if the stock don't want water, the housewife has to do without it also. Let your menfolk know you will not tolerate such treatment. Tell them you don't want to end up in your old age with a bad heart from drag-ging water over long distances."

The ICA's energetic campaign had the desired effect of boosting demand for water on tap over the 1960s. Group schemes were set up in many areas and were supported by government and local authorities. These schemes shared the costs of drilling for water and installation of the pump. In some cases, these costs were quite substantial. Despite the country's abundant rainfall, it is often necessary to drill 200–300 meters to reach the water table. By 1969, a survey showed that 49% of rural families had water on tap in their homes.

Growth in the Use of Electric Appliances

Reducing the need to carry water from wells and pumps was not the only way electricity reduced domestic drudgery. Though most rural households opted first for lighting, the purchase of labor-saving electric appliances also grew over time. Table 11-2 demonstrates how the ownership of domestic electric appli-ances grew in Ireland during the period 1960–1985. At the end of the period, the gap between ownership rates in the country's urban and rural areas had greatly narrowed.

Support for Rural Development

Throughout the rural electrification program, the ESB continued to promote the use of electricity on the farm. The availability of reliable rural electricity supplies gradually acquired greater importance and underpinned the rapid development that eventually transformed the Irish countryside.

In the late 1940s, when the pay from farming was still low and the value of electricity and electrical equipment had yet to be proven, the ESB's emphasis was on simple and inexpensive improvements that would appeal to an unsophisticated and inexperienced audience. As electricity supplies became more widely available, a variety of small industrial enterprises emerged. These were very often extensions of existing businesses. The local garage or blacksmith might, for example, install a welder, lathe, grinder, drills, and perhaps a power hacksaw, and thus develop a general engineering works. The greatest impact on farm use from electricity was in dairy farming. The number of milking machines in 1946 was 1,000; in 1960 it had grown to 10,000. This progress virtually eliminated hand milking, one of the most burdensome chores of rural life.

These advances were, however, mainly confined to the richer eastern and south-ern areas. Elsewhere, there was little appreciable growth in agricultural productivity

Table 11-2. *Growth in Electric Appliance Ownership in Ireland, 1960–1985 (percentage household penetration)*

Appliance	1960	1965	1970	1975	1980	1985
Urban consumers						
Electric cooker	25	30	36	42	50	56
Washing machine	15	35	47	57	70	87
Refrigerator or freezer[a]	9	25	51	72	98	111
Dishwasher	—	1	3	4	6	9
Vacuum cleaner	33	30	50	60	75	86
Electric kettle	37	39	46	52	68	86
Rural consumers						
Electric cooker	18	20	25	31	37	40
Washing machine	12	22	32	43	60	69
Refrigerator or freezer[a]	5	12	21	48	100	114
Dishwasher	—	—	1	3	5	7
Vacuum cleaner	10	12	16	23	47	63
Electric kettle	40	48	55	58	70	78

[a] Many people have both refrigerators and freezers, so the row for these appliances can be as high as 200 percent (if everyone had both appliances).

Source: Shiel 1988.

or in the average rural dweller's standard of living. Farming remained dependent on cattle grazing, and there was little inclination or incentive to use electricity for other than domestic purposes.

Change, however, began to take place in the 1960s, when the government launched an economic expansion program. This program encouraged industrial development, especially in the rural areas. It also vigorously promoted tourism, providing incentives for hotel and guesthouse owners. Agricultural output gradually increased and the volume output of livestock and livestock products doubled during the period 1951–1970, but the economic benefits were limited by the fact that Britain, with its cheap food policy, remained the only significant market outlet for Irish agricultural produce.

It was the country's entry in 1971 into the European Union (then called the European Economic Community) that transformed the rural picture, opening up the huge European market to the Irish farmer. Rural electrification, which by then was available to more than 300,000 farmers, enabled the more progressive to make rapid increases in production and quality to take advantage of the new opportunities. The availability of electricity was not, of course, the only factor involved. Much work had been done in the areas of research, development, and education in modern agricultural technology, but without electricity, the Irish farming community would not have been able to respond as effectively as it did.

There was also an enormous expansion of industrial activity in the rural areas. Many were of the indigenous type such as fish processing, vegetable freezing, spinning and weaving, and, of course, processing of dairy products. During the 1970s and 1980s, however, more sophisticated industries, such as computer services, manufacture of optical lens and liquid crystal displays, and a wide variety of others were established.

In the late 1960s and 1970s, there was also a surge of new farmhouse building. Modern, well-equipped houses replaced the old picturesque but uncomfortable thatched cabins. This change has been so widespread as to become a matter of controversy; the flood of new buildings is regarded by many as changing the traditional character of the Irish countryside.

Lessons from Ireland's Experience

Since the post-World War II era, rural electrification has played an integral role in the transformation of Irish rural life. Although no one would claim that electricity alone brought about all the changes that have occurred over the past half century, none would deny that, without rural electrification, most of these changes could not have happened. In hindsight, it is possible to see the main reasons that the Irish rural electrification scheme was such a success.

Government Commitment

Despite financial hardships and conflict-related deprivations, the Irish government wholeheartedly committed itself to improving the quality of life of the country's impoverished rural people. In addition, it recognized electricity's potential role in increasing farm productivity and, in turn, elevating the country's overall socioeconomic and export position.

Visionary Leadership

That the truly crucial, underlying aspects of Ireland's rural electrification program were evident at the outset to William Roe and others concerned with the early planning is a tribute to their vision and good sense. Roe was the first to recognize that a finely honed construction organization was insufficient to guarantee program success. Achieving social and economic objectives, as well as financial viability, depended on the rural community's enthusiastic response. Moreover, Roe believed that the task of rural electrification could only be carried out by mobilizing staff enthusiasm and involving the rural community extensively in program planning and execution.

The ESB's Role in Maximizing Local Partnerships

The ESB realized that the rural electrification program could not succeed, financially or socially, without an effective partnership between itself and the local community. As a result, the board committed itself to an active role in promoting electricity use in rural areas, instead of acting simply as the provider of an electricity distribution system. In its role, the ESB became involved in the process of rural development, in collaboration with government departments, grassroots organizations, and local communities.

The ESB deliberately planned, controlled, and implemented the program to create maximum local involvement. It continued to educate rural people and promote the use of electricity after initial connections had been made. In return, it demanded that local communities make a serious commitment to the program before any electrification work was begun in their areas. By helping to mobilize and stimulate rural demand in this way, the ESB not only improved program finances, it also maximized the program's social and economic effect.

Autonomous Decisionmaking, High Performance, and Capital Subsidies

While creating a clear policy framework, the government allowed the ESB complete independence in its policies and judgments on staffing, technical matters, and relationships with other organizations and the community. The ESB studied accounts of rural electrification programs in North American and Scandinavian countries and considered various technical and institutional options, which it then customized to meet rural Ireland's unique needs. An enthusiastic, well-trained management team was recruited by the REO and assigned a high level of responsibility and accountability. By initially excluding the 30% of rural households living in the most remote rural areas, where investment costs were highest, and by contributing 50% of the capital cost in the early, difficult years, the government afforded a lifeline to the ESB when it was most needed.

Enlightened Community Leadership and Social Awareness

The major rural grassroots organizations supported the government—indeed, they were ahead of it in recognizing the need for rural infrastructure development—and gave the rural electrification program their whole-hearted, highly effective support. They introduced scientific farming methods to young, progressive farmers who were anxious to innovate because they had become the absolute owners (as opposed to tenants) of their holdings. From the outset, national, county, and local community leaders (including NGOs and parish priests) agreed on the urgent need for widespread rural development, which ensured strong political support for the program. Government agencies, rural grassroots organizations, and the ESB worked together closely with the common objectives of promoting higher living standards, greater productivity, and higher quality farm produce. Moreover, they recognized that farmers needed convincing of the benefits of electrification, which led to the development of effective communications and community outreach. Emphasis was placed on building good relationships with local communities and motivating homeowners to sign up for connections. The area organizer, in particular, played a critical role in bridging communications between the ESB and the local community and in recruiting local construction workers, thereby forging a strong sense of local ownership.

Conclusion

The rural electrification program in Ireland points toward common features in providing electricity service to rural people. With hindsight, it is possible to see why the rural electrification scheme was such a success. Early on, the realization that the program could not succeed—financially or socially—without an effective partnership between the ESB and the local community was decisive. As a result, the ESB, instead of acting simply as a supplier of electricity, committed itself to an active role in promoting the use of electricity in rural areas. It became involved with government departments, NGOs, and local communities in the process of rural development.

This involvement was not mere public relations or window dressing. The ESB deliberately went about the planning, control, and implementation of the program to create maximum local involvement; it also continued to educate people and promote the use of electricity when the initial connections had been made. In return it demanded that local communities make a serious commitment to the program before any electrification work was begun in their areas. By helping to mobilize and stimulate rural demand in this way, the ESB not only improved the finances of the program, it also maximized its social and economic effect. In this way, the goal of the government, to change and improve the way of life of its rural population, was achieved well beyond its early expectations.

Acknowledgments

This chapter is primarily based on the author's experience, research, and analysis carried out for the book *The Quiet Revolution* published in 1984. The Panos Institute in London published an earlier version of this case study by the same author in 1988 entitled *Rural Electrification in Ireland,* edited by Gerald Foley.

CHAPTER 12

Meeting the Challenge of Rural Electrification

Douglas F. Barnes

THE DILEMMA OF RURAL ELECTRIFICATION for developing nations is that those who already have electricity are fairly well off and demand that their service be continued at the lowest possible cost. The populations that do not have electricity are poor people living in regions where it is expensive for distribution companies to serve. The electricity companies have an almost thankless task to accomplish. This involves keeping their existing customers happy with the highest quality service while simultaneously making the expensive investments to provide electricity for the remaining people in the country.

The contrasts in approaches to this problem are almost as stark as the problems themselves. Some countries take on rural electrification with a missionary zeal. For them, electricity is viewed as a public good that must be provided to their citizens. Some of these countries have succeeded, and their stories are told in this study; unfortunately, many other committed countries have experienced frustration and wasted efforts. Finally, some others are worried about how the financial stress of serving poor rural people will affect their growing electricity businesses, and they are cautious about rural electrification.

Compounding the dilemma is that many development theorists think in terms of the electricity provision models. Thus, countries are advised extensively on what models of electricity distribution to follow, such as public, private or cooperative. Unfortunately, such approaches have proved to be problematic. For instance, in some countries rural electric cooperatives have been successful, whereas other countries have done well with public or private approaches to rural electrification. Likewise, the same models have succeeded in some countries and suffered problems in others. Thus, it would appear that the institutional model is not the critical element in successful rural electrification programs. This finding matches well with the discussion highlighted in Chapter 1: that successful rural electrification programs seem to be a part of the social, cultural, economic, and political fabric of a society (Sen 2000; Jechoutek 2005). It appears that a combination of factors consistent with a national resolve can lead to the success of rural electrification programs.

To be sure, privatization has brought many benefits and has done remarkably well in providing electricity to rural people in some countries (Estache et al. 2002; Irwin and Brook 2003; Komives et al. 2005; Besant-Jones 2006). Private companies have been improving the reliability of electric power services and in general have been improving the efficiency of distribution. The transfer of responsibilities that comes with privatization has also, to a certain degree, taken some of the politics out of the industry, but the private sector cannot be expected to serve poor populations in remote areas without some form of public policy and subsidy support. The perception that privatization would lift the burden of rural electrification from governments has not proved to be true (World Bank 2004; Covindasamy et al. 2005; Eberhardt et al. 2005; Victor 2005).

So what is the reason for all these problems? Why is rural electrification regarded as such a hard program to implement by power or distribution companies in developing countries (Zomers 2001)? Why have so many projects gone sour while others have succeeded? These questions are not easy to answer. However, the answers are important because rural electrification is a necessary condition for development. Some regions may not be ready to take advantage of the myriad development benefits of rural electricity, but no one would argue that there can be development without it. So the real question is how to approach rural electrification at a pace that is neither too fast nor too slow and is both patient and sustained, allowing for revenues to catch up with investment costs. Also, there is a fine balance between providing incentives or subsidies in such a way that it encourages independence rather than dependence on public funds.

The Challenge of Rural Electrification

Successful rural electrification is a process of solving problems over long periods of time. The nature of the problems actually changes as the programs evolve and mature. As some solutions in the early stages of electrification later turn into problems, rigid prescriptions on accomplishing rural electrification are probably more harmful than good.

Despite this dynamic process, some principles can guide programs. These principles can be applied to the many different forms of rural electrification. In a sense, the reliance on principles rather than rigid institutional formulas is a comforting idea (Table 12-1). The table illustrates that rigid formulas often do not work well across diverse cultures and societies. Some countries are strong on cost reduction but weak in other areas. Other countries are weak in pricing electricity but strong on supporting appropriate forms of subsidies. Also, some countries adopted effective policies that were adapted over time. Principles, therefore, prove more malleable and can be formed and reformed as the problems change. Thus, we can reject the notion that rural electrification is too formidable a task for developing countries. This does not mean that it will be easy, however. Addressing the varied

Table 12-1. Summary of Country Case Studies and Rural Electrification Principles for Successful Programs

Principle	Costa Rica	Philippines	Bangladesh	Thailand	Mexico	Tunisia	Chile	China
Institution of rural electrification and program management	** Support from U.S.-based National Rural Electric Cooperative Association	**** National Electrification Administration (NEA)	**** Rural Electrification Board (REB)	**** Office of Rural Electrification in PEA (ORE)	*** Central support from National Utility	*** Implemented by STEG, the national utility	**** Rural Electrification Program (PER)	** Multiple Central Agency TA to local county companies
Political interference	**** Planning used to avoid problems	* Significant problems in different stages of the program	*** Planning used to avoid problems	**** Planning and program support from the king assisted in avoiding problems	**** Planning and progressive decentralization of social infrastructure funds helped avoid problems	**** Planning used to avoid problems	**** Transparent process of project selection by PER	** Many regional problems
Nature of subsidies	**** Initial loan with subsidy and grace period	*** Subsidized loans and grace periods for new capital cost construction and in some cases bulk power subsidies	**** Subsidized loans and grace periods for new capital cost construction and in some cases bulk power subsidies	**** Cross-subsidies from regional consumers and bulk power subsidies; some local contribution of social fund subsidies	** Social infrastructure funds to provide subsidies for capital construction costs of new service	**** Direct budget transfer of subsidies from the government and presidential fund for political connections	**** Agency set up to award and provide capital cost subsidies to cooperatives and private electricity companies in ways that would not affect their tariffs	**** Government provided free technical assistance and subsidized loans for microhydroelectricity generation

(continued)

Table 12-1. *Summary of Country Case Studies and Rural Electrification Principles for Successful Programs (continued)*

Principle	Costa Rica	Philippines	Bangladesh	Thailand	Mexico	Tunisia	Chile	China
Pricing of electricity	*** Cooperative charges full price based on regulatory process	**** Cooperative charges full price after subsidies based on regulatory process	** Cooperative charges full price after subsidies based on REB regulation	**** National tariff based on cost recovery	** Regional tariffs subsidized from central government budget	**** National tariffs based on cost recovery after subsidies	**** Cooperative and private utility charge full prices based on the regulatory process	*** Regional electricity companies use cost recovery prices based on state guidelines and self regulation
Lowering barriers to obtaining supply	** Connection fees based on connection costs for the whole community in the initial phase; loans for house wiring and appliances	*** Low connection fees partially subsidized through existing price	*** Low connection fees partially subsidized through existing price	**** Low connection fees and house wiring loans in initial program	**** Low connection fees partially subsidized through existing price	**** Low connection fees partially subsidized through existing price	**** Low connection fees partially subsidized through existing price	*** Low connection fees with significant variation among counties
Reducing construction and operating costs	**** U.S. based single-phase	**** U.S. based single-phase	**** U.S. based single-phase modified to local conditions	*** Three-phase but reduced costs	*** U.S. based single-phase for residential, and three-phase for irrigation	**** Single-phase and single wire earth return modified to local grid system	* Low cost system design, but due to subsidy and adverse mountainous terrain initial program costs were high	** Initially many different low-cost techniques that were later standardized

Table 12-1. *Summary of Country Case Studies and Rural Electrification Principles for Successful Programs (continued)*

Principle	Costa Rica	Philippines	Bangladesh	Thailand	Mexico	Tunisia	Chile	China
Community involvement	**** Ratepayers are the owners of the cooperative	**** Ratepayers are owners of the cooperative	**** Ratepayers are owners of the cooperative	** Extensive community consultation	**** Regional boards require community contribution through social infrastructure funds	** Extensive community consultation	** Community expresses need for service and may contribute financing as necessary, but utility takes over after initial request	*** Decentralized distribution and generating companies based on counties with quite varied community involvement
Planning for rural expansion	**** Estimated cost and expected revenue based on expansion plans taking least expensive and highest revenue potential areas	**** Estimated cost and expected revenue based on expansion plans taking least expensive and highest revenue potential areas	**** Estimated cost and expected revenue based on expansion plans taking least expensive and highest revenue potential areas	**** Estimated cost and expected revenue based on expansion plans taking least expensive and highest revenue potential areas	*** Cost based on expansion plans from communities requests and support from infrastructure funds	**** Estimated cost and expected revenue based on expansion plans based on subsidy received for expansion	**** PER evaluates subsidy contribution based on costs, benefits, and regional price and awards grants for new area expansion	** Early planning had many difficulties, with mix of cost based but politically based priorities

Note: The stars indicate the author's evaluation of how closely or how weakly the programs followed the main principles governing successful rural electrification. Four stars means that the programs have quite closely followed the principles deemed necessary for a successful program and one star means either a weak or no association with the principles.

problems of rural electrification generally continues throughout the life of the program until reaching near universal service levels.

The process of applying the principles of rural electrification can try the patience of both developing country governments and international donors, who want quick solutions to the problems. How can either group have patience and a long-term view of development in the face of a steady stream of crises and demands for instant solutions? Rural electrification is indeed a long-term process that requires the government to give up financial resources and to stay at arm's length from the planning and implementation of the program. In the next section, we discuss the main features of successful programs as they relate to the country studies in this book.

Necessity of Sustained Government Commitment

In all of the successful countries, there was a national commitment, not only to make the program happen, but also to ensure that it survives the natural ebb and flow of political support. The sustained government commitment is most obvious in the development of new institutions to deal with the program. This was true whether there was private or public involvement in the electricity sector. In Thailand, the task of rural electrification was given to a public-sector company called the Provincial Electricity Authority. Within this company, a coordinating unit named the Office of Rural Electrification was set up and given remarkable freedom to coordinate and plan the program. By contrast, in Chile the government established a rural electrification program to provide capital cost subsidies used to supply service to communities without electricity. This program started with existing private-sector and cooperative electricity grid distribution companies in the early stages of their program, and the coverage in recent years was expanded to nongovernmental and other organizations serving off-grid rural consumers.

Partly because large-scale, grid-based rural electrification is a relatively complex business, an effective implementing agency is virtually a requirement for enabling programs. The exact institutional structure, however, does not appear to be critical because a variety of approaches have been successful. In addition to the examples given above, a Rural Electrification Board was set up in Bangladesh to supervise the rural electric cooperatives. The U.S.-based National Rural Electrification Cooperative Association provided technical assistance for the development of rural electric cooperatives in Costa Rica. In Tunisia, responsibilities were delegated to regional offices of the utility; this delegation was combined with strong program and technical support from the main national power company.

Although no one institutional model appears unquestionably superior, there are common factors among those that have worked well. A high degree of operating autonomy in which the implementing agency can pursue rural electrification as its primary objective seems to be essential. However, with autonomy comes responsibility as well. A typical example was Ireland, where the rural electrification agency had its own budget and control over access to materials and labor and worked within its own realistically drawn up expansion plan. For this to be implemented, a basic requirement was the availability of stable financial resources made available annually to the company. In turn, the company also was held accountable for meeting its targets every year.

Less tangible but even more important, implementing agencies need dynamic leadership with a capacity to motivate staff and bring a sense of dedication to the task of rural electrification. In Thailand and other countries with successful programs, the staff of the implementing agencies felt that they were laying the foundation for the development and advancement of their countries. A sense of security and clear career prospects within the implementing agency can contribute significantly to building such attitudes among staff.

Thus, there typically exists a strong management group or institution supporting the process of rural electrification. These groups often take different forms, such as a separate administrative unit within a public utility, an independent administrative unit, or a specialized part of an existing government ministry. They all share the common feature of having a degree of independence in creating standards and evaluating proposals for rural electrification investments.

Unfortunately, strong government support for rural electrification is a necessary but not a sufficient condition for a program's success. In many cases, strong government support has led to a squandering of resources and heightened dissatisfaction among rural consumers. One such case is India, where the government has poured resources into the program for decades. Although progress has occurred, even today rural people complain about lack of access to electricity and about the unreliability of the electricity supply. Some of the reasons for this are highlighted in the sections below.

Effective Prioritization and Planning of Rural Electrification

The Achilles heel of many rural electrification programs is that there is significant political interference in the way the established electricity companies carry out their business. The necessary sustained political commitment can turn into sustained interference in planning and development of the rural electrification grid. The cost of service in rural areas is quite high even among the best planned systems. The effect of political pressure to extend the system too far and too fast can lead many companies to the brink of bankruptcy or failure or, in some cases, extensive dependence on government subsidies for operating costs. For public companies that are not allowed to fail, it may mean that they lack the necessary resources to maintain system reliability, which has the adverse effect of making their customers unhappy and less likely to pay their electricity bills because of poor service. This problem further aggravates the downward spiral away from financial sustainability.

The Importance of Transparent Criteria for Rural Electrification

The adherence to specific project-selection criteria is a common feature of successful large-scale rural electrification programs, including those administered by public companies. Successful programs typically are based on multiyear electrification expansion plans that take into consideration the financial viability of the investments and their economic effect on the region. Project-selection methodologies are critical both for ensuring the financial viability of the electricity company and for minimizing the interference of politicians in the selection process. The many financial problems encountered in implementing rural electrification programs have resulted from political pressure to expand too fast into areas with low or marginal electricity demand.

Successful rural electrification programs have all developed their own systems for ranking or prioritizing areas to receive electricity. Capital investment costs, level of local contributions, numbers and density of consumers, and the likely demand for electricity are among the factors usually taken into account. In Costa Rica, the ranking of communities was based on their population density, level of commercial development, and potential electricity load. Thailand developed a numerical ranking system that weighed a variety of factors, such as level of income, the number of existing commercial enterprises, and the government's plans for other infrastructure investments in the area.

Another rationale for good planning is that providing an electricity supply typically will only make a significant contribution to sustainable rural development when the other necessary conditions are present. Security of land tenure, availability of agricultural inputs, access to health and educational services, reliable water supply, and adequate dwellings are among the more obvious of these conditions. If farmers are to invest in increased agricultural production, they must have access to markets where they can obtain fair prices for their higher outputs. Families must have a level of disposable income that allows them to pay for improved lighting and ownership of television sets and other electric appliances.

Another common feature of many of the successful country studies in this book is that the rural electrification agencies often coordinate their expansion activities with other organizations that are involved in rural development. This was especially true for the program in Tunisia, where rural electrification was made one of the pillars of the country's development strategy, along with education and health. Periodic meetings take place in Tunisia to discuss coordination of electricity, education, and health programs. Whether the meetings are regular or only periodic, it makes quite a bit of sense to extend the electricity grid to communities where there will be investments in education, water supply, or other rural development programs. There is evidence that these programs work better together, rather than independently (World Bank 1999a). Schools with electricity work better than those without it. Water pumping can be done by hand, but it is much easier and more effective to use an electric pump. Thus, many of the most successful programs have reaped benefits in terms of greater electricity sales by coordinating with other rural development efforts.

Dealing with the Political Dimension

Before turning to other important factors for rural electrification, one must recognize that because of public funding these programs often are subject to a great deal of politics, especially attempts to distort the planning process. Politicians intervene in rural electrification planning in an attempt to favor their constituencies.

Well-meaning politicians regard public funding as giving them rights to interfere in project planning by insisting that their home communities receive electricity on a priority basis, but experience shows that such requests can be damaging to the long-term success of the program. Once technical and financial decisionmaking in the implementing agency becomes based on political string pulling, professional discipline is destroyed and the organizational structure is undermined. Waste of resources, low staff morale, and operational ineffectiveness are the characteristics of rural electrification programs suffering from a high degree of political interference.

Sometimes this political desire for providing electricity to constituencies can be turned into a positive force, as was the case in Thailand. In this program, local politicians were encouraged to raise and contribute funds so that their constituents could receive electricity before the planned time. The company required a contribution of 30% of the capital cost that would be necessary to provide the connection of the village to the grid, so that the village would move up in the expansion plan. Generally, such communities would receive electricity within a year of making such a contribution. Many rural communities had social development funds available to them, and they were permitted to use these funds to pay for the partial capital costs of connection to the grid. Thus, successful programs often use clearly defined criteria to rank areas in order of priority for electrification, so that the decisionmaking is clearly seen by all to be transparent and fair; this system can take the political pressure off the company to expand to remote areas too quickly.

Reducing Construction and Operating Costs

In addition to system planning, there are major opportunities for the reduction of construction and operating costs of rural electrification in most countries. In many cases, careful attention to system design enables construction costs to be reduced by up to 20–30% (World Bank 2000), contributing significantly to the pace and scope of the rural electrification program.

Where the main use of electricity is expected to be for lights and small appliances—a pattern typical of many rural areas—there is no reason to apply the design standards used for more heavily used urban systems (World Bank 1975; Foley 1992). The rural distribution system can be designed for the actual loads, often in the range of 15 to at most 75 kilowatt-hours per month, imposed on it by rural households. In most newly constructed systems, consumption usually grows at a fairly slow pace, so it is important to start off with low-cost systems that can be relatively cheaply upgraded later.

Each country will have its own cost-saving opportunities. In Thailand, materials were standardized and manufactured locally, reducing procurement, materials handling, and purchasing expenses. In Costa Rica, the Philippines, and Bangladesh, they adopted the well-proven single-phase distribution system based on standards developed for the U.S. rural electrification program. This use of the single-phase MALT system was a major feature for the cost-saving program in Tunisia. After extensive technical studies, the internally controversial decision was made to use a three-phase backbone combined with a single-phase distribution system. Estimates are that distribution costs were reduced as much as 30%. This was not an easy decision for the electricity company in Tunisia to make because it faced opposition from many of its engineers, who were more accustomed to designing systems for urban areas. Thus, the country studies show that careful and critical analysis of design assumptions and implementation practices reveals potential for significant cost savings.

Sustainable Financing of Rural Electrification

Cost recovery through customer billing is probably the most important factor determining the long-term effectiveness of rural electrification programs. When cost recovery is pursued, many of the other program elements fall into place. All the successful programs reviewed in the country studies placed a strong emphasis on covering their costs, although there is a wide variation in how it was approached.

Electricity supply organizations that depend on operational subsidies are critically vulnerable to any downturn in the availability of subsidies. When the subsidy is reduced, as inevitably happens during such long-term programs, the resulting losses or at the least financial stress can create significant disincentives to extend electricity to new customers, and especially to poor ones. An example of the problems of short-term financing is the country of Kenya, where in the past the rural electrification program was almost entirely dependent on the availability of grant funds from donors. The result is that progress in rural electrification has been slow and intermittent. Thus, such financial fits and starts can be quite destructive to a program that requires 20–30 years to reach maturity. This section describes some of the successful ways other countries have dealt with the issue of sustainable financing, mostly through effective subsidies and charging a fair price for electricity.

Development of Effective Subsidies

Most of the world's rural electrification programs were supported by governments to promote development and to get people out of poverty. Passing along the full cost of electrification in many cases can have a negative effect on social equity. Thus, all rural electrification programs worldwide have involved some form of subsidy. However, subsidies must be efficient, effective, and equitable. That is, they should be fairly easy to administer, they should have an effect on the desired population, and they should reach the poorest of society.

The rural electrification programs reviewed in this study pass most of these tests. However, the types of subsidies for rural electrification have been quite varied in successful programs. There have been capital subsidies, capital investment subsidy funds based on principles of output-based aid, bulk power subsidies, and others. Thus, the idea that any one form of subsidy is good or bad seems to belie the evidence from the country studies.

Capital subsidies for rural electrification have worked in several countries. The concept of providing partial capital subsidies for new electricity connections and requiring the distribution company to cover the operating costs of service through revenue collection has proven to be the most common way to make rural electrification programs financially viable and sustainable over the long term (World Bank 1994). When properly administered, such subsidies can make rural electrification a profitable business.

Capital investment subsidies can be implemented in several ways. In most programs, a substantial proportion of the capital has been obtained either through loans with concessionary interest rates or from a blending of loans with grants. The program in Costa Rica started with low-interest loans from the U.S. Agency for International Development. In Ireland, a proportion of the investment costs, which varied depending on the state of the national exchequer, were covered by government grants. In Thailand, capital contributions from donors and subsidized loans from international banks helped lower the barriers to providing electricity to its rural population. Provided it is used wisely and operating costs are covered, having access to such concessionary capital need have no ill effects on the implementing agency or the rural electrification program. However, concessionary capital should never be provided to organizations that are not covering their operating and maintenance costs (World Bank 1996) because it will simply worsen their financial position.

Another form of subsidy that has been used in many rural electrification programs is the reduction in the price that rural distribution utilities must pay for bulk electricity compared to companies serving urban regions or large industry or commercial enterprises. Although the use of bulk-power subsidies is not a good idea from a strictly economic point of view (Komives et al. 2005; Besant-Jones 2006), they have been used successfully in several countries. In Thailand, the Provincial Electricity Authority pays less for bulk electricity than the distribution company serving the Bangkok region, but in this case the bulk-power supplier has not been financially affected. This is because the company supplying the electricity is required to be financially viable, so the bulk electricity subsidy is really a cross-subsidy from Bangkok to the rest of the nation. In the United States, the cheap power from the public dams has been given to the rural cooperatives on a priority basis because of their role in providing affordable electricity to people in rural areas.

In an era of increasing privatization of electricity distribution, the idea of creating a subsidy fund to encourage distribution companies to expand supply systems and connect new, poorer consumers is a worthwhile objective. This approach to subsidies is often called public–private partnership. The country of Chile is a case in point. Chile established a rural electrification fund for the purpose of promoting new investments in electricity distribution. This fund was open to mostly existing private and cooperative distribution companies. The fund required that the companies must serve new consumers and demonstrate that they can do this in a financially viable way. The rules and regulations governing qualification for the funds are administered by an institution that was created especially for that purpose. At present, for most projects the fund subsidizes approximately 75% of the capital costs of new line extension to rural communities in Chile, but to receive the funds there is a rigorous test of both financial and economic viability of the investment.

Lifeline rates and cross-subsidies have been an effective policy for encouraging rural consumers to adopt and use electricity in countries where electricity companies serve both urban and rural consumers (Barnes et al. 2005). Although lifeline rates can be a useful way to make electricity affordable to rural consumers, they also are often abused in many countries, where there are political pressures to keep electricity rates low for everyone. In the poorly performing rural electrification programs, the lifeline rate often is set so high that it covers almost all consumers and therefore compromises the financial viability of the companies serving rural consumers. However, lifeline rates that are properly related to the poverty profiles of consumers have been able to easily absorb the subsidies for the poorest consumers without causing financial difficulties for the companies. For instance, in Thailand the electricity distribution adjusts the increasing block rate used for electricity pricing to its records of consumer demand for electricity. The poorest households use little electricity; therefore, allowing them to pay a lower price does not cause significant problems for the company. The problems arise when the lifeline rates are not based on the demand by the poorest households. In such a case the subsidies become poorly targeted, lead to problems for the electricity companies, and prove wasteful for the country.

Thus, the form of the subsidy might not be as important as how such subsidies are administered. One form of subsidy that is acceptable in one country may be unfitting in another one. The type of subsidy that is suitable is often an implicit part of a country's culture.

A final caveat on subsidies is that the programs in which rural electrification has been problematic generally have not attained the balance between too little or too much subsidy. The programs with too little subsidy generally are able to provide electricity access to wealthy households, and they refrain from extending it to the poor. The programs with too much subsidy often become so dependent on them that they do not have an incentive to serve paying rural customers. Achieving the right balance is not an easy task to achieve, so it is instructive to see how it has been achieved by the countries with relatively successful programs reviewed in this book. Countries just beginning the journey toward providing electricity to their rural populations can use the lessons learned.

Charging the Right Price for Electricity

There is some belief in developing country governments that electricity tariffs need to be extremely low, often well below their true supply costs, if rural electrification is to be affordable to rural people. By contrast, many international development agencies stress the principle of full or near full cost recovery. Several things must be taken into consideration in deciding the "right price" of rural electrification.

Rural electrification tariffs set at realistic levels allow people to improve their lives vastly, even if they spend a bit more of their income on energy. Charging a fair price allows the electricity company to provide an electricity supply in an effective, reliable, and sustainable manner to an increasing number of satisfied consumers. In Costa Rica, the price of electricity is set through a regulatory process, but it is high enough for the cooperatives to make a modest profit. Recent surveys in regions without electricity indicate that there is a willingness to pay for electricity services. In some cases, rural people may spend as much as US$5 per month on other energy sources, even in remote areas (Barnes and Floor 1996; World Bank 1999a, 2000, 2002a, 2002b; Barnes and Toman 2006). Private suppliers sometimes find a ready market among wealthy households and rural enterprises for electricity as high as US$1.00 per kilowatt-hour. However, charging such full cost for electricity service would inevitably mean that there will be slow uptake in electricity by both the middle class and the poor households.

There also is an issue of changing prices. In its early stages, rural electrification mainly makes sense in areas where there is already a demand for electricity-using services, such as lighting, television, refrigeration, and motive power. In the absence of a grid supply, these services are obtained by spending money on kerosene, liquid petroleum gas, dry-cell batteries, car battery recharging, and small power units, all of which are expensive per unit of electricity supplied (Foley 1992). Thus, when the grid arrives, there is a ready customer base to adopt electricity. As the program moves further into rural areas, some adjustment in the capital cost subsidy mechanisms may be necessary to keep prices at moderate levels and thus to make it affordable by the generally poorer households in such areas. This is the rationale for the subsidies in Chile, where the rural electrification fund adjusts the capital cost subsidies to allow the distribution companies to have the same price as their established customer base.

Many countries have a regulator who sets the price of electricity. The rules of electricity pricing followed by many developing country regulators often are geared mainly toward urban areas. It is usually wise not to overregulate rural utilities because their unique operating environments must be taken into consideration (Reiche et al. 2006).

In summary, the price that distribution companies charge for electricity in rural areas is country specific, but several principles seem to apply. The first is that some kind of subsidy is necessary to lower the effective price that rural people must pay for electricity so that electricity will be adopted more broadly by the rural population. The second is that depending on the customer base, lifelines rates to encourage poor consumers to gain the benefits of electricity are justified mainly in distribution companies that balance the costs of serving the poor by charging marginally higher prices to others. Finally, the willingness to pay for electricity charges by rural people should not be underestimated. In many cases, they do not connect to electricity service because of the barriers to obtaining an electricity supply rather than because of the monthly energy charge for electricity. Finally, it is preferable for a company's main source of revenue to come from its customers rather than from subsidies, as this inevitably focuses the attention of the business on providing quality service.

Customer Focus by Effective Distribution Companies

The relationship between rural electrification distribution companies and their customers is often somewhat overlooked by international development agencies and

utility engineers carrying out programs of rural electrification in developing countries. Everyone knows that effective customer service is desirable, but the emphasis on investments in wires and poles sometimes overshadows many of the softer
requirements of having sustainable programs, such as community involvement and
customer service (Cernea 1985; World Bank 1992, 1996; Barnes et al. 2005).

Billing and Lowering the Barriers to Obtaining a Supply

Most of the effective rural electrification programs insist that customers pay their
bills. The ability to collect monthly bills may seem like an elementary part of a
utility's business, but in many countries bill payment rates are at or below 50%.
This situation points to the issue of perceived fairness of the electricity distribution company and its relations with its customers.

Rural electrification utilities are often conscious about keeping the costs of
service low. The initial connection charges demanded by the utility are often far
greater barriers to rural families than is the monthly electricity bill. In Bolivia, for
example, a small local grid, despite charging 25 to 30 U.S. cents per kilowatt-hour,
immediately doubled its number of consumers when it offered them the option
of paying for the connection cost over five years (World Bank 1991). By contrast,
in Malawi, where the electricity company charges virtually the full cost of line
extension—a 30-year investment—to new customers, the rural electrification rate
is less than 10% (World Bank 2005b, 28). Reducing these charges, or spreading
them over several years, even if it means charging more per unit of electricity,
allows larger numbers of low-income rural families to obtain electricity.

Benefits of Community Involvement

Traditional thinking in many utilities often ignores the importance of local
community involvement. Rural electrification is seen simply as a technical matter of stringing lines to grateful consumers eager to receive the benefits of
electricity. The country studies show clearly that rural electrification programs
can benefit greatly from the involvement of local communities—or can suffer
because of its absence.

For instance, in Ireland, the parish rural electrification committee that was set
up to represent the local community did much to smooth the implementation
of the program. The committee played a crucial role in helping assess the level of
demand, educating consumers in advance, encouraging them to sign up for a supply, and promoting the wider use of electricity. The efforts made by parish rural
electrification committees to recruit customers ensured that the utility received an
adequate return on its investment and contributed to the rapid implementation
of the country's rural electrification program. In Bangladesh, consumer meetings
were held before the arrival of the electricity supply, helping to avoid costly and
time-consuming disputes over rights of way and construction damage. Finally, in
Thailand, community contributions in cash or kind were often the decisive factor
in bringing electricity to rural communities earlier than might have been possible
without such assistance.

The benefits of rural electrification are well understood by rural people. Poorly run and managed programs that fail to provide adequate service are bound to meet the disdain of their customers. On the other hand, the lesson from the case studies in this book is that those companies that do proper planning, promote productive uses, provide electricity for essential community services, and work with their customers to solve problems that arise gain the respect of both customers and community leaders. This respect can then be passed on through the political process that will support the program over the long term without making distorting political demands that inevitably cause problems.

The Way Forward

Well-planned, carefully targeted, and effectively implemented rural electrification programs provide enormous benefits to rural people. Indeed, once an area has reached a certain level of development, further progress in raising standards of living to socially and politically acceptable levels will depend on the availability of an electricity supply. As restructuring of national power utilities gathers pace around the developing world, it is essential to keep in mind that appropriate institutional frameworks and incentives are created to ensure that rural electrification does not get lost in the transition.

This book has dealt mainly with grid-based rural electrification, which is often portrayed as being in competition with alternatives, especially photovoltaic systems. This is a mistake because there is little conflict between the two and, in fact, they can complement one another. Providing electricity to people in rural areas is a daunting task for organizations, whether it involves connecting communities to the grid or developing off-grid approaches to rural electrification.

Although the task is challenging, the main message from the countries that have gone through the experience is quite encouraging. There are major opportunities for increasing the pace and widening the scope of rural electrification. If these opportunities are grasped, it will enable large numbers of new consumers to enjoy the benefits of an electricity supply at acceptable costs and without burdening national governments and power utilities with unsustainable subsidies.

We should not return to the days of thinking that electricity will solve all development problems, but we also should not be so naive as to think that poor people in developing countries can take full advantage of other forms of development assistance without electricity. The effect that a well-managed program can have for people in developing countries should not be underestimated. Some countries have already made great strides in meeting the challenge of rural electrification, and now they can provide guidance for others willing to do the same.

Acknowledgment

I thank Gerald Foley, who provided significant input into this chapter at an earlier stage of its development.

References

ADB (African Development Bank). 1995a. *The African Development Bank Experience with Rural Electrification: The Case of Tunisia.* Workshop Proceedings on Financing Photovoltaic Rural Electrification in Developing Countries, March 5–7. Rome: ENEA.

———. 1995b. *Tunisie: Rapport d'Achèvement, Projet d'Electrification Rurale (Electricite IV), Departement Infrastructure et Industrie, Region Nord.* Abidjan, Tunisia: ADB.

———. 1999. *Rapport d'Evaluation du Projet d'Electrification Rurale (Electricite VI): Tunisie, Departement par pays Region Nord.* Abidjan, Tunisia: ADB.

AME and GTZ. 1999. *Solar Rural Electrification in Tunisia: Approach and Practical Experience.* (Volumes 1 and 2). Tunis, Tunisia: Agence pour la Maîtrise de l'Energie.

Anderson, Douglas D. 1980. State Regulation of Electric Utilities. In *The Politics of Regulation,* edited by James Q. Wilson. New York: Basic Books, 3–41.

Arcia, Gustavo. 2000. *Education and Poverty in Nicaragua. Consulting Report, Nicaragua Poverty Assessment.* Washington, DC: World Bank.

Asian Institute of Technology. 1992. *Rural Electrification Guidebook for Asia and Pacific.* Bangkok, Thailand: Asian Institute of Technology.

Atkinson, Edward. 1887. The Relative Strength and Weakness of Nations. *The Century* 33(3): 422–36 and 33(4):613–22. http://cdl.library.cornell.edu/cgi-bin/moa/moa-cgi?notisid=ABP2287-0033-77 and http://cdl.library.cornell.edu/cgi-bin/moa/moa-cgi?notisid=ABP2287-0033-120 (accessed December 8, 2006).

Bangladesh Bureau of Statistics. 1998b. *Yearbook of Agricultural Statistics of Bangladesh.* Dhaka, Bangladesh: Government of Bangladesh.

———. April 1999. *Statistical Bulletin Bangladesh.* Dhaka Bangladesh: Government of Bangladesh.

Barkat, Abdul, M. Rahman, S. Zaman, A. Podder, S. Halim, N. Ratna, M. Majid, A. Maksud, A. Karim, and S. Islam. 2002. *Economic and Social Impact Evaluation Study of the Rural Electrification Program in Bangladesh.* Report to the National Rural Electric Cooperative Association International. Dhaka, Bangladesh: NRECA International.

Barnes, Douglas. 1988. *Electric Power for Rural Growth: How Electricity Affects Rural Life in Developing Countries.* Boulder, CO: Westview Press.

Barnes, Douglas, and Willem Floor. 1996. Rural Energy and Developing Countries: A Challenge for Economic Development. *Annual Review of Energy and Environment* 21: 497–530.

Barnes, Douglas, and Jonathan Halpern. 2000. The Role of Energy Subsidies. In *Energy Services for the World's Poor.* World Bank Energy and Development Report. Washington, DC: World Bank.

Barnes, Douglas, and Michael Toman. 2006. Energy, Equity and Economic Development. In *Economic Development and Environmental Sustainability: New Policy Options,* edited by Ramon Lopez and Michael Toman. Oxford: Oxford University Press.

Barnes, Douglas, Kerry Krutilla, and William Hyde. 2005. *The Urban Household Energy Transition: Social and Environmental Impacts in the Developing World.* Washington, DC: RFF Press.

Barnes, Douglas, Henry Peskin, and Kevin Fitzgerald. 2003. *The Benefits of Rural Electrification in India: Implications for Education, Household Lighting, and Irrigation.* Draft paper prepared for South Asia Energy and Infrastructure. Washington, DC: World Bank.

Belkhiria, Hachemi. 1996. *Etude d'un Projet Pilote pour l'Electrification Rurale en 4.16-kV Mono-phasé*. Tunis, Tunisia: Tunisian Electricity and Gas Company.

————. 2001. *Etude de l'Alimentation des Zones de Pompage par un Réseau Basse Tension à 3 Fils et 2 Niveaux de Tension*. Tunis, Tunisia: Tunisian Electricity and Gas Company.

Benjamin, G.R. 1964. *Estudio de Ingeniería y Factibilidad Económica para Tres Proyectos Pilotos de Cooperativas de Electrificación Rural en Guanacaste, Tres Amigos, y Los Santos*. Washington, DC: National Rural Electric Cooperative Association.

Besant-Jones, John. 2006. Reforming Power Markets in Developing Countries: Wgat Gave We Learned?, Energy and Mining Sector, Board Discussion Paper. Washington, DC: World Bank.

Bose, Sarmila. 1994. *Money, Energy, and Welfare: The State and the Household in India's Rural Electrification Policy*. New York: Oxford University Press.

Bouraoui, Soukaina. 2001. Droits des Femmes et Evolution des Rapports Hommes–Femmes depuis l'Indépendance. In *Population et Développement en Tunisie*, edited by Jacques Vallin and Thérèse Locoh. Tunis, Tunisia: Centre d'Etudes et de Recherches Economiques et Sociales (CERES).

Bradford, Ernst. 1925. The Influence of Cheap Power on Factory Location and on Farming. *Annals of the American Academy of Political and Social Science* 118 (March): 91–95.

Brodman, Janice. 1982. *Rural Electrification and the Commercial Sector in Indonesia*. Resources for the Future Discussion Paper D–73L. Washington, DC: Resources for the Future.

Brown, Ashly, Jon Stern, Bernard Tenenbaum, and Defne Gencer. 2006. *Handbook for Evaluating Infrastructure Regulatory Systems*. Washington, DC: World Bank.

Brown, D. Clayton. 1980. *Electricity for Rural America: The Fight for the REA*. Contributions in Economic History No. 29. Westport, CT: Greenwood Press.

Bustamante Castañeda, Cesar. 1978. *La Electrificación Rural, Factor de Desarrollo Económico en el Estado de Guerrero*. Tesis de Licenciatura, Universidad Nacional Autónoma de México. Mexico City, México.

Butler, Edward, Karen Poe, and Judith Tendler. 1980. *Bolivia: Rural Electrification*. U.S. Agency for International Development (USAID) project impact evaluation. Washington, DC: USAID.

Cabraal, Anil, and Douglas Barnes. 2006. Productive Uses of Energy for Rural Development. *Annual Review of Environment and Resources* 30: 117-144.

Cabrera, R.E. 1992. Rural Electrification in the Philippines. In *Rural Electrification Guidebook for Asia and Pacific*, edited by G. Saunier. Bangkok, Thailand: ESCAP.

Cernea, Michael. 1985. *Putting People First: Sociological Variables in Rural Development*. New York: Oxford University.

Chaib, Sawsen, Moncef Aissa, and Ahmed Ounalli. 2001. *Enquête sur l'Impact Socio-économique de l'Electrification Rurale: Méthodologie et Résultats*. Internal report.

Chandler, Alfred D. 1990. *Scale and Scope: The Dynamics of Industrial Capitalism*. Cambridge, MA: Harvard University Press.

CFE (Comisión Federal de Electricidad). 1988. *Electrificación Rural y Recursos Forestales*. Mexico City, México: CFE.

————. 1997. *Conexión, 1937–1997*. 60th anniversary of CFE. El plan para la próxima década, 1997–2007. Mexico City, México: CFE, August.

————. From 1985 to 2006. Informe de Labores. Mexico City, México: CFE.

————. From 1990 to 2005. Grado de Electrificación. Sudirección de Distribución, Unidad de Electrificación. Mexico City, México: CFE.

————. From 1990 to 2005. Obra realizada, Subdirección de Distribución, Unidad de Electrificación. Mexico City, México: CFE.

Comisión Nacional para el Ahorro de Energía. 1999. Energías Renovables y Ahorro de Energía. *Instituto de Investigaciones Estéticas*. Mexico City, México, September–October.

Commission on Country Life. 1909. *Report of the Country Life Commission*. U.S. Senate 60th Cong., 2d sess., Document No. 705. Reprint. Spokane, WA: Chamber of Commerce.

Cooke, Morris L. 1948. The Early Days of the Rural Electrification Idea, 1914–1936. *American Political Science Review* 42 (June): 431–47.

COOPESANTOS. 1994. *25 Años Generando Progreso*. Costa Rica: San Marcos de Tarrazú.

————. 1996. XXXVII Asamblea General de Delegados Abril 14 de 1996. Costa Rica: San Marcos de Tarrazú.

Covindasamy, Ananda, Daizo Oda, and Yabei Zhang. 2005. *Analysis of Power Projects with Private Participation under Stress*. ESMAP Report 311/05. Washington, DC: World Bank.

Danbom, David B. 1979. *The Resisted Revolution: Urban America and the Industrialization of Agriculture, 1900–1930*. Ames, IA: Iowa State University Press.

Das, Kumar Bar. 1990. *Electrical Energy and Economic Development of Rural India*. New Delhi: Ashis.

Denton, Frank H. 1979. *Lighting up the Countryside: The Story of Electric Cooperatives in the Philippines*. Development Academy of the Philippines. Manila: Academy Press.

DuBoff, Richard B. 1979. *Electric Power in American Manufacturing, 1889–1958*. New York: Arno.

Eberhardt, Anton, Alix Clark, Njeri Wamukonya, and Katherine Gratwick. 2005. *Power Sector in Africa: Assessing the Impact on Poor People*. ESMAP Report 306/05. Washington, DC: World Bank.

EGAT (Electric Generating Authority of Thailand). 1992–2001. Annual Report. Bangkok, Thailand: EGAT.

EIA (Energy Information Administration). 2000. Country Analysis Briefs: Chile. http://www.eia.doe.gov/emeu/cabs/chile.html (accessed December 8, 2006).

ESMAP (Energy Sector Management Assistance Programme). 1990. *Tunisia: Interfuel Substitution Study: A Joint Report*. ESMAP Report 114/90. Washington, DC: World Bank.

————. 1996. *Tunisie: Les Potentiels de Valorisation des Energies Renouvelables*. ESMAP Report 190A/96 (Volumes 1 and 2). Washington, DC: World Bank.

Essebaa, Hachemi. 1994. *Le Système MALT: Un Système de Distribution Adapté*. Tunis, Tunisia: Tunisian Electricity and Gas Company.

Estache, Antonio, Vivien Foster, and Quentin Wodon. 2002. *Accounting for Poverty in Infrastructure Reform: Learning from Latin America's Experience*. World Bank Institute Development Studies. Washington, DC: World Bank.

Filmer, Deon, and Lant Pritchett. 1998. *The Effect of Household Wealth on Educational Attainment around the World: Demographic and Health Survey Evidence*. Washington, DC: World Bank.

Foley, Gerald. 1992. Rural Electrification in the Developing World. *Energy Policy* 20 (2): 145–152.

FAO (Food and Agriculture Organization of the United Nations). 1993. *Evaluation du Programme Rural Intégré (PDRI)*. Rome: FAO.

Fluitman, Fred. 1983. *The Socioeconomic Impact of Rural Electrification in Developing Countries: A Review of Evidence*. International Labour Organisation (ILO) working paper. Geneva: ILO.

Friedlander, Amy. 1996. *Power and Light: Electricity in the U.S. Energy Infrastructure, 1870–1940*. Reston, VA: Corporation for National Research Initiatives.

Funigiello, Philip. 1973. *Toward a National Power Policy*. Pittsburgh: University of Pittsburgh Press.

Girer, R.G. 1986. Rural Electrification in Costa Rica: Membership Participation and Distribution of Benefits. M.Sc. Thesis, University of Pennsylvania.

Glaser, Leah S. 2002. Rural Electrification in Multiethnic Arizona: A Study in Power, Urbanization, and Change. Ph.D. dissertation, Arizona State University, Tempe, AZ. Glendale, AZ: Southwest Museum of Engineering, Communications and Computation. http://www.smecc.org/rural_electrification_in_arizona.htm (accessed December 8, 2006).

Goddard, Paul, Gustavo Gomez, Polly Harrison, and George Hoover. 1981. *The Product Is Progress: Rural Electrification in Costa Rica*. Project Impact Evaluation No. 22. Washington, DC: USAID.

Goodwyn, Lawrence. 1976. *Democratic Promise: The Populist Movement in America*. New York: Oxford University Press.

Gordon, Adele. 1997. *Facilitating Education in Rural Areas of South Africa: The Role of Electricity and Other Sources of Energy*. University of Cape Town Energy and Development Research Centre (EDRC) Report Series. Cape Town: EDRC.

Harrington, Michael. 1962. *The Other America: Poverty in the United States*. New York: Macmillan.

Hellman, Richard. 1972. *Government Competition in the Electric Utility Industry*. New York: Praeger.

Hernández-Trillo, Fausto. 2004. Analisis de Aportaciones Federales para Infraestructura. Mexico City, México: Centro de Investigación y Docencia Económicas (CIDE).

Herrin, Alexandro. 1979. Rural Electrification and Fertility Change in the Southern Philippines. *Population and Development Review* 5(1): 61–87.

Hicks, Jack, Allen Inversin, Mounir Majdoub, and Seyoum Solomon. 1993. *Rural Electrification Evaluation Study: Tunisia, November 1993.* Washington, DC: National Rural Electrification Cooperative Association.

Hirsh, Richard F. 2002. Emergence of Electrical Utilities in America. Smithsonian Institution Web site, Powering a Generation of Change. http://americanhistory.si.edu/powering/past/h1main.htm (accessed December 8, 2006).

Huacuz-Villamar, Jorge M. 1995. Renewable Energy Rural Electrification. Sustainability Aspects of the Mexican Programme in Practice. Natural Resource Forum 19(3).

Hughes, Thomas P. 1983. *Networks of Power: Electrification in Western Society, 1880–1930.* Baltimore, MD: Johns Hopkins University Press.

Hyman, Leonard S. 1995. *America's Electric Utilities: Past, Present and Future,* 5th ed. Vienna, Virginia: Public Utilities Reports, Inc.

IEA (International Energy Agency). 2002. *World Energy Outlook.* Paris: IEA.

———. 2004. *World Energy Outlook.* Paris: IEA.

INEGI (Instituto Nacional de Estadística, Geografía e Informática). Aguascalientes, México: Sistema Municipal de Bases de Datos (SIMBAD).

International Herald Tribune. 2001. Contemporary Tunisia: Women in Business. August 13.

Irwin, Timothy, and Penelope Brook. 2003. Private Infrastructure and the Poor. In *Infrastructure for Poor People: Public Policy for Private Provision,* edited by P. Brook and T. Irwin. Washington, DC: World Bank.

Jechoutek, Karl. 2005. Developing Through Deliberation: Democratisation and Culture in Transition Polities. Paper prepared for New Directions in Democratic Theory, University of Cape Town.

Kahn, Alfred E. 1988. *The Economics of Regulation: Principles and Institutions,* Vol. 2. Cambridge, MA: MIT Press.

Kazin, Michael. 1995. *The Populist Persuasion: An American History.* New York: Basic Books.

Keun, Odette. 1937. *A Foreigner Looks at TVA.* New York: Longmans, Green. http://newdeal.feri.org/tva/tva12.htm (accessed December 8, 2006).

Khandker, Shahidur. 1996. *Education Achievements and School Efficiency in Rural Bangladesh.* World Bank Discussion Paper No. 319. Washington, DC: World Bank.

Khandker, Shahidur, Victor Lavy, and Deon Filmer. 1994. *Schooling and Cognitive Achievements of Children in Morocco: Can the Government Improve Outcomes?* World Bank Discussion Paper No. 264. Washington, DC: World Bank.

Knittel, Christopher R. 2006. The Adoption of State Electricity Regulation: The Role of Interest Groups. *Journal of Industrial Economics* 54(2): 201–21.

Komives, Kristin, Quentin Wodon, Vivien Foster, and Jonathan Halpern. 2005. *Water, Electricity, and the Poor: Who Benefits from Utility Subsidies?* Directions in Development. Washington, DC: World Bank.

Kulkarni, Veena, and Douglas Barnes. 2004. The Impact of Electricity on School Participation in Rural Nicaragua. Working paper, University of Maryland, College Park, MD.

Lara-Beautell, Cristóbal. 1953. La Industria de Energía Eléctrica. Fondo de Cultura Económica. Mexico City, México.

Lay, J.D. 1976a. *Evaluation Report of COOPELESCA.* Washington, DC: NRECA.

———. 1976b. *Evaluation Report of COOPEGUANACASTE.* Washington, DC: NRECA.

Lay, J.D., et al. 1984. *Analysis of the Rural Electric Cooperative Movement in Costa Rica and Recommendations for Training and Technical Assistance.* Washington, DC: NRECA.

Lay, James, and Joan H. Hood. 1976. *Interim Evaluation Report: Rural Electric Cooperative of Guanacaste, R.L. and Rural Electric Cooperative of San Carlos, R.L.* Report to International Program Division. Washington, DC: NRECA.

Madigan, Francis, Alejandro N. Herrin, and William F. Mulcahy. 1976. *Evaluation Study of the MISA-MIS Oriental Rural Electric Service Cooperative (MORESCO).* Report. Washington, DC: USAID.

Mandel, David, Peter Allgeier, Gary Wasserman, Gerald Hickey, Robert Salazar, and Josephine Alviar. 1980. *The Philippines Rural Electrification.* Project Impact Evaluation Report No. 15. Washington, DC: USAID.

Masmoudi, Radhouane. 1997. *L'Electrification Rurale.* Revue Tunisienne de l'Energie, No. 47, Fourth Trimester. Tunis, Tunisia: Ministère de l'Industrie.

McGuire, Patrick, and Mark Granovetter. 1998. Business and Bias in Public Policy Formation: The National Civic Federation and Social Construction of Electric Utility Regulation, 1905–1907. Paper presented to the American Sociological Association, August 1998.

McPherson, Ian. 2002. The Rochdale Principles of Co-operation. The North American Students of Cooperation, Faculty of Arts & Science, University of Victoria, Canada.

MEA (Metropolitan Electricity Authority). 1982–2002. Annual Report. Bangkok, Thailand: MEA.

Middle West Service Company. 1973. National Plan for Thailand Accelerated Rural Electrification. Paper prepared for the Provincial Electricity Company of Thailand under USAID Contract No. AID/SA/IR–207. Washington, DC: USAID.

Ministerio de Industria, Energía, y Minas. 1986. Plan Nacional de Energía, 1986–2005. San José: Government of Costa Rica.

Modi, Vijay, Susan McDade, Dominique Lallement, and Jamal Saghir. 2006. *Energy Services for the Millennium Development Goals.* New York: World Bank and United Nations Development Program.

Muller, Frederick William. 1944. *Public Rural Electrification.* Washington, DC: American Council on Public Affairs.

Muñoz, J.V., and R.A. Gonzalez. 1995. *Cooperativas de Electrificación Rural de Costa Rica.* San Marcos de Tarrazú, Costa Rica: COOPESANTOS.

National Agency for the Management of Energy (ANME). 1996. *Stratégie de Développement des Energies Renouvelables.* Tunis, Tunisia: National Agency for Renewable Energy.

———. 1998. *Project d'Electrification de 200 Foyers Ruraux, d'une Ecole Primaire et d'un Dispensaire au Gouvernorat de Siliana.* Coopération Tuniso-Espagnole, Rapport de Synthèse. Tunis, Tunisia: National Agency for Renewable Energy.

———. 1999. *Rapport Annuel d'Activité.* Tunis, Tunisia: National Agency for Renewable Energy.

NCF (National Civic Federation). 1907. *Municipal and Private Operation of Public Utilities: Report to the National Civic Federation Commission on Public Ownership and Operation,* 3 vols. New York: NCF. Commission on Public Ownership and Operation.

NEA (National Electrification Administration). 1978. *Nationwide Survey on Socio-Economic Impact of Rural Electrification.* Prepared for the U.S. Agency for International Development. Quezon City, Philippines: NEA.

———. 1981. *Vital Documents on the Philippine Rural Electrification Program.* Quezon City, Philippines: NEA.

———. 1990. *Functions and Responsibilities of the REC Board of Directors.* Quezon City, Philippines, NEA, NEA Bulletin No. 35.

———. 1994. *Criteria for Categorization of Electric Cooperatives (ECs).* Quezon City, Philippines, National Electrification Administration, memorandum to all electric cooperatives, June 15

NRECA (National Rural Electric Cooperative Association). 1967. *Victoria–Manapla–Cadiz Rural Electric Cooperative: Engineering, Feasibility and Loan Application Report.* Washington, DC, NRECA.

———. 1978. Rural Electrification Feasibility Study: Study prepared for Bangladesh Power Development Board. Washington, DC: NRECA.

———. 1980. *Rural Electrification in Costa Rica: Viability Concepts and Evaluation.* Washington, DC: NRECA.

———. 1981. *Report on the Philippine Rural Electrification Impact Survey.* Washington, DC: NRECA.

NRECA International. 1999. *Reducing the Cost of Grid Extension for Rural Electrification.* Washington, DC: Energy Sector Management Assistance Programme (ESMAP), World Bank.

Nieuwenhout, F., P. Van de Rijt, and E. Wiggelinkhuizen. 1998. *Rural Lighting Services.* Paper prepared for the World Bank. Petten, Netherlands: Netherlands Energy Research Foundation.

Nordic Consulting Group. 1997. *Etude d'Impact de Secteur de l'Energie, Tunisie: Aide Memoire.* Abidjan, Tunisia: ADB.

Norris, George W. 1945. *Fighting Liberal: The Autobiography of George W. Norris.* New York: Macmillan.

Nye, David F. 1990. *Electrifying America: Social Meanings of a New Technology.* Cambridge, MA: MIT Press.

Pence, Richard A. (ed.). 1984. *The Next Greatest Thing: 50 Years of Rural Electrification in America*. Washington, DC: NRECA.

Prokaushali Sangsad Ltd. 2000. *Off-Grid Rural Electrification Study: Feasibility Study for a Solar Home Systems Project within the Context of Alternative Options for Rural Electrification*. Dhaka, Bangladesh: World Bank.

PEA (Provincial Electricity Authority). 1986. *Report on Monitoring and Evaluation of Accelerated Rural Electrification Project: First Stage*. Prepared by Project Evaluation Division, Office of Rural Electrification. Bangkok, Thailand: PEA. In Thai.

———. 1989. *Load Promotion Technical Manual for Rice Mills*. Prepared by Load Promotion Unit, Office of Rural Electrification. Bangkok, Thailand: PEA. In Thai.

———. 1990. *Report on Monitoring and Evaluation of Accelerated Rural Electrification Project: Second Stage*. Prepared by Project Evaluation Division, Office of Rural Electrification. Bangkok, Thailand: PEA. In Thai.

Reiche, Kilian, Bernard Tenenbaum, and Clemencia Torres. 2006. *Promoting Electrification: Regulatory Principles and a Model Law*. ESMAP Report. Washington, DC: World Bank.

———. *Rapport National sur le Développement Humain en Tunisie: 1999*. Tunis, Tunisia: Ministry of Economic Development, National Institute of Statistics.

Roddis, Suzanne. 2000. *Poverty Reduction and Energy: The Links Between Electricity and Education*. Washington, DC: World Bank. Processed.

Rodríguez-Rodríguez, Guillermo. 1994. Evolución de la Industria Eléctrica en México in *El Sector Eléctrico de México*. Comisión Federal de Electricidad-Fondo de Cultura Económica. Mexico City, México.

Rose, Mark H. 1984. Urban Environments and Technological Innovation: Energy Choices in Denver and Kansas City, 1900–1940. *Technology and Culture* 25 (July): 509.

Ross, James E. 1972. *Cooperative Rural Electrification*. New York: Praeger Press.

Sachs, Jeffrey. 2005. *The End of Poverty: Economic Possibilities for Our Time*. New York: Penguin Press.

Samanta, B.B., and A.K. Sundaram. 1983. *Socioeconomic Impact of Rural Electrification in India*. Discussion Paper D–730 prepared by the Operations Research Group. Washington, DC: Resources for the Future and USAID.

Sathaye, Jayant. 1987. Issues of Rural Electrification Demand and Supply in the Philippines. *Energy Policy* 29: 715–724.

Saunders, John, J. Michael Davis, Galen C. Moses, and James E. Ross. 1978. *Rural Electrification and Development: Social and Economic Impact in Costa Rica and Colombia*. Boulder, CO: Westview Press.

Schuller, Tom. 1985. *Democracy at Work*. New York: Oxford University Press.

Sen, Amartya. 1995. *Inequality Re-examined*. Cambridge: Harvard University Press.

———. 2000. *Development as Freedom*. New York: Anchor Books.

SENER (Secretaria de Energía). 2001. Programa sectorial de energía 2001–2006 (PROSENER). Mexico City, México: SENER.

Shiel, Michael J. 1984. *The Quiet Revolution: The Electrification of Rural Ireland*. Dublin, Ireland: O'Brien Press.

———. 1988. *Rural Electrification in Ireland*. Case Study No. 1, Panos Rural Electrification Programme. London: Panos Insttitute.

Sklar, Martin J. 1988. *The Corporate Reconstruction of American Capitalism, 1890–1916: The Market, the Law, and Politics*. New York: Cambridge University Press.

———. 1992. *The United States as a Developing Country: Studies in U.S. History in the Progressive Era and the 1920s*. New York: Cambridge University Press.

Sterrett, Joseph E., and Joseph S. Davis, 1928. *The Fiscal and Economic Condition of Mexico*. Report submitted to the International Committee of Bankers on Mexico, New York.

Taylor, Carl C. 1942. Rural Life. *American Journal of Sociology* 47(6): 842–53.

Tennessee Valley Authority (TVA). 2002. The Father of Public Power [David E. Lilienthal]. http://www.tva.com/heritage/lilienthal/ (accessed December 5, 2006).

Tunisian Electricity & Gas Company (STEG). 1996. *L'Electrification Rurale par la Méthode SWER: Rapport d'Evaluation de l'Expérience Tunisienne*. Tunis, Tunisia: STEG.

———. *Rapports Annuels d'Activité: 1980–1999*. Tunis, Tunisia: STEG.

————. 2000. *Statistiques Rétrospectives 1989–1999*. Direction des Etudes et de la Planification. Tunis, Tunisia: STEG.

United Nations. 1990. *Power Systems in Asia and the Pacific, with Emphasis on Rural Electrification*. New York: United Nations.

————. 2003. *Human Development Report 2003—Millennium Development Goals: A Compact Among Nations to End Human Poverty*, edited by Sakiko Fukuda-Parr. New York: Oxford University Press.

Unnayan, Shamannay. 1996. *A Socioeconomic Impact Evaluation of the Rural Electric Program in Bangladesh*. Report Prepared for USAID, Dhaka, Bangladesh.

USAID (U.S. Agency for International Development). 1980. *The Philippines: Rural Electrification*. Washington, DC: USAID.

————. 1981. *Power to the People: Rural Electrification Sector Summary Report*. Draft summary report. Washington, DC: USAID.

————. 1988. *Project Grant Agreement between the Republic of the Philippines and the United States of America for the Rural Electrification Project*. Washington, DC: USAID.

U.S. Census Bureau. 1934. *Fifteenth Census of the United States*. U.S. Bureau of the Census. Washington, DC: Government Printing Office.

————. 1976. *Historical Statistics of the United States, Colonial Times to 1970*. Prepared by the U.S. Bureau of the Census. New York: Basic Books.

————. 1980. *Nationwide Survey of Rural Electrification in the Philippines, 1979*. Washington, DC: Bureau of the Census.

U.S. Department of Commerce. 1915. *Special Report on Central Electric Light and Power Stations 1912*. Washington, DC: Government Printing Office.

U.S. Department of Energy, Energy Information Administration. 1996. Appendix A: History of the U.S. Electric Power Industry, 1882–1991. In *The Changing Structure of the Electric Power Industry: An Update*. Office of Coal, Nuclear, Electric and Alternate Fuels. DOE/EIA-0562(96). Washington, DC: Government Printing Office. http://www.eia.doe.gov/cneaf/electricity/page/electric_kid/append_a.html#N_1_ (accessed December 5, 2006).

von-der-Fehr, Nils-Henrik M., and Jaime Millán. 2003. Power Sector Reform: Lessons Learned. In *Keeping the Lights On: Power Sector Reform in Latin America*, edited by Jaime Millán and Nils-Hendrik M. von-der-Fehr. Washington, DC: Inter-American Development Bank.

Van der Plas, Robert, and A. de Graaff. 1988. *A Comparison of Lamps for Domestic Lighting in Developing Countries*. World Bank Industry and Energy Department Working Paper (Energy Series) No. 6. Washington, DC: World Bank.

Velez, Eduardo, Carlos Becerra, and Alberto Carrasquilla. 1983. *Rural Electrification in Colombia*. Report by Instituto SER de Investigacion for Resources for the Future and USAID. Washington, DC: Resources for the Future.

Victor, David, 2005. *The Effects of Power Sector Reform on Energy Services for the Poor*. Paper produced for the Department of Economic and Social Affairs. New York: United Nations.

Wasserman, G., and A. Davenport. 1983. *Power to the People: Rural Electrification Sector Summary Report*. USAID Program Evaluation Report No. 11. Washington, DC: USAID.

Webb, Michael, and David Pearce. 1985. *Economic Benefits of Power Supply*. World Bank Energy Department Paper No. 25. Washington, DC: World Bank.

Wolman, Paul. 1992. *Most Favored Nation: The Republican Revisionists and U.S. Tariff Policy, 1896–1912*. Chapel Hill: University of North Carolina Press.

World Bank. 1975. *Rural Electrification*. Washington, DC: World Bank.

————. 1984. Costa Rica: Issues and Options in the Energy Sector. Energy Strategy Management Assistance Program Report 4655-CR. Washington, DC: World Bank.

————. 1989. *Philippines Rural Electrification Sector Study: An Integrated Program to Revitalize the Sector*. Sector Study Report. Washington, DC: World Bank.

————. 1991. *Bolivia Prefeasibility Evaluation of Rural Electrification and Demand Assessment*. Washington, DC: World Bank. ESMAP Report 129/91.

————. 1992. *Effective Implementation: Key to Development Impact*. Report of the Portfolio Management Task Force. Washington, DC: World Bank.

————. 1994. *Rural Electrification in Asia: A Review of Bank Experience.* Operation Evaluation Department Report No. 13291. Washington, DC: World Bank.

————. 1996. *Rural Energy and Development: Improving Energy Supplies for Two Billion People.* Development in Practice. Washington, DC: World Bank.

————. 1999a. *Lao PDR: Institutional Development for Off-grid Electrification.* ESMAP Report 215/99. Washington, DC: World Bank.

————. 1999b. *Poverty and Social Development in Peru, 1994–1997.* World Bank Country Study. Washington, DC: World Bank.

————. 2000. *Reducing the Cost of Grid Extension.* ESMAP Report 227/00. Washington, DC: World Bank.

————. 2000a. *Republic of Tunisia, Social and Structural Review 2000: Integrating into the World Economy and Sustaining Economic and Social Progress.* Washington, DC: World Bank.

————. 2000b. Memorandum from IBRD President to Executive Directors on a CAS of the World Bank Group for the Republic of Tunisia, Maghreb Country Management Unit, Middle East and North Africa Region.

————. 2001. Tunisia: Country Brief. Available at http://www.worldbank.org.

————. 2002a. *Energy Strategy for Rural India: Evidence from Six States.* ESMAP Report 358/02. Washington, DC: World Bank.

————. 2002b. *Rural Electrification and Development in the Philippines: Measuring the Social and Economic Benefits.* ESMAP Report 255/02. Washington, DC: World Bank.

————. 2003. Mexico Public Expenditure Review. Washington, DC: World Bank.

————. 2004a. *The Impact of Energy on Women's Lives in Rural India.* ESMAP Report 276/04. Washington, DC: World Bank.

————. 2004b. *Power Sector Reform and the Rural Poor in Central America.* ESMAP Report 297/05. World Bank, Washington, DC.

————. 2005a. *Improving the World Bank's Development Effectiveness: What Does Evaluation Show?* Operations Evaluation Department. Washington, DC: World Bank.

————. 2005b. *Malawi Rural Energy and Institutional Development.* ESMAP Technical Paper 66/05. Washington, DC: World Bank.

————. 2006. *Costa Rica: The Challenge of Sustainable Growth.* Country Economic Memorandum. World Bank Report No 36180-CR. Washington, DC: World Bank.

Yergin, Daniel. 1991. *The Prize: The Epic Quest for Oil, Money, and Power.* New York: Simon & Schuster.

Zomers, Adriaan. 2001. *Rural Electrification: Utilities' Chafe or Challenge?* Enschede, Netherlands: Twente University Press.

Index